PLANE TRIGONOMETRY WITH TABLES

FOURTH EDITION

PLANE TRIGONOMETRY WITH TABLES

E. RICHARD HEINEMAN

Texas Tech University

McGRAW-HILL BOOK COMPANY

New York   St. Louis   San Francisco   Düsseldorf
Johannesburg   Kuala Lumpur   London   Mexico
Montreal   New Delhi   Panama   Paris
São Paulo   Singapore   Sydney   Tokyo   Toronto

Library of Congress Cataloging in Publication Data

Heineman, Ellis Richard.
  Plane trigonometry with tables.

  1. Trigonometry, Plane.
QA533.H47   1974   516'.24   74-2039
ISBN 0-07-027931-4

    7 8 9 0   KPKP   7 9 8

This book was set in Baskerville by York Graphic Services, Inc.
The editors were A. Anthony Arthur and Carol First;
the designer was Nicholas Krenitsky;
the production supervisor was Leroy A. Young.
Kingsport Press, Inc., was printer and binder.

CONTENTS

PREFACE

This fourth edition of "Plane Trigonometry" continues the emphasis of the third edition on the analytic aspects of the subject. By the use of timely comments and analogies, an attempt has been made to deepen the student's concepts of the principles of algebra.

Since an increasing majority of calculus books now use the notation Sin^{-1} rather than Arcsin, the inverse sine symbol is used. The new chapter on complex numbers contains a mathematical induction proof of De Moivre's theorem.

Instructors can save time in making problem selections by noticing the two following features:

1. The problems in each exercise are so arranged that by assigning numbers 1, 5, 9, etc., or similar sets beginning with 2, 3, or 4,

the instructor can obtain balanced coverage of all points involved without undue emphasis on some principles at the expense of others. For example, in the solution of right triangles without logarithms in Exercise 11, each of the four sets of problems includes problems involving the use of the sine, the cosine, and the tangent; each set contains problems involving the angle of elevation, the angle of depression, and the concept of bearing; each set contains approximately the same number of problems in which the unknown is an acute angle (or a leg, or the hypotenuse). This does not mean that the problem lists consist of sets of four problems that are identical except for numerical quantities. Wherever possible, the author has tried to make each problem different in some way, other than numerically, from all other problems in that exercise.

2. Answers to three-fourths of the problems are given at the back of the book. Answers to problems numbered 4, 8, 12, etc., appear in the Instructor's Manual.

Completely new problem lists appear in this edition. Additional features retained from the third edition (1964) include:

3. Many of the exercises contain true-false questions to test the student's ability to avoid pitfalls and to detect camouflaged truths. An effort has been made to thwart the development of such false notions as "In logarithms, division is replaced by subtraction" and "To find the square root of a number, divide its logarithm by 2." The duty of the instructor is not only to teach correct methods but also to convince the student of the error in the false methods.

4. Definite instructions are given for proving identities and solving trigonometric equations. The subject of identities is approached gradually with practice in algebraic operations with the trigonometric functions.

5. A careful explanation of approximations and significant figures is given early in the text. The principle of accuracy in figures is adhered to throughout the book.

6. All problem sets are carefully graded and contain an abundance of simple problems that involve nothing more than the principles being discussed. There is also an ample supply of problems of medium difficulty and some "head-scratchers."

7. In addition to a discussion (Chap. 6) of the graphs of the trigonometric functions of θ, there is a body of material (Chap. 9)
   on graphical methods. Included are the graphs of $a \sin (bx + c)$,
   $a \sin x + b \cos x$, and $\sin^n x$. The two chapters on graphing are
   intentionally separated by Chap. 7 (Functions of Two Angles)
   and Chap. 8 (Trigonometric Equations), in the belief that this
   arrangement will result in better comprehension by the student.

8. There is a discussion of logarithmic and exponential functions,
   the trigonometric functions of numbers, the circular functions,
   and Euler's formula.

9. Miscellaneous points include (a) a note to the student, (b) problems that are encountered in calculus, (c) a careful explanation
   of the concept of infinity, (d) memory schemes, (e) the uses of
   the sine and cosine curves, and (f) interesting applied problems.

In a 45-lesson course the following outline (allowing four periods
for examinations) could be used.

Chapter	1	2	3	4	5	6	7	8	9	10	11	12	13	14
No. of Lessons	4	4	$3\frac{1}{2}$	$1\frac{1}{2}$	$1\frac{1}{2}$	$1\frac{1}{2}$	5	2	2	$5\frac{1}{2}$	$1\frac{1}{2}$	4	2	3

In a 30-lesson course designed for students who have studied at least
the computational aspects of trigonometry in high school, the following outline (allowing three periods for examinations) is reasonable.

Chapter	1	2	3	4	5	6	7	8	9	10	11	12	13	14
No. of Lessons	$1\frac{1}{2}$	$1\frac{1}{2}$	$2\frac{1}{2}$	1	1	1	4	2	2	$2\frac{1}{2}$	1	2	2	3

E. Richard Heineman

NOTE TO THE STUDENT

A mastery of the subject of trigonometry requires (1) a certain amount of memory work, (2) a great deal of practice and drill in order to acquire experience and skill in the application of the memory work, and (3) an insight and understanding of "what it is all about." Your instructor is a "trouble-shooter" who attempts to prevent you from going astray, supplies missing links in your mathematical background, and tries to indicate the "common sense" approach to the problem.

The memory work in any course is one thing that the student can and should perform by himself. The least you can do for your instructor and yourself is to *commit to memory each definition and theorem as soon as you contact it*. This can be accomplished most rapidly, not by reading, but by writing the definition or theorem until you can reproduce it without the aid of the text.

In working the problems, do not continually refer back to the illustrative examples. Study the examples so thoroughly (by writing them) that you can reproduce them with your text closed. Only after the examples are entirely clear and have been completely mastered should you attempt the unsolved problems. These problems should be worked *without referring to the text*.

Bear in mind, too, that *memory* and technical *skill* are aided by *understanding;* therefore, as the course develops you should review the definitions and theorems from time to time, always seeking a deeper insight into them.

E. Richard Heineman

PLANE
TRIGONOMETRY
WITH
TABLES

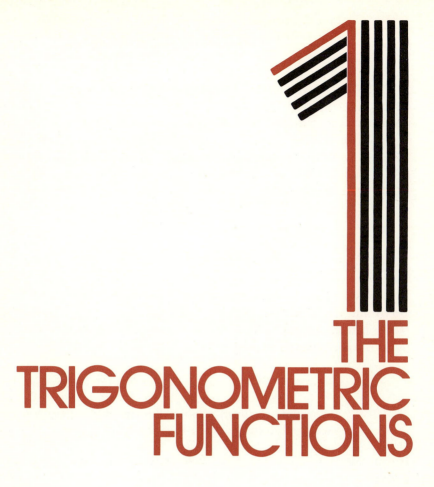

THE TRIGONOMETRIC FUNCTIONS

1 TRIGONOMETRY Trigonometry is that branch of mathematics which deals primarily with six ratios called the trigonometric functions. These ratios are important for two reasons. First, they are the basis of a theory which is used in other branches of mathematics as well as in physics and engineering. Second, they are used in solving triangles. From geometry we recall that two sides and the included angle of a triangle suffice to fix its size and shape. It will be shown later that the length of the third side and the size of the remaining angles can be computed by means of trigonometry.

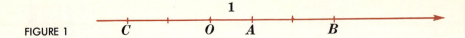

FIGURE 1

2 DIRECTED SEGMENTS   A directed line is a line upon which one direction is considered positive; the other, negative. Thus in Figure 1 the arrowhead indicates that all segments measured from left to right are positive. Hence if $OA = 1$ unit of length, then $OB = 3$, and $BC = -5$. Observe that since the line is directed, CB is not equal to BC. However, $BC = -CB$; or $CB = -BC$. Also note that $OB + BC + CO = 0$.

3 THE RECTANGULAR COORDINATE SYSTEM   A rectangular (or Cartesian) coordinate system consists of two perpendicular *directed* lines. It is conventional to draw and direct these lines as in Figure 2. The **x-axis** and the **y-axis** are called the **coordinate axes;** their intersection O is called the **origin.** The position of any point in the plane is fixed by its distances from the axes.

> The **x-coordinate*** (or **x**) of point **P** is the directed segment **NP** (or **OM**) measured from the **y**-axis to point **P**. The **y-coordinate*** (or **y**) of point **P** is the directed segment **MP**, measured from the **x**-axis to point **P**.

It is necessary to remember that each coordinate is measured *from axis to point.* Thus the x of P is NP (not PN); the y of P is MP (not PM).

* The x-coordinate and y-coordinate are also called the *abscissa* and *ordinate,* respectively.

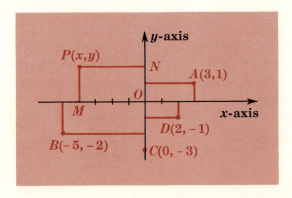

FIGURE 2

The point P, with x-coordinate x and y-coordinate y, is denoted by $P(x,y)$. It follows that the x of any point to the right of the y-axis is positive; to the left, negative. Also the y of any point above the x-axis is positive; below, negative.

To **plot** a point means to locate and indicate its position on a coordinate system. Several points are plotted in Figure 2.

The distance from the origin O to point P is called the **radius vector** (or r) of P. This distance r is not directed and *is always positive** by agreement. Hence with each point of the plane we can associate three coordinates: x, y, and r. The radius vector r can be found by using the Pythagorean† relation $x^2 + y^2 = r^2$ (see Figure 3).

The coordinate axes divide the plane into four parts called **quadrants** as indicated in Figure 4. We shall sometimes denote these as Q I, Q II, Q III, Q IV, respectively.

ILLUSTRATION 1   To find r for the point $(5, -12)$, use

$$r^2 = 5^2 + (-12)^2 = 169 \qquad r = 13$$

ILLUSTRATION 2   If $x = 15$ and $r = 17$, we obtain y by using

$$x^2 + y^2 = r^2$$

Hence $(15)^2 + y^2 = (17)^2$; $225 + y^2 = 289$; $y = \pm 8$. If the point is in quadrant I, $y = 8$; if the point is in quadrant IV, $y = -8$. (Since x is positive, the point cannot lie in either quadrant II or quadrant III.)

---

* Or zero (for the origin O).
† *Pythagorean theorem: The square of the hypotenuse of a right triangle equals the sum of the squares of its legs.*

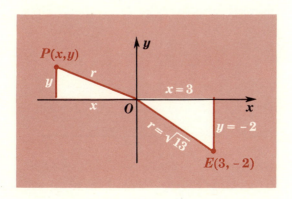

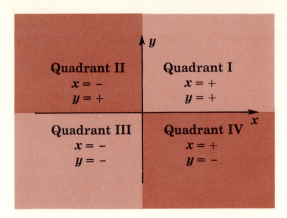

FIGURE 4

4   THE DISTANCE FORMULA   Let $P_1(x_1, y_1)$ and $P_2(x_2, y_2)$ be any two points in the xy-plane (Figure 5). We shall use the Pythagorean theorem to express the distance P_1P_2 in terms of the coordinates of the points. Through P_1, draw a line parallel to the x-axis. Through P_2, draw a line parallel to the y-axis. These lines meet at $Q(x_2, y_1)$. The length of the positive segment P_1Q is $x_2 - x_1$ (that is, the *right x* minus the *left x**). And the length of the positive segment QP_2 is $y_2 - y_1$ (that is, the *upper y* minus the *lower y*). Since $(P_1P_2)^2 = (P_1Q)^2 + (QP_2)^2$,

$$P_1P_2 = \sqrt{(x_2 - x_1)^2 + (y_2 - y_1)^2}$$

* More precisely, the x of the point on the right (the large x) minus the x of the point on the left (the small x). For $P_1(-2,1)$ and $Q(5,1)$, $P_1Q = 5 - (-2) = 7$.

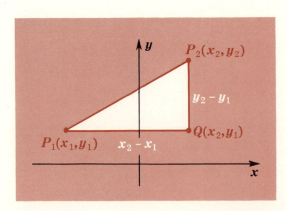

FIGURE 5

This formula holds for all positions of P_1 and P_2. If, for example, P_2 lies below P_1, then $QP_2 = y_2 - y_1$, which is a negative quantity. But the square of $(y_2 - y_1)$ is equal to the square of the positive quantity $(y_1 - y_2)$; that is, $(y_2 - y_1)^2 = (y_1 - y_2)^2$. Hence

the distance d between $P_1(x_1, y_1)$ and $P_2(x_2, y_2)$ is

$$d = \sqrt{(x_2 - x_1)^2 + (y_2 - y_1)^2}$$

ILLUSTRATION   The distance between $A(-1,5)$ and $B(3, -2)$ is $d = \sqrt{(-1 - 3)^2 + (5 + 2)^2} = \sqrt{16 + 49} = \sqrt{65}$. Either A or B can be designated as P_1.

EXERCISE 1

1 Plot on coordinate paper and then find the value of r for each of the following points: $(-6, -8)$, $(-5,0)$, $(3, -2)$, $(-\sqrt{11},5)$.

2 Plot on coordinate paper and then find the value of r: $(-12,5)$, $(8,0)$, $(1,6)$, $(-2, -2\sqrt{3})$.

3 Plot on coordinate paper and then find the value of r: $(24,7)$, $(0,2)$, $(-3, -5)$, $(6, -\sqrt{13})$.

4 Plot on coordinate paper and then find the value of r: $(8, -15)$, $(0, -4)$, $(-8,3)$, $(\sqrt{21},2)$.

5 Use the Pythagorean theorem to find the missing coordinate and then plot the point:

(a) $y = 15$, $r = 17$, point is in Q I.

(b) $x = -3$, $r = \sqrt{10}$, point is in Q III.

(c) $y = -9$, $r = 9$.

6 Find the missing coordinate and then plot the point:

(a) $x = 4$, $r = 5$, point is in Q IV.

(b) $y = 2$, $r = \sqrt{29}$, point is in Q II.

(c) $x = 0$, $r = 4$, y is negative.

7 Find the missing coordinate and then plot the point:

(a) $x = -5$, $r = 13$, point is in Q III.

(b) $y = 4$, $r = 7$, point is in Q I.

(c) $y = 0$, $r = 8$, x is positive.

8 Find the missing coordinate and then plot the point:

(a) $y = 24$, $r = 25$, point is in Q II.

(b) $x = 6$, $r = 7$, point is in Q IV.

(c) $x = -5$, $r = 5$.

9 In which quadrants is the following ratio positive?

(a) $\dfrac{y}{r}$ (b) $\dfrac{x}{r}$ (c) $\dfrac{y}{x}$

10 In which quadrants is the following ratio negative?

(a) $\dfrac{y}{r}$ (b) $\dfrac{x}{r}$ (c) $\dfrac{y}{x}$

11 What is the x of all points on the y-axis? What is the y of all points on the x-axis?

12 Without plotting, identify the quadrant in which each of the following points lies if s is a negative number: $L(3,s)$, $M(-1, -s)$, $N(s, -s^2)$, $Q(-s^3,s^2)$.

13 Use the distance formula to find the distance between the points $(3, -2)$ and $(-1,5)$.

14 Use the distance formula to find the distance between the points $(-4,6)$ and $(3,a)$.

5 TRIGONOMETRIC ANGLES   In geometry, you may have thought of an angle as the "opening" between two rays* which form the sides of the angle and which emerge from a point called the vertex of the angle.

> **A trigonometric angle is an amount of rotation used in moving a ray from one position to another.**

A *positive* angle is generated by *counterclockwise* rotation; a *negative* angle, by *clockwise* rotation.   Figure 6 illustrates the terms used and shows an angle of 200°.   Figure 7 shows angles of 500° and −420°.   The −420° angle may be thought of as the amount of rotation effected by the minute hand of a clock between 12:15 and 1:25.   *To specify a trigonometric angle, we need,* in addition to its sides, *a curved arrow extending from its initial side to its terminal side.*

---

* A *ray*, or half-line, is the part of a line extending in one direction from a point on the line.

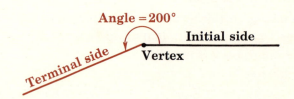

FIGURE 6

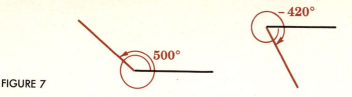

FIGURE 7

6 STANDARD POSITION OF AN ANGLE

An angle is said to be in standard position if its vertex is at the origin and its initial side coincides with the positive *x*-axis.

An angle is said to be in a certain quadrant if its terminal side lies in that quadrant *when the angle is in standard position.* For example, 600° is in the third quadrant; or, −70° is a fourth-quadrant angle.

Angles are said to be **coterminal** if their terminal sides coincide when the angles are in standard position. For example, 200°, 560°, −160° are coterminal angles. From a trigonometric viewpoint these angles are not equal; they are merely coterminal.

EXERCISE 2

Place each of the following angles in standard position; draw a curved arrow to indicate the rotation. Draw and find the size of the two other angles, one positive and one negative, that are coterminal with the given angle.

1	70°	**2**	340°	**3**	420°	**4**	560°
5	540°	**6**	620°	**7**	170°	**8**	280°

Each of the following points is on the terminal side of a positive angle in standard position. Plot the point; draw the terminal side of the angle; indicate the angle by a curved arrow; use a protractor to find to the nearest degree the size of the angle.

9	(−1,7)	**10**	(3, −2)	**11**	(5,6)	**12**	(−4, −9)
13	(−5, −2)	**14**	(−4,5)	**15**	($\sqrt{14}$, −1)	**16**	($\sqrt{3}$,1)

17 A flywheel makes 1100 revolutions per minute. Through how many degrees does it move in 1 sec.?

7 DEFINITIONS OF THE TRIGONOMETRIC FUNCTIONS OF A GENERAL ANGLE
The whole subject of trigonometry is based upon the six **trigonometric functions.** The names of these functions, with their abbreviations in parentheses, are: sine **(sin),** cosine **(cos),** tangent

(tan), cotangent **(cot)**, secant **(sec)**, cosecant **(csc)**. In a certain sense, the following definitions are the most important in this book.

A COMPLETE DEFINITION OF THE TRIGONOMETRIC FUNCTIONS OF ANY ANGLE θ

1. *Place the angle θ* in standard position.*
2. *Choose any point P† on the terminal side of θ.*
3. *Drop a perpendicular from P to the x-axis, thus forming a triangle of reference for θ.*
4. *The point P has three coordinates x, y, r, in terms of which we define the following trigonometric functions:*

$$\sin \theta = \frac{y}{r}$$

$$\cos \theta = \frac{x}{r}$$

$$\tan \theta = \frac{y}{x}$$

$$\cot \theta = \frac{x}{y}$$

$$\sec \theta = \frac{r}{x}$$

$$\csc \theta = \frac{r}{y}$$

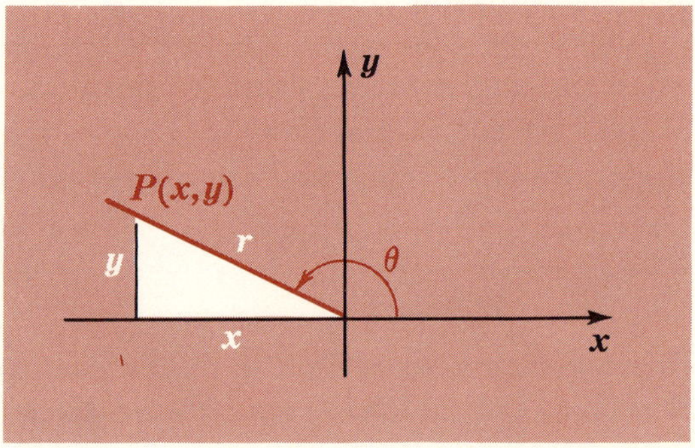

FIGURE 8

These six ratios are called *functions,* in accordance with the modern definition of the word, because they give rise to *ordered pairs,* for example, $(\theta, \sin \theta)$, with only one value—sin θ in our example—corresponding to a given θ.

The *domain* of each function consists of all θ for which the corresponding denominator is not 0; thus tan θ and sec θ are not defined when $x = 0$, and cot θ and csc θ are not defined when $y = 0$. You will recall that *division by zero is impossible.* The definition of division states that $a/b = c$ if and only if $bc = a$, provided c is a unique number. If $\frac{1}{0} = a$, then $(0)(a)$ must equal 1. No such number

* See Greek alphabet on front endpaper.
† Other than the origin O.

a exists.   Another explanation: When we write $\frac{12}{3}$, we are asking, "How many 3's add up to 12?"   Consequently $\frac{1}{0}$ means "How many zeros will add up to 1?"   Such a question is obviously absurd.

The *range* of each trigonometric function will be discussed later. The range of sin θ, for example, is by definition the set of all values taken on by sin θ as θ varies over its domain.

While a function is defined to be a set of ordered pairs, such as $(\theta, \sin \theta)$, it is customary to speak somewhat loosely of the second member as a function of the first.   Thus we say that sin θ is a function of θ.   In general, we say that *a function of θ is a quantity whose value is determined whenever a value in the domain of θ is assigned to θ.*   For example, $3\theta^2 + 1$ is a function of θ for all numbers θ.   If θ has the value 5, then $3\theta^2 + 1$ has the value 76.   If $\theta = -4$, then $3\theta^2 + 1 = 49$.   Likewise, $\theta^3 + 7$ and 8θ are functions of θ for all θ.   Also, $(\theta - 2)/(\theta - 3)$ is a function of θ for all θ except $\theta = 3$.

In order to prove that sin θ is a function of θ, we must show that the value of sin θ is independent of the choice of point P on the terminal side of θ.   Let P' (x', y') be any other point on OP (see Figure 9).   Then, using the coordinates of P', we have $\sin \theta = y'/r'$.   Since triangles $OP'M'$ and OPM are similar, it follows that

$$\frac{y'}{r'} = \frac{y}{r}$$

and the value of sin θ is the same whether it is obtained by using P or by using P'.   Since the value of sin θ is determined by the value of θ,

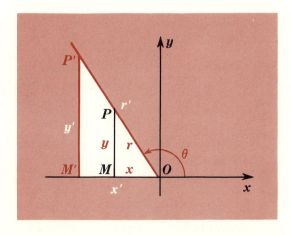

FIGURE 9

and is independent of the choice of P, we can say that $\sin \theta$ is a function of θ. *The values of the trigonometric functions* of θ depend solely upon the value of θ.*

8   CONSEQUENCES OF THE DEFINITIONS

(a) The reciprocal of the number a is $\dfrac{1}{a}$.

Hence the reciprocal of 3 is $\frac{1}{3}$; the reciprocal of $-\frac{1}{4}$ is -4; $\frac{2}{5}$ and $\frac{5}{2}$ are reciprocals.   Since $\dfrac{x}{y} = \dfrac{1}{y/x}$, we can say, for values of θ for which these functions are defined, that

$$\cot \theta = \frac{1}{\tan \theta}$$

Similarly, $$\sec \theta = \frac{1}{\cos \theta}$$

and $$\csc \theta = \frac{1}{\sin \theta}$$

Multiplying both sides of the last equation by $\sin \theta$, we get

$$\sin \theta \csc \theta = 1$$

Dividing both sides of this equation by $\csc \theta$, we obtain

$$\sin \theta = \frac{1}{\csc \theta}$$

Hence *$\sin \theta$ and $\csc \theta$ are reciprocals;* also *$\cos \theta$ and $\sec \theta$ are reciprocals;* and *$\tan \theta$ and $\cot \theta$ are reciprocals.*   The following table indicates the reciprocal functions:

$$
\left.
\begin{array}{l}
\sin \theta \\
\cos \theta \\
\left.\begin{array}{l}\tan \theta \\ \cot \theta\end{array}\right\} \\
\sec \theta \\
\csc \theta
\end{array}
\right\}\ \text{Reciprocals}
$$

---

* Three other functions sometimes used are versed sine, coversed sine, and haversine: vers $\theta = 1 - \cos \theta$; covers $\theta = 1 - \sin \theta$; hav $\theta = \frac{1}{2}(1 - \cos \theta)$.   We shall not use these functions in this book.

Caution. The symbol *cos* in itself has no meaning.* To have interpretation, it must be followed by some angle. Write $\cos\theta$, not *cos*. Notice that $\sin\theta\csc\theta = 1$ means that the sine of any angle times the cosecant of the *same* angle equals unity.

(b) Any trigonometric function of an angle is equal to the same function of all angles coterminal with it.

This follows directly from Section 7. Thus

$$\sin 370° = \sin (370° - 360°) = \sin 10°$$
$$\cos (-100°) = \cos (-100° + 360°) = \cos 260°$$
$$\tan 900° = \tan (900° - 720°) = \tan 180°$$

(c) The sine is positive for angles in the top quadrants; the cosine is positive for angles in the right-hand quadrants.

Since r is always positive, $\sin\theta$ is positive whenever y is positive, i.e., in the upper quadrants, I and II. Similarly $\sin\theta$ is negative in the lower quadrants, III and IV. Also $\sin\theta$ is 0 when $y = 0$, which occurs when θ is coterminal with 0° or 180° (see Figure 10).

Likewise, $\cos\theta$ has the same sign as x. Hence $\cos\theta$ is positive in the right-hand quadrants, I and IV. Also $\cos\theta$ is negative in the left-hand quadrants, II and III. And $\cos\theta$ is 0 whenever $x = 0$, which occurs when θ is coterminal with 90° or 270°.

Moreover, $\tan\theta = y/x$ is positive when y and x have the same sign, namely in quadrants I and III. Also, $\tan\theta$ is negative when y and x have opposite signs, namely in quadrants II and IV.

The sign of each of the three remaining functions is the same as the sign of its reciprocal. Thus $\csc\theta$ is positive in the upper quadrants.

* We can, however, speak of *the cosine function,* referring to the entire set of ordered pairs $(\theta, \cos\theta)$.

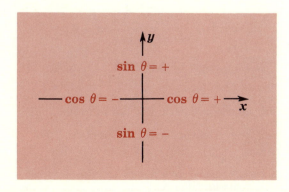

FIGURE 10

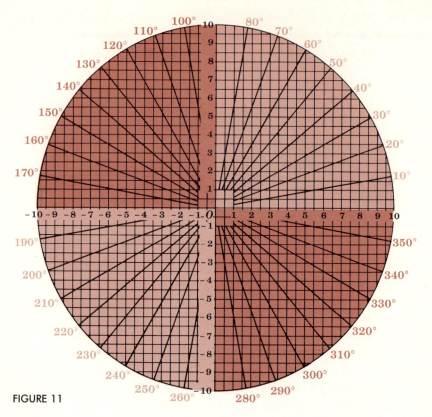

FIGURE 11

EXAMPLE 1   Find the values of sin 160°, cos 160°, tan 160°.

Solution   In Figure 11 the angle 160° is in standard position. For convenience, on the terminal side of 160° choose point P so that $r = 10$. In forming a triangle of reference we find that $x = -9.4, y = 3.4$* (see Figure 12). Hence

$$\sin 160° = \frac{y}{r} = \frac{3.4}{10} = 0.34$$

$$\cos 160° = \frac{x}{r} = \frac{-9.4}{10} = -0.94$$

$$\tan 160° = \frac{y}{x} = \frac{3.4}{-9.4} = -0.36$$

EXAMPLE 2   Find the trigonometric functions of 180°.

Solution   Place the angle in standard position. For point P let us choose $(-1,0)$. Since r is always positive, $r = 1$. The tri-

---

* Since this number was obtained from a drawing, it is merely an approximation. Equally acceptable would be 3.3 or 3.5.

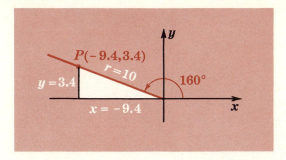

FIGURE 12

angle of reference has "collapsed," but P does have the coordinates x, y, r (see Figure 13). Then

$$\sin 180° = \frac{y}{r} = \frac{0}{1} = 0$$

$$\cos 180° = \frac{x}{r} = \frac{-1}{1} = -1$$

$$\tan 180° = \frac{y}{x} = \frac{0}{-1} = 0$$

$$\cot 180° = \frac{x}{y} = \frac{-1}{0}, \quad \textit{which does not exist}$$

$$\sec 180° = \frac{r}{x} = \frac{1}{-1} = -1$$

$$\csc 180° = \frac{r}{y} = \frac{1}{0}, \quad \textit{which does not exist}$$

EXAMPLE 3 Assuming angle θ is in standard position, compute the trigonometric functions of θ if point $P(-2, -3)$ is on its terminal side.

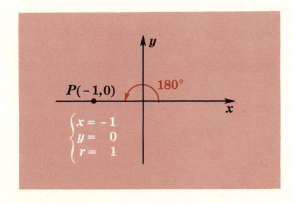

FIGURE 13

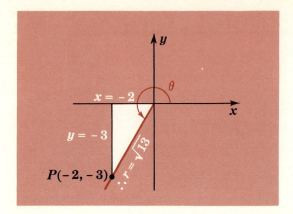

FIGURE 14

Solution   The Pythagorean theorem gives us $r = \sqrt{13}$. Angle θ and its triangle of reference are shown in Figure 14.   Then

$$\sin \theta = \frac{y}{r} = \frac{-3}{\sqrt{13}} = -\frac{3\sqrt{13}*}{13}$$

$$\cos \theta = \frac{x}{r} = \frac{-2}{\sqrt{13}} = -\frac{2\sqrt{13}*}{13}$$

$$\tan \theta = \frac{y}{x} = \frac{-3}{-2} = \frac{3}{2}$$

$$\cot \theta = \frac{x}{y} = \frac{-2}{-3} = \frac{2}{3}$$

$$\sec \theta = \frac{r}{x} = \frac{\sqrt{13}}{-2} = -\frac{\sqrt{13}}{2}$$

$$\csc \theta = \frac{r}{y} = \frac{\sqrt{13}}{-3} = -\frac{\sqrt{13}}{3}$$

EXERCISE 3

Place each of the following angles in standard position, using a curved arrow to indicate the rotation.   Use Figure 11 to label the sides of the triangle of reference.   Choose P so that $r = 10$.   Then approximate the values of the sine, cosine, and tangent of each angle.

1 110°	**2** 310°	**3** 80°	**4** 200°
5 220°	**6** 160°	**7** 340°	**8** 50°
9 10°	**10** 260°	**11** 140°	**12** 290°

* Rationalizing the denominator by multiplying top and bottom by $\sqrt{13}$.

13 Using $r = 10$, read from Figure 11 the sine of each of the following angles: $0°$, $10°$, $20°$, $30°$, $40°$, $50°$, $60°$, $70°$, $80°$, $90°$.

14 Using $r = 10$, read from Figure 11 the cosine of each of the following angles: $0°$, $10°$, $20°$, $30°$, $40°$, $50°$, $60°$, $70°$, $80°$, $90°$.

15 Draw a figure and compute the trigonometric functions of $0°$.

16 Draw a figure and compute the trigonometric functions of $90°$.

17 Draw a figure and compute the trigonometric functions of $270°$.

18 Draw a figure and compute the trigonometric functions of $-180°$.

Each of the following points is on the terminal side of an angle θ, in standard position. Use a curved arrow to specify θ. Construct and label the sides of the triangle of reference as in Figure 14. Find the six trigonometric functions of θ.

19 $(-15, -8)$ 20 $(-4,3)$ 21 $(5, -12)$ 22 $(7,24)$

23 $(-1,2)$ 24 $(-5, -6)$ 25 $(7,3)$ 26 $(9, -4)$

27 $(5, -\sqrt{11})$ 28 $(2\sqrt{2},1)$ 29 $(-\sqrt{2},\sqrt{7})$ 30 $(-\sqrt{6}, -\sqrt{10})$

Copy the following statements and identify the quadrant in which θ must be in order to satisfy each set of conditions.

31 $\sin \theta = -$ and $\cos \theta = -$ 32 $\sin \theta = -$ and $\sec \theta = +$

33 $\cos \theta = -$ and $\tan \theta = +$ 34 $\cos \theta = -$ and $\csc \theta = +$

35 $\sin \theta = +$ and $\tan \theta = -$ 36 $\tan \theta = +$ and $\csc \theta = -$

37 $\cot \theta = -$ and $\csc \theta = -$ 38 $\cot \theta = -$ and $\sec \theta = +$

Copy the following statements and identify each as possible or impossible.

39 $\cos \theta = 3$ 40 $\cos \theta = 0$

41 $\tan \theta = 200$ 42 $\tan \theta = 0.03$

43 $\sin \theta = 0$ 44 $\sin \theta = -2$

45 $\cos \theta = \frac{1}{3}$ and $\sec \theta = -3$ 46 $\tan \theta = 5$ and $\sec \theta = 4$

Copy the following statements and in each case state whether θ is close to $0°$ or close to $90°$.

47 $\tan \theta = 0.01$ 48 $\tan \theta = 100$

49 $\sin \theta = 0.01$ 50 $\sin \theta = 0.99$

51 $\cos \theta = 0.01$ 52 $\cos \theta = 0.99$

Refer to Figure P1 on page 16 in answering the following questions.

53 In which quadrant is α?   54 In which quadrant is β?

55 In which quadrant is γ?   56 In which quadrant is δ?

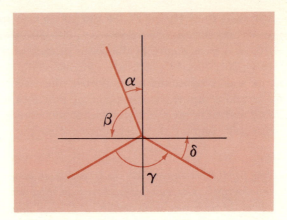

FIGURE P1

9 GIVEN ONE TRIGONOMETRIC FUNCTION OF AN ANGLE, TO DRAW THE ANGLE AND FIND THE OTHER FUNCTIONS

When we know (1) the quadrant in which an angle lies and (2) the value of one trigonometric function of this angle, it is possible, by using the Pythagorean theorem and the general definition, to draw the angle and find its other five trigonometric functions.

EXAMPLE   Given $\cos \theta = \frac{2}{5}$ and θ is not in Q I, to draw θ and find its other functions.

Solution   Since $\cos \theta$ is positive in the two right-hand quadrants and since Q I is ruled out, θ must lie in Q IV.   Remembering that for all angles $\cos \theta = x/r$ and that in this case $\cos \theta = \frac{2}{5}$, we can use $x = 2$ and $r = 5$.*   By means of $x^2 + y^2 = r^2$, we find $y = -\sqrt{21}$, the negative sign being chosen because $P(x,y)$ is in Q IV (see Figure 15).   Then

$$\sin \theta = \frac{y}{r} = \frac{-\sqrt{21}}{5} = -\frac{\sqrt{21}}{5}$$

$$\tan \theta = \frac{y}{x} = \frac{-\sqrt{21}}{2} = -\frac{\sqrt{21}}{2}$$

$$\cot \theta = \frac{x}{y} = \frac{2}{-\sqrt{21}} = -\frac{2\sqrt{21}}{21}$$

$$\sec \theta = \frac{r}{x} = \frac{5}{2}$$

$$\csc \theta = \frac{r}{y} = \frac{5}{-\sqrt{21}} = -\frac{5\sqrt{21}}{21}$$

* Equally correct but not so convenient would be $x = 6$, $r = 15$ or $x = 1$, $r = \frac{5}{2}$.

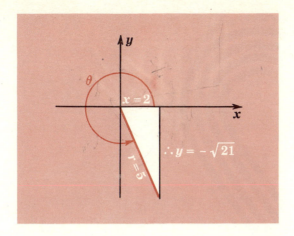

FIGURE 15

Construct and label the sides of the triangle of reference. Use a curved arrow to indicate θ. Find the remaining trigonometric functions.

1 $\sin \theta = \frac{3}{4}$, θ not in Q II.
2 $\cos \theta = -\frac{1}{4}$, θ not in Q III.

3 $\tan \theta = -\frac{3}{2}$, θ not in Q II.
4 $\sin \theta = -\frac{5}{8}$, θ not in Q IV.

5 $\cos \theta = \frac{24}{25}$, $\sin \theta < 0$.*
6 $\tan \theta = \frac{15}{8}$, $\cos \theta > 0$.†

7 $\sin \theta = \frac{4}{5}$, $\tan \theta < 0$.*
8 $\cos \theta = \frac{5}{13}$, $\sin \theta > 0$.†

9 $\tan \theta = \dfrac{2\sqrt{5}}{5}$, θ not in Q I. *Hint:* $\dfrac{2\sqrt{5}}{5} = \dfrac{2}{\sqrt{5}} = \dfrac{-2}{-\sqrt{5}}$

10 $\sin \theta = -\dfrac{2\sqrt{85}}{85}$, θ not in Q III. *Hint:* $-\dfrac{2\sqrt{85}}{85} = \dfrac{-2}{\sqrt{85}}$

11 $\cos \theta = -\dfrac{\sqrt{15}}{8}$, θ not in Q II.

12 $\tan \theta = -\frac{1}{8}$, θ not in Q IV.

13 $\sec \theta = -3$, θ not in Q III.
14 $\cot \theta = 5$, θ not in Q I.

15 $\csc \theta = a$, θ is in Q I.
16 $\cot \theta = b$, θ is in Q IV.

* The symbol $<$ is read "is less than."
† The symbol $>$ is read "is greater than."

2

TRIGONOMETRIC FUNCTIONS OF AN ACUTE ANGLE

10 TRIGONOMETRIC FUNCTIONS OF AN ACUTE ANGLE   Let θ be an acute angle of a right triangle (Figure 16).   If θ were in standard position, this right triangle would be its triangle of reference.*   The hypotenuse (*hyp.*) would be the r of some point P on the terminal side of θ; the side opposite θ (*opp.*) would be the y of P; and the side adjacent θ (*adj.*) would be the x of P.   Then the general definitions (Section 7) involving x, y, and r would become special definitions involving *adj.*, *opp.*, and *hyp.*

We conclude that

* In some cases it may be necessary to turn the triangle over.

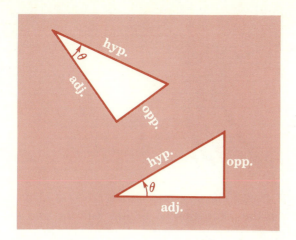

FIGURE 16

for any **acute** *angle θ lying in a right triangle,*

$$\sin \theta = \frac{\text{opp.}}{\text{hyp.}}$$

$$\cos \theta = \frac{\text{adj.}}{\text{hyp.}}$$

$$\tan \theta = \frac{\text{opp.}}{\text{adj.}}$$

The other three functions can be obtained through their reciprocals.

EXERCISE 5

For each of the following right triangles write the sine, cosine, and tangent of each acute angle. Leave results in fractional form.

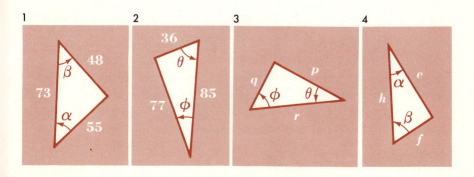

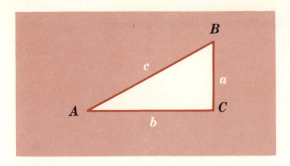

11 COFUNCTIONS   The sine and cosine are said to be *cofunctions;* *
i.e., the cosine is the cofunction of the sine and the sine is the cofunction of the cosine.   Similarly, the tangent and cotangent are cofunctions, and the secant and cosecant are cofunctions.

If A and B are the acute angles in right triangle ABC above, then

$$\sin A = \frac{a}{c} = \cos B$$

$$\cos A = \frac{b}{c} = \sin B$$

Similarly, $\tan A = \cot B$ $\cot A = \tan B$
               $\sec A = \csc B$ $\csc A = \sec B$

Hence we have the following:

THEOREM   **Any trigonometric function of an acute angle is equal to the
cofunction of its complementary angle.**

Thus $\sin 70° = \cos 20°$, $\cos 80° = \sin 10°$, $\tan 50° = \cot 40°$.

12 VARIATION OF THE FUNCTIONS OF AN ACUTE ANGLE   If r is
fixed and if θ increases from $0°$ to $90°$ (Figure 11), then y increases
and x decreases.   It follows that *as an acute angle increases, its sine,
tangent, and secant increase* while their cofunctions, the cosine, cotangent,
and cosecant, decrease.   Since neither leg of a right triangle can equal
the hypotenuse, the sine and cosine of an acute angle must always be
less than 1; the secant and cosecant must be greater than 1.

---

* Not to be confused with *reciprocal* functions.

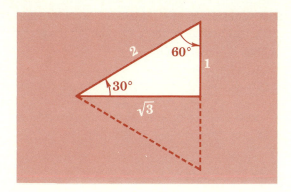

FIGURE 17

13　THE TRIGONOMETRIC FUNCTIONS OF 30°, 45°, 60°　Consider an equilateral triangle of side 2.　The bisector of one of the 60° angles will also bisect the opposite side (Figure 17).　By the Pythagorean theorem, the length of the bisector is $\sqrt{3}$.　Using Section 10, we find

$$\sin 30° = \frac{1}{2} \qquad\qquad \sin 60° = \frac{\sqrt{3}}{2}$$

$$\cos 30° = \frac{\sqrt{3}}{2} \qquad\qquad \cos 60° = \frac{1}{2}$$

$$\tan 30° = \frac{1}{\sqrt{3}} = \frac{\sqrt{3}}{3} \qquad \tan 60° = \sqrt{3}$$

The 30°-60°-90° triangle can be easily remembered if we note that the largest side (the hypotenuse) is twice the shortest side.

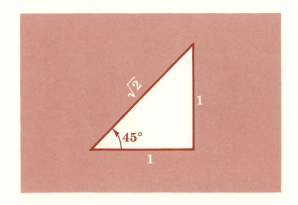

FIGURE 18

To compute the functions of 45°, draw an isosceles right triangle of leg 1 (Figure 18). The hypotenuse, by the Pythagorean theorem, must be $\sqrt{2}$. Then, by Section 10,

$$\sin 45° = \frac{1}{\sqrt{2}} = \frac{\sqrt{2}}{2}$$

$$\cos 45° = \frac{1}{\sqrt{2}} = \frac{\sqrt{2}}{2}$$

$$\tan 45° = \frac{1}{1} = 1$$

Because of their frequent occurrence, the following should be memorized, along with the figures from which they are derived:

$$\sin 30° = \cos 60° = \frac{1}{2}$$

$$\sin 45° = \cos 45° = \frac{\sqrt{2}}{2}$$

$$\sin 60° = \cos 30° = \frac{\sqrt{3}}{2}$$

As we shall see later, the tangent of any angle can be obtained by dividing its sine by its cosine. The three remaining functions can be obtained through their reciprocals.

The expression $\sin^2 \theta$ is a shorter way of writing $(\sin \theta)^2$. Since $\sin \theta$ is merely a number (the ratio of two lengths), we can speak of the square of this number and call it $\sin^2 \theta$. Thus

$$\sin^2 60° = \left(\frac{\sqrt{3}}{2}\right)^2 = \frac{3}{4} \qquad \sin^3 30° = \left(\frac{1}{2}\right)^3 = \frac{1}{8}$$

$$\sec^2 30° = \left(\frac{1}{\cos 30°}\right)^2 = \frac{1}{\cos^2 30°}$$

$$= \frac{1}{\left(\frac{\sqrt{3}}{2}\right)^2} = \frac{1}{\frac{3}{4}} = \frac{4}{3}$$

If we recall (Prob. 15, page 15) that $\sin 0° = 0$, $\cos 0° = 1$, $\sin 90° = 1$, $\cos 90° = 0$, we can easily remember the sine and cosine of special first-quadrant angles by forming a mental picture of Figure 19.

	0°	30°	45°	60°	90°
sin	$\dfrac{\sqrt{0}}{2}$	$\dfrac{\sqrt{1}}{2}$	$\dfrac{\sqrt{2}}{2}$	$\dfrac{\sqrt{3}}{2}$	$\dfrac{\sqrt{4}}{2}$
cos	$\dfrac{\sqrt{4}}{2}$	$\dfrac{\sqrt{3}}{2}$	$\dfrac{\sqrt{2}}{2}$	$\dfrac{\sqrt{1}}{2}$	$\dfrac{\sqrt{0}}{2}$

FIGURE 19

EXERCISE 6

Compute the value of each of the following expressions. Leave results in fractional form.

1   $8 \sin 30° - \cos^4 45° - \sin^2 60°$
2   $\sqrt{2} \sin 45° - \cos^4 30° + \cos 60°$
3   $14 \sin 60° - 4 \cos 30° + \sin^4 45° \sin 0°$
4   $\sin^3 30° + \sin 60° \cos 90° + \cos^6 45°$
5   $\sin 90° + 2\sqrt{2} \sin 45° - 3 \cos 60° - \cos^2 30°$
6   $\cos 0° - \sqrt{3} \sin^3 60° + \sin 30° \cos^2 45°$
7   $\sqrt{2} \cos 45° - 8 \sin^5 30° + \cos^3 60°$
8   $6 \sin^2 45° - 20 \cos^4 60° - 4 \cos^6 30°$

Identify as true or false and give reasons.

9   $\tan 10° < \tan 20°$
10   $\csc 40° < \csc 50°$
11   $\sin 55° = \cos 35°$

12   $\sin 15° = \dfrac{1}{\sec 15°}$

13   $\sin 30° + \sin 60° = \sin 90°$
14   $\cos (80° - \theta) = \sin (10° + \theta)$

15   $\sin 80° = \dfrac{1}{\sec 10°}$

16   $\csc^2 40° = \dfrac{1}{\cos^2 50°}$

14   TABLES OF TRIGONOMETRIC FUNCTIONS   In Table 1 (pages 244–248 in the tables) there are listed, to four decimal places, the sine, cosine, tangent, and cotangent for acute angles at intervals of 10′.   You will recall that there are 60 minutes in 1 degree ($60' = 1°$).   For angles less than 45°, find the name of the function at the *top* of the column, then read *down* until the angle is found at the *left*.   For angles greater than 45°, find the name of the function at the *bottom* of the column, then read *up* until the angle is found at the *right*.

The two problems we shall need to consider are:

1. Given an angle, to find one of its trigonometric functions.
2. Given a trigonometric function of some angle, to find the angle.

15   GIVEN AN ANGLE, TO FIND ONE OF ITS FUNCTIONS

EXAMPLE 1   Find tan 8° 20′.

Solution   On page 244, in the column with *tan* at its *head,* come down to the number in line with 8° 20′. Thus tan 8° 20′ = 0.1465.

EXAMPLE 2   Find sin 74° 50′.

Solution   On page 245, in the column with *sin* at its *foot,* move up to the number in line with 74° 50′. Hence sin 74° 50′ = 0.9652.

16   GIVEN A FUNCTION OF AN ANGLE, TO FIND THE ANGLE

EXAMPLE 1   Find θ if sin θ = 0.9387.

Solution   Since sines are found in column three reading *down* and in column six reading *up,* we must search through these two columns until we find the number 0.9387. It appears in the sixth column which has *sin* at its *foot.* This column contains the sines of the angles in the *right* column. On a line with 0.9387, we find in the *right* column the angle 69° 50′. Hence,

if   sin θ = 0.9387

then θ = 69° 50′

EXAMPLE 2   Find θ if cot θ = 1.288.

Solution   We search the two cotangent columns, the fourth going up and the fifth going down, and find 1.288 in the fifth column. Since this column has *cot* at its *head,* we associate this number with the angle at the *left,* 37° 50′. Hence, if $\qquad$ $\cot \theta = 1.288$

then* $\qquad$ $\theta = 37° 50'$

EXERCISE 7

Use a four-place table (Table 1) to find the value of each of the following.

1	$\cos 63° 20'$	2	$\sin 78° 10'$
3	$\tan 18° 40'$	4	$\cot 59° 30'$
5	$\sin 27° 10'$	6	$\cos 34° 0'$
7	$\cot 80° 20'$	8	$\tan 46° 50'$
9	$\cot 8° 40'$	10	$\tan 41° 30'$
11	$\sin 55° 10'$	12	$\cos 17° 0'$

Use a four-place table (Table 1) to find θ from the following function of θ.

13	$\sin \theta = 0.8450$	14	$\cos \theta = 0.1880$
15	$\cot \theta = 2.050$	16	$\tan \theta = 0.6830$
17	$\cos \theta = 0.9863$	18	$\tan \theta = 11.06$
19	$\sin \theta = 0.6517$	20	$\cos \theta = 0.6134$
21	$\tan \theta = 2.246$	22	$\cot \theta = 1.060$
23	$\cos \theta = 0.0436$	24	$\sin \theta = 0.3228$

17 INTERPOLATION When a sports announcer says, "The ball is on the 27-yard line," most football fans realize that the announcer estimates that the ball is $\frac{2}{5}$ of the way from the 25-yard line to the 30-yard line. This process of literally "reading between the lines" is called interpolation. Another example is, "Interpolate to approximate the value of $\sqrt{8}$." Knowing $\sqrt{4} = 2$ and $\sqrt{9} = 3$, we conclude that $\sqrt{8}$ is a number between 2 and 3.† Moreover, 8 is $\frac{4}{5}$ of the way from 4 to 9. Assume that for a small increase in a number N, the change in $\sqrt{N}$ is proportional to the change in N.‡ Then $\sqrt{8}$ would lie $\frac{4}{5}$ of the way from 2 to 3. Since $\frac{4}{5}$ of 1 is 0.8, we conclude that $\sqrt{8}$ is approximately 2.8. This result is correct to only one decimal place.

---

* The student should guard against writing $\cot \theta = 1.288 = 37° 50'$. The second equality sign is incorrectly used because 1.288 does *not* equal $37° 50'$; and $\cot \theta$ does *not* equal $37° 50'$.

† The square root of a number increases when the number increases.

‡ Not strictly true, but a good approximation if the change is small.

The process of interpolation is important in all work involving the use of tables.

We already know that the trigonometric functions do not change uniformly with the change in the angle (if an angle is doubled, its sine does not double). But if the angle is changed by only a few minutes, the change in the function is very nearly proportional to the change in the angle.

EXAMPLE 1   Find sin 56° 14′.

Solution   Here we must interpolate between 56° 10′ and 56° 20′.

$$
\begin{aligned}
\sin 56° \ 10′ &= 0.8307 \\
\sin 56° \ 14′ &= \\
\sin 56° \ 20′ &= 0.8323
\end{aligned}
$$

As the angle increases 10′ (from 56° 10′ to 56° 20′), its sine increases 16 ten-thousandths. Our angle is $\frac{4}{10}$ of the way from 56° 10′ to 56° 20′. Hence the sine of our angle is $\frac{4}{10}$ of the way from 0.8307 to 0.8323. But $\frac{4}{10}(16) = 6\frac{2}{5} \rightarrow 6$. (Round off to 6 because $6\frac{2}{5}$ is closer to 6 than it is to 7.) Since the sine is increasing, *add* the 6 to 0.8307 to get

$$\sin 56° \ 14′ = 0.8313$$

EXAMPLE 2   Find cos 31° 17′.

Solution
$$
\begin{aligned}
\cos 31° \ 10′ &= 0.8557 \\
\cos 31° \ 17′ &= \\
\cos 31° \ 20′ &= 0.8542
\end{aligned}
$$

An *increase* of 10′ in the angle causes a *decrease* of 15 in the cosine. Our angle is $\frac{7}{10}$ of the way from 31° 10′ to 31° 20′. Hence we want $\frac{7}{10}$ of the decrease of 15. But $\frac{7}{10}(15) = 10\frac{1}{2} \rightarrow 11.*$ *Subtracting* this number from 0.8557 gives

$$\cos 31° \ 17′ = 0.8546$$

---

* In "rounding off" a number that is *exactly halfway,* it is conventional to choose the number that makes the *final result even* rather than odd. In this case the final result is written as 0.8546 rather than 0.8547. This procedure will be followed throughout this book.

EXAMPLE 3   Find θ if $\tan \theta = 0.4934$.

Solution

$$\left.\begin{array}{l} \tan 26°\ 10' \\ \\ \tan\quad \theta \\ \\ \tan 26°\ 20' \end{array}\right)_{10'} \begin{array}{l} = 0.4913 \\ \\ = 0.4934 \\ \\ = 0.4950 \end{array} \left.\vphantom{\begin{array}{l}a\\b\\c\end{array}}\right)_{37}^{21}$$

Our number 0.4934 is $\frac{21}{37}$ of the way from 0.4913 to 0.4950. Hence θ should be $\frac{21}{37}$ of the way from 26° 10′ to 26° 20′. But $\frac{21}{37}(10') = 5\frac{25}{37}' \to 6'$. Hence

$$\theta = 26°\ 16'$$

EXAMPLE 4   Find θ if $\cos \theta = 0.2581$.

Solution

$$\left.\begin{array}{l} \cos 75°\ 0' \\ \\ \cos\quad \theta \\ \\ \cos 75°\ 10' \end{array}\right)_{10'} \begin{array}{l} = 0.2588 \\ \\ = 0.2581 \\ \\ = 0.2560 \end{array} \left.\vphantom{\begin{array}{l}a\\b\\c\end{array}}\right)_{28}^{7}$$

Our number is $\frac{7}{28}$ of "the way down." Hence θ is $\frac{7}{28}(10') = \frac{10'}{4} = 2\frac{1}{2}' \to 2^*$ away from 75° 0′. Therefore

$$\theta = 75°\ 2'$$

It is to be noted that, in this three-line method of interpolation, *the small angle is always written on top. All differences are measured from the small angle and its function.*

EXERCISE 8

Use a four-place table (Table 1). Interpolate to find the value of each of the following.

1	sin 58° 1′	2	cos 37° 51′
3	tan 66° 14′	4	cot 31° 27′
5	tan 20° 36′	6	cot 48° 6′
7	sin 15° 17′	8	cos 79° 22′
9	cot 16° 45′	10	sin 60° 9′
11	cos 9° 57′	12	tan 70° 13′
13	cos 83° 58′	14	tan 28° 38′
15	cot 51° 46′	16	sin 4° 24′

* Number is exactly halfway. Make result even.

Interpolate to find θ.

17	$\cos\theta = 0.9768$		18	$\tan\theta = 2.000$
19	$\cot\theta = 2.599$		20	$\sin\theta = 0.8256$
21	$\sin\theta = 0.2468$		22	$\cos\theta = 0.0620$
23	$\tan\theta = 0.1152$		24	$\cot\theta = 0.8765$
25	$\tan\theta = 3.116$		26	$\cot\theta = 1.248$
27	$\sin\theta = 0.9072$		28	$\cos\theta = 0.9887$
29	$\cot\theta = 0.4466$		30	$\sin\theta = 0.6000$
31	$\cos\theta = 0.2472$		32	$\tan\theta = 0.8772$

18   APPROXIMATIONS AND SIGNIFICANT FIGURES

If a given distance is *measured* and if its length is expressed in decimal form, it is conventional to write no more digits than are correct (or probably correct). Thus, if we say that the measured distance between points A and B is 17 ft., we mean that the result is given to the nearest foot; i.e., the true distance is closer to 17 ft. than it is to 16 ft. or 18 ft. This is an example of two-figure accuracy. If we say that the measured distance AB is 17.0 ft., we mean that the true distance is given to three significant figures; i.e., it is closer to 17.0 than it is to 16.9 or 17.1. This implies that the true distance is somewhere between 16.95 and 17.05. Notice that 17 and 17.0 do not mean the same thing when they represent approximate values.

The number of significant digits in a number is obtained by counting the digits from left to right, beginning with the first nonzero digit and ending with the rightmost digit. * Thus, 0.078060 has five significant digits, 70.00 has four, and 0.790 has only three. Notice that the number of significant digits does not depend on the position of the decimal point.

Results computed by multiplication or division from approximate data will usually have no higher degree of accuracy than that of the data used. *We agree to round off the result so that it will have as many sig-*

---

* Ambiguity may result if the number in question is an integer ending in one or more 0's. For example, if the radius of the earth is given as 4000 miles, we may not know how many 0's are significant. If, however, the number 4000 was obtained from 3960 by rounding it off to the nearest multiple of 100 miles, then the first 0 is significant; the other two are not. In a case of this kind we sometimes use **scientific notation.** In this notation the number is expressed as a product. The first factor is the number formed by the significant digits with a decimal point placed after the first digit; the second factor is an integral power of 10. Thus, $4.0(10^3)$ indicates the number 4000, in which only the first 0 is significant. Also, $0.0123 = 1.23(10^{-2})$.

nificant figures as there are in the least accurate number in the data. If a field is measured and found to be 11.3 rods long and 10.7 rods wide, we would be tempted to say that its area is $(11.3)(10.7) = 120.91$ square rods. To do so would be to claim false accuracy. The result should be rounded off to three significant figures (the same as in the given data) to obtain 121 square rods. The first two figures in this result are correct, but the third is only a good approximation because the true area is somewhere between $(11.25)(10.65) = 119.8125$ square rods and $(11.35)(10.75) = 122.0125$ square rods.

Since nearly all the angles listed in Table 1 have trigonometric functions that are nonending decimals, the numbers appearing in the body of this table are merely four-figure approximations. Hence most of the results obtained by the use of this table will be approximations and should be considered as such.

Recall that, in rounding off a number that is exactly halfway, it is conventional to choose the number that makes the final result even rather than odd.

In solving triangles, we agree to set up the following correspondence between accuracy in sides and angles.

ACCURACY IN SIDES	ACCURACY IN ANGLES
2-figure	Nearest degree
3-figure	Nearest multiple of 10 minutes
4-figure	Nearest minute
5-figure	Nearest tenth of a minute

Hence within each of the following sets of data the same degree of accuracy prevails:

23, 42, 62°
0.461, 61° 10′, 44° 50′, 74° 0′
8.624, 82° 33′, 64° 00′
0.078247, 23° 19.7′, 48° 0.0′

If the data include a side with two-figure accuracy and another side with three-figure accuracy, then the computed parts should be written with only two-figure accuracy, which means that computed angles should be taken to the nearest degree. In general, *our results can be no more accurate than the least accurate item of the data.* If the given data include a number whose degree of accuracy is doubtful, we shall (in this book) *assume the maximum degree of accuracy.* For example, with no information to the contrary, the side 700 ft. will be treated as a three-figure number; the angle 46° 20′ will be considered as having four-figure accuracy.

EXERCISE 9

Round off the following numbers and angles to (a) four-figure accuracy, (b) three-figure accuracy, and (c) two-figure accuracy.

1	37.419	2	0.028065
3	0.53715	4	9.2613
5	72° 48.5′	6	16° 26.9′
7	85° 18.3′	8	66° 53.5′

The number 77.4 is the best three-figure approximation for all numbers from 77.35 to 77.45 *inclusive.** *What range of numbers is covered by each of the following approximations?*

9	61.3	10	8.40
11	926	12	0.0579
13	0.4828	14	3691
15	73.15	16	902.4

19   THE SOLUTION OF RIGHT TRIANGLES   To solve a triangle means to find from the given parts the values of the remaining parts. A right triangle is determined by

(*a*) Two of its sides, or
(*b*) One side and an acute angle.

In either case it is possible to find the remaining parts by using the special definitions in Section 10 together with the fact that the acute angles of a right triangle are complementary. For convenience we list here again these special definitions.

---

* See footnote on p. 26.

For any acute angle θ lying in a right triangle:

$$\sin \theta = \frac{\text{opp.}}{\text{hyp.}}$$

$$\cos \theta = \frac{\text{adj.}}{\text{hyp.}}$$

$$\tan \theta = \frac{\text{opp.}}{\text{adj.}}$$

For any triangle we shall use the small letters a, b, and c to denote the lengths of the sides that are opposite the angles A, B, and C, respectively. In a right triangle we shall always reserve the letter c for the hypotenuse.

EXAMPLE 1   Solve the right triangle having an acute angle of 38° 50′, the side adjacent to this angle being 311.

Solution   We first draw the triangle to scale and label numerically the parts that are known (Figure 20). Then

(1) $B = 90° - 38° 50′$
            $= 51° 10′$

(2) To find a, we observe that the given side and the required side are related to the given angle by the equation

$$\tan 38° 50′ = \frac{a}{311}$$

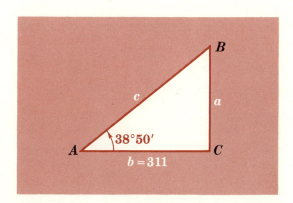

FIGURE 20

Multiply both sides of the equation by 311:

$$311 \tan 38° 50' = a$$

Hence $\qquad\qquad a = 311(0.8050) \doteq 250$

(3) To find c, we notice that the given parts, 38° 50' and 311, are related to the required part through the cosine of the angle:

$$\cos 38° 50' = \frac{311}{c}$$

Multiply both sides by c:

$$c \cos 38° 50' = 311$$

Divide both sides by $\cos 38° 50'$:

$$c = \frac{311}{\cos 38° 50'} = \frac{311}{0.7790} \doteq 399$$

This problem illustrates three-figure accuracy in the data and the computed results.

EXAMPLE 2   Solve the right triangle whose hypotenuse is 20.00 and one of whose legs is 16.40.

Solution   Draw the triangle (Figure 21) and label numerically the given parts.

(1) Since the hypotenuse and the side opposite A are given,

$$\sin A = \frac{16.40}{20.00} = 0.8200$$

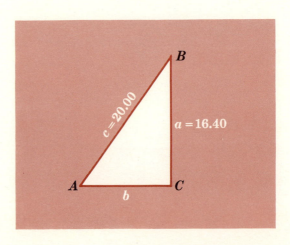

FIGURE 21

Four-place accuracy in the given sides means that the angles should be found to the nearest minute. Interpolating, we find

$$A = 55° 5'$$

(2) $B = 90° - 55° 5'$
        $= 34° 55'$

(3) To find b, use

$$\cos A = \frac{b}{20.00}$$

Multiply both sides by 20.00:

$$20.00 \cos A = b$$

Hence $b = (20.00) \cos 55° 5' = (20.00)(0.5724) = 11.45$

The value of b could have been found by using

$$\tan A = \frac{16.40}{b} \qquad b = \frac{16.40}{\tan 55° 5'} = \frac{16.40}{1.432}$$

$$= 11.45$$

This serves as a partial check on the solution.
It may be easier to use

$$b = \frac{16.40}{\tan 55° 5'} = 16.40 \cot 55° 5' = (16.40)(0.6980)$$

$$= 11.4472 \rightarrow 11.45$$

This problem illustrates four-figure accuracy in the data and the results.

In solving a right triangle by use of trigonometric functions it is desirable to find as many as possible of the required parts directly from the given parts. Why?

EXERCISE 10

Use a four-place table (Table 1) to solve the following right triangles.

1   $A = 31° 20'$, $a = 208$
2   $B = 24° 50'$, $b = 3.78$
3   $B = 70°$, $a = 2.6$
4   $A = 7°$, $b = 88$
5   $B = 45° 51'$, $c = 8008$
6   $A = 72°$, $c = 99$

7   $a = 0.4280, b = 0.6000$
8   $a = 2.000, b = 1.822$
9   $b = 3.07, c = 7.70$
10 $a = 1975, c = 5005$
11 $a = 33.1, c = 300$
12 $b = 365, c = 500$
13 $a = 80, b = 48$
14 $a = 0.393, b = 0.500$
15 $A = 29° 50', c = 515$
16 $B = 36° 10', c = 0.700$
17 $A = 67° 40', b = 0.760$
18 $b = 19, c = 100$
19 $b = 2730, c = 3000$
20 $B = 15° 40', b = 2430$

20   ANGLES OF ELEVATION AND DEPRESSION; BEARING OF A

LINE   The *angle of* $\begin{Bmatrix} elevation \\ depression \end{Bmatrix}$ of a point P as seen by an observer O is the vertical angle measured from the horizontal line through O $\begin{Bmatrix} upward \\ downward \end{Bmatrix}$ to the line of sight OP (see Figure 22).

The *bearing* of a line in a horizontal plane is the *acute* angle made by this line with a north-south line.   In giving the bearing of a line, write first the letter N or S, then the angle of deviation from north or south, then the letter E or W.   Thus, in Figure 23 the bearing of line OA is N 70° E; or the bearing of point A from point O is N 70° E.

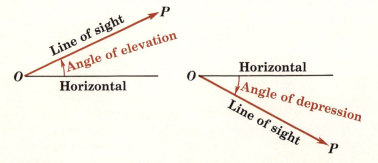

FIGURE 22

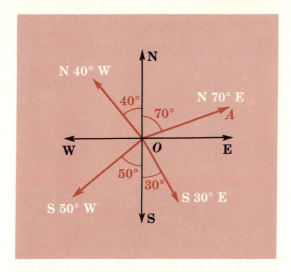

FIGURE 23

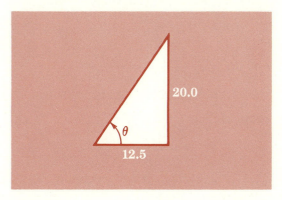

FIGURE 24

EXAMPLE 1   A vertical stake 20.0 in. high casts a horizontal shadow 12.5 in. long. What time is it if the sun rose at 6:00 A.M. and will be directly overhead at noon?

Solution   The angle of elevation of the sun (Figure 24) is found by

$$\tan \theta = \frac{20.0}{12.5} = 1.600$$

$$\theta = 58° \ 0'$$

It takes the earth 6 hr. to rotate through 90°. Since this rotation is uniform, each degree of elevation of the sun will correspond to $\frac{6}{90}$ of an hour or 4 min. Consequently a rotation through 58° 0' will require (58)(4 min.) = 232 min. = 3 hr. and 52 min. Hence the time is 9:52 A.M.

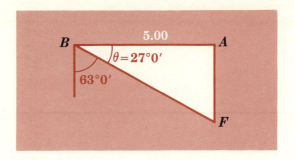

FIGURE 25

EXAMPLE 2   From a lookout tower A a column of smoke is sighted due south. From a second tower B, 5.00 miles west of A, the smoke is observed in the direction S 63° 0′ E. How far is the fire from B? From A? (See Figure 25.)

Solution   Angle $FBA = \theta = 90° - 63° 0′ = 27° 0′$. To get BF, use

$$\cos 27° 0′ = \frac{5.00}{BF}$$

Hence $BF = \dfrac{5.00}{\cos 27° 0′} = \dfrac{5.00}{0.8910}$

$= 5.61$ miles

To obtain AF, use

$$\tan 27° 0′ = \frac{AF}{5.00}$$

Hence $AF = 5.00 \tan 27° 0′ = (5.00)(0.5095)$
$= 2.55$ miles

EXERCISE 11

1   The angle of elevation of the top of Gateway Arch, St. Louis, from a point on the ground 208 ft. from its base is 72° 0′. Find the height of the arch.

2   Find the height of the Sears Tower in Chicago if the angle of elevation of its top from a point on the ground 1177 ft. from its base is 51° 0′.

3   A 20-ft. ladder rests against a vertical wall with its foot on level ground and 11 ft. from the base of the wall. What is the angle of elevation of the ladder?

4   Find the height of a tree if it casts a shadow of 137 ft. when the angle of elevation of the sun is 25°.

5   The Royal Gorge Bridge crosses the Arkansas River near Canon City, Colo. From a point 900.0 ft. from, and in the same horizontal plane

with, the bridge, the angle of depression of the river flowing beneath the bridge is 49° 29'. What is the distance from the bridge down to the river?

6   A lovesick swain looks out of his office window, 40 ft. above the ground, and sees, with an angle of depression of 35°, his girlfriend standing on the sidewalk, talking to another suitor. How far apart are the lovers?

7   A balloon hovers 718 ft. above one end of a bridge that spans the Mississippi River at New Orleans. The angle of depression of the other end of the bridge from the balloon is 24° 30'. How long is the bridge?

8   From the top of a building on level ground, the angle of depression of the bottom of a nearby television tower is 23° 30'. From the bottom of the building, the angle of elevation of the top of the tower is 64° 40'. If the height of the building is 200 ft., find the height of the tower.

9   Akron, Ohio, is 190 miles due north of Charleston, W.Va. Stamford, Conn., is due east of Akron and is 460 miles from Charleston. What is the bearing of Stamford from Charleston? Charleston from Stamford? *

10 Evanston, Ill., is 300 miles due east of Ames, Iowa. Sheboygan, Wis., is 120 miles due north of Evanston. What is the bearing of Sheboygan from Ames? Ames from Sheboygan? *

11 Chapel Hill, N.C., is due south of Staunton, Va. Frankfort, Ky., is due west of Staunton and is 350 miles N 63° W from Chapel Hill. How far is Staunton from Frankfort? From Chapel Hill? *

12 Philadelphia is 420 miles due east of Columbus, Ohio. Detroit is due north of Columbus and is N 68° W from Philadelphia. How far is Detroit from Philadelphia? From Columbus? *

13 An airplane moves in a straight line that makes an angle of 9°10' with the horizontal. How many feet does the plane rise while traveling 800 ft. in the air?

14 At 3:00 P.M. a lady on a motorboat traveling north at 20.0 mph noticed a television tower with bearing N 65° E. At 3:06 P.M. the tower was due east of the boat. How far was the tower from the lady when she first noticed it?

15 In order to determine the width CA of a river, a surveyor laid off a distance of 100 ft. on the bank CB at right angles to CA. Using a transit, he found angle ABC to be 70° 50'. How wide is the river?

16 Pittsburg, Kans., is 117 miles due south of Kansas City, Kans. A plane leaves Kansas City at noon and flies 200 mph in the direction S 21° 30' W. When (to the nearest minute) will the plane be due west of Pittsburg?

17 A plane leaves an airport at 9:00 A.M. and flies 200 mph southwest. A second plane leaves the airport at 9:30 A.M. and flies 320 mph north-

*Ignore the curvature of the earth and assume only two-place accuracy. Get angles to the nearest degree. Get distances to the nearest multiple of 10 miles.

west. Find the bearing of the second plane from the first one at 10:00 A.M.

18 Pottstown, Pa., is 47 miles due west of Trenton, N.J. A helicopter leaves Trenton at noon and travels 50 mph in the direction N 72° W. When (to the nearest minute) will it be closest to Pottstown?

19 A flagpole 32.0 ft. high casts a horizontal shadow 25.0 ft. long. What time is it if the sun rose at 6:10 A.M. and will be directly overhead at 12:10 P.M.?

20 A vertical stake 30.3 in. high casts a horizontal shadow 75.0 in. long. What time is it if the sun was directly overhead at 11:50 A.M. and will set at 5:50 P.M.?

21 Find the radius of a circle in which a chord of 20.0 in. subtends a central angle of 38° 40′.

22 The sides of an isosceles triangle are 8.00, 8.00, and 6.00. Find the angles of the triangle.

23 Find the perimeter of a regular polygon of 60 sides inscribed in a circle of radius 1.000 ft. Compare this number with the circumference of the circle.

24 Show that each of the acute angles formed by the intersection of two diagonals of a cube is 70° 32′.

25 In order to find the north-south width AB of a swamp, a surveyor measures 600 ft. from B in the direction N 37° 0′ E to a point C. He then notices that the bearing of CA is N 53° 0′ W. Find the distance AB.

26 From the top of a lighthouse a ft. above sea level at high tide, the angle of depression of a buoy is θ at high tide and ϕ at low tide. Show that the height of the tide is $a(\cot \theta \tan \phi - 1)$.

27 From a window in a building across the street from a skyscraper, the angle of elevation of the top of the skyscraper is θ and the angle of depression of the bottom is ϕ. If the window is a ft. above the street level, express the height of the skyscraper in terms of a, θ, and ϕ.

28 Let ABC be any triangle with acute angles A and B and included side c. If CD is the perpendicular from C to AB, show that $CD = \dfrac{c}{\cot A + \cot B}$.

Hint: Show that $AD = CD \cot A$ and $DB = CD \cot B$. Then add these equations.

29 From an observation tower 70.0 ft. above the level of a lake, the angle of depression of a point on the near shore is 48° 0′ and that of a point directly beyond on the far shore is 4° 0′. Find the width of the lake.

30 From point A, the angle of elevation of a mountain peak is 27° 0′. From point B, in the same horizontal plane with A and 1320 ft. closer to the base of the mountain, the angle of elevation of the peak is 34° 0′. Find the height of the peak above the level of A and B. *Hint:* See the figure on page 39. Show that $1320 + x = y \cot 27° 0′$, $x = y \cot 34° 0′$. Eliminate x by subtracting one equation from the other.

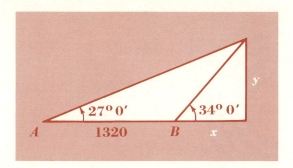

31 An observer at A looks due north and sees a meteor with an angle of elevation of 60°. At the same instant, another observer, 20 miles east of A, sees the same meteor and approximates its position as N 40° W but fails to note its angle of elevation. Find the height of the meteor and its distance from A.

32 A ladder 30.0 ft. long stands on level ground and leans against a vertical wall. The angle of elevation of the ladder is 58° 0′.

(a) Find the distance from the foot of the ladder to the wall.

(b) How far is the top of the ladder above the ground?

(c) If the foot of the ladder is moved 4.0 ft. closer to the wall, how much does the top of the ladder rise?

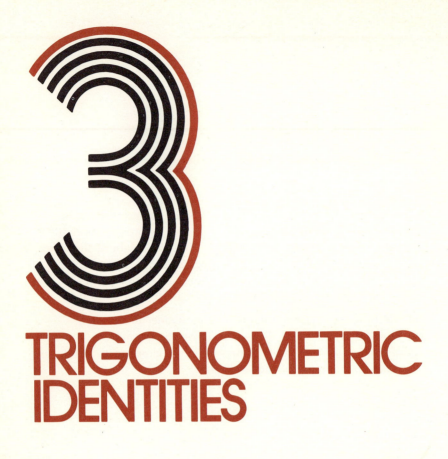

3
TRIGONOMETRIC IDENTITIES

21 THE FUNDAMENTAL RELATIONS In Section 9 we discussed the problem of determining all the trigonometric functions of an angle if one of them is given. This was a geometric process involving the construction of a triangle of reference for the angle. We shall now consider purely analytic relations among the functions themselves. These relations are of considerable importance in other branches of mathematics as well as in engineering and physics. Very often, the key to the solution of some important problem will be your ability to replace a given mathematical expression with an equivalent one that is more readily usable. Many such problems will involve the trigonometric functions. Consequently, the basic transformations which you will learn to use in this chapter are among the most important things that you should carry with you from this course into more advanced work in mathematics and science.

For any* angle θ, the following eight fundamental relations are true:

$$\csc\ \theta = \frac{1}{\sin \theta} \tag{1}$$

$$\sec\ \theta = \frac{1}{\cos \theta} \tag{2}$$

$$\cot\ \theta = \frac{1}{\tan \theta} \tag{3}$$

$$\tan\ \theta = \frac{\sin \theta}{\cos \theta} \tag{4}$$

$$\cot\ \theta = \frac{\cos \theta}{\sin \theta} \tag{5}$$

$$\sin^2 \theta + \cos^2 \theta = 1 \tag{6}$$

$$1 + \tan^2 \theta = \sec^2 \theta \tag{7}$$

$$1 + \cot^2 \theta = \csc^2 \theta \tag{8}$$

These relations are invaluable for many considerations that follow in this book and should be memorized immediately. The first three relations have already been discussed (Section 8) and used. We shall prove only (1):

$$\frac{1}{\sin \theta} = \frac{1}{y/r} = \frac{r}{y} = \csc \theta$$

The proofs of (4) and (5) are similar. To prove (4):

$$\frac{\sin \theta}{\cos \theta} = \frac{y/r}{x/r} = \frac{y}{x} = \tan \theta$$

---

* Strictly speaking, for every angle for which the functions actually exist. For example, (4) has no meaning when $\theta = 90°$ because $\cos 90° = 0$; hence $\tan 90° = \dfrac{\sin 90°}{\cos 90°} = \dfrac{1}{0}$, which does not exist. Likewise (8) has no meaning for $\theta = 180°$ because $\cot 180°$ and $\csc 180°$ do not exist (see Section 8). Exceptions like these *can* occur only for $\theta = 0°, 90°, 180°, 270°$, and angles coterminal with them. Notice, however, that (6) does hold for $\theta = 180°$: $\sin^2 180° + \cos^2 180° = 0^2 + (-1)^2 = 1$.

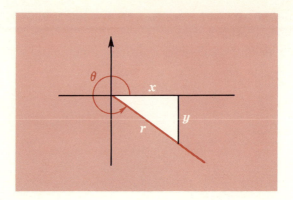

FIGURE 26

In order to prove (6), (7), (8), we recall that for any angle θ in standard position, the coordinates of point P on its terminal side (see Figure 26) are related by the equation

$$x^2 + y^2 = r^2 \tag{9}$$

Dividing both sides by r^2, we get

$$\frac{x^2}{r^2} + \frac{y^2}{r^2} = \frac{r^2}{r^2}$$

$$\left(\frac{x}{r}\right)^2 + \left(\frac{y}{r}\right)^2 = 1$$

$$\cos^2 \theta + \sin^2 \theta = 1$$

Dividing Equation (9) in turn by x^2 and y^2, we get (7) and (8), respectively.

The student should be able to recognize these eight fundamental relations in other forms.   For example,

(1) may be reduced to $\sin \theta \csc \theta = 1$

(2) may be reduced to $\cos \theta = \dfrac{1}{\sec \theta}$

(6) may be reduced to $\sin \theta = \pm \sqrt{1 - \cos^2 \theta}$

(6) may be reduced to $\cos \theta = \pm \sqrt{1 - \sin^2 \theta}$

(7) may be reduced to $\sec \theta = \pm \sqrt{1 + \tan^2 \theta}$*

---

* The student should realize from his study of algebra that $\sqrt{1 + \tan^2 \theta}$ is *not* equal to $1 + \tan \theta$.

The sign that should be chosen in the last three equations is determined by the quadrant in which θ lies. In Q I and Q IV, $\cos \theta = \sqrt{1 - \sin^2 \theta}$; but in Q II and Q III, $\cos \theta = -\sqrt{1 - \sin^2 \theta}$.

Again the student is warned against the careless habit of writing *sin* instead of *sin* θ. Equation (6) says that the square of the sine of any angle plus the square of the cosine of *that same angle* is equal to 1. It could just as well have been written

$$\sin^2 A + \cos^2 A = 1$$
or
$$\sin^2 7B + \cos^2 7B = 1$$

EXAMPLE 1   Prove or disprove:

$$\sin^4 5A + 2 \sin^2 5A \cos^2 5A + \cos^4 5A = 1$$

Solution   Since the given equation involves only sines and cosines, we shall attempt to derive it from (6), which says that for all values of θ

$$\sin^2 \theta + \cos^2 \theta = 1$$

Square both sides of the equation:

$$(\sin^2 \theta + \cos^2 \theta)^2 = 1^2$$
$$\sin^4 \theta + 2 \sin^2 \theta \cos^2 \theta + \cos^4 \theta = 1$$

Now let $\theta = 5A$:

$$\sin^4 5A + 2 \sin^2 5A \cos^2 5A + \cos^4 5A = 1$$

This proves the statement is true for all values of A.

EXAMPLE 2   Prove or disprove:

$$\sin A + \cos A = 1$$

Solution   We can demonstrate that this equation is not generally true by setting A equal to some specific angle and then showing that the two sides of the equation are not numerically equal. Choosing $A = 30°$, we have

$$\sin 30° + \cos 30° = 1$$
$$0.5 + 0.866 = 1$$
$$1.366 = 1 \qquad \text{False}$$

This proves conclusively that the given equation is not true for all values of A. It is, however, true for some values of A; e.g., if $A = 90°$, we have

$$\sin 90° + \cos 90° = 1$$
$$1 + 0 = 1 \qquad \text{True}$$

EXAMPLE 3   Simplify $\sqrt{\csc^2 3B - 1}$.

Solution   Since $1 + \cot^2 3B = \csc^2 3B$, it follows that $\cot 3B = \pm\sqrt{\csc^2 3B - 1}$. This implies that $\sqrt{\csc^2 3B - 1} = \pm\cot 3B$. If $3B$ is an angle in Q I or Q III, then $\cot 3B > 0$. In this case $\sqrt{\csc^2 3B - 1} = \cot 3B$. But if $3B$ is in Q II or Q IV, then $\cot 3B < 0$. By definition, $\sqrt{\csc^2 3B - 1}$ means the *principal* square root of $\csc^2 3B - 1$. It is positive or zero but never negative. In this case $\sqrt{\csc^2 3B - 1} = -\cot 3B$. Hence

$$\sqrt{\csc^2 3B - 1} = \cot 3B \quad \text{if } 3B \text{ is in Q I or Q III}$$

and $\quad \sqrt{\csc^2 3B - 1} = -\cot 3B \quad \text{if } 3B \text{ is in Q II or Q IV}$

The student should note carefully that a general statement can be *disproved* by citing one instance in which it is not true. Such an instance is called a *counterexample*. But a general statement cannot be proved by merely showing that it is true for one special case. It must be proved for all cases.

EXERCISE 12

Use the eight fundamental relations* to write each of the following expressions as a single trigonometric function of some angle.

1 $\dfrac{\sin 8A}{\cos 8A}$

2 $\dfrac{\cos 320°}{\sin 320°}$

3 $\dfrac{1}{\sec 4A}$

4 $\dfrac{1}{\tan 200°}$

5 $\sqrt{\csc^2 A - 1}$

6 $-\sqrt{1 - \sin^2 140°}$

7 $\sqrt{1 - \cos^2 100°}$

8 $\sqrt{\sec^2 200° - 1}$

9 $-\sqrt{1 + \tan^2 260°}$

10 $\tan 48° \cos 48°$

Decide which of the following statements are valid consequences of the eight fundamental relations. If the statement is true, cite proof; if false, correct it.

11 $\tan^3 5\theta = \dfrac{\sin^3 5\theta}{(\cos 5\theta)^3}$

12 $\dfrac{\cos 6A}{\sin 6A} = \cot A$

13 $\dfrac{1}{\cos A} = \sec$

14 $(5 \csc \theta)^2 = \dfrac{25}{\sin^2 \theta}$

*Do not express in terms of x, y, r.

15 $\cot A = \dfrac{1}{\tan B}$ **16** $\sin A = \dfrac{1}{\cos A}$

17 $\sin A \csc B = 1$ **18** $\cos^3 \theta \sec^3 \theta = 1$

19 $(3 \sec \theta)^2 - (3 \tan \theta)^2 = 9$ **20** $\tan^4 B + 2 \tan^2 B + 1 = \sec^4 B$

21 $(3 \cot A)^4 = \dfrac{81 \cos^4 A}{\sin^4 A}$ **22** $\cot^2 \theta - \csc^2 \theta = -1$

23 $\sin^4 \theta + \cos^4 \theta = 1$ **24** $(7 \sin \theta)^2 + (7 \cos \theta)^2 = 49$

25 $\sin \theta = \sqrt{1 - \cos^2 \theta}$ holds for θ in Q I and Q II.

26 $\tan \theta = -\sqrt{\sec^2 \theta - 1}$ holds for θ in Q II and Q III.

27 $\csc \theta = \sqrt{\cot^2 \theta - 1}$ holds for θ in Q I and Q III.

28 $\cos \theta = \sqrt{1 - \sin^2 \theta}$ holds for θ in Q I and Q II.

29 $\sqrt{\cos^2 350°} = \cos 350°$ **30** $\sqrt{\cos^2 100°} = \cos 100°$

31 $\cot 60° = \dfrac{\cos 60°}{\sin 60°} = \dfrac{\frac{1}{2}}{\frac{1}{2}\sqrt{3}} = \dfrac{1}{\sqrt{3}} = \dfrac{1}{\sqrt{3}} \cdot \dfrac{\sqrt{3}}{\sqrt{3}} = \dfrac{\sqrt{3}}{3}$

32 $\sec 45° = \dfrac{1}{\cos 45°} = \dfrac{1}{\frac{1}{2}\sqrt{2}} = \dfrac{2}{\sqrt{2}} = \sqrt{2}$

22 ALGEBRAIC OPERATIONS WITH THE TRIGONOMETRIC FUNC-TIONS

The expression sin θ, meaning the sine of angle θ, is an abstract number. It is the ratio of two distances such as

$$\frac{2 \text{ ft.}}{3 \text{ ft.}} = \frac{2}{3} \qquad \text{or} \qquad \frac{2 \text{ in.}}{3 \text{ in.}} = \frac{2}{3}$$

For this reason it can be treated in the same way that we deal with numbers and letters (representing numbers) in algebra. For example, $\sin^3 \theta + \cos^3 \theta$ may be expressed as the sum of two cubes.* Since

$$a^3 + b^3 = (a + b)(a^2 - ab + b^2),$$
$$\sin^3 \theta + \cos^3 \theta = (\sin \theta + \cos \theta)(\sin^2 \theta - \sin \theta \cos \theta + \cos^2 \theta)$$
$$= (\sin \theta + \cos \theta)(1 - \sin \theta \cos \theta)$$

Also $(\sec \theta + \tan \theta)^2$ may be expanded as the square of a binomial to equal "the square of the first plus twice the product plus the square of the last":

$$(\sec \theta + \tan \theta)^2 = \sec^2 \theta + 2 \sec \theta \tan \theta + \tan^2 \theta$$

A glance at the fundamental relations reveals that the functions occurring most often are sin θ and cos θ. Equations (1), (2), (4), (5)

---

* Recall that $\sin^3 \theta$ is a short way of writing $(\sin \theta)^3$.

express each of the other functions directly in terms of sin θ and cos θ. Also, sin θ and cos θ are the only trigonometric functions that are defined for all θ. For these reasons it is advantageous in many problems to reduce an expression to sines and cosines.

EXAMPLE 1   Express $\dfrac{3 \csc \theta}{5 \csc \theta - 6 \cot^2 \theta}$ in terms of sin θ and cos θ.

Solution

$$\dfrac{3 \csc \theta}{5 \csc \theta - 6 \cot^2 \theta}$$

$$= \dfrac{3\left(\dfrac{1}{\sin \theta}\right)}{5\left(\dfrac{1}{\sin \theta}\right) - 6\left(\dfrac{\cos^2 \theta}{\sin^2 \theta}\right)}$$

$$= \dfrac{\dfrac{3}{\sin \theta}}{\dfrac{5 \sin \theta - 6 \cos^2 \theta}{\sin^2 \theta}}$$
Getting a common denominator for the bottom

$$= \dfrac{3}{\sin \theta} \cdot \dfrac{\sin^2 \theta}{5 \sin \theta - 6 \cos^2 \theta}$$
Inverting the denominator and multiplying

$$= \dfrac{3 \sin \theta}{5 \sin \theta - 6 \cos^2 \theta}$$
Simplifying the fraction

In order to express this quantity in terms of just sin θ, replace $\cos^2 \theta$ with $(1 - \sin^2 \theta)$ to obtain $\dfrac{3 \sin \theta}{5 \sin \theta - 6 + 6 \sin^2 \theta}$.

EXAMPLE 2   Express each of the other trigonometric functions of θ in terms of sin θ.

Solution 1

$$\cos \theta = \pm \sqrt{1 - \sin^2 \theta} \qquad\qquad \text{using (6)}$$

$$\tan \theta = \frac{\sin \theta}{\cos \theta} = \frac{\sin \theta}{\pm \sqrt{1 - \sin^2 \theta}} \qquad \text{using (4), (6)}$$

$$\cot \theta = \frac{\cos \theta}{\sin \theta} = \frac{\pm \sqrt{1 - \sin^2 \theta}}{\sin \theta} \qquad \text{using (5), (6)}$$

$$\sec \theta = \frac{1}{\cos \theta} = \frac{\pm 1}{\sqrt{1 - \sin^2 \theta}} \qquad \text{using (2), (6)}$$

$$\csc \theta = \frac{1}{\sin \theta} \qquad\qquad\qquad\qquad \text{using (1)}$$

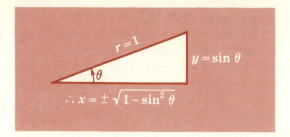

FIGURE 27

Solution 2   Place θ in standard position, as in Figure 27.   In order to make $y/r = \sin\ \theta$, let $y = \sin\ \theta$ and $r = 1$.   The Pythagorean theorem gives $x = \pm\sqrt{1 - \sin^2\theta}$.   Then

$$\cos\theta = \frac{x}{r} = \pm\sqrt{1 - \sin^2\theta}$$

$$\tan\theta = \frac{y}{x} = \frac{\sin\theta}{\pm\sqrt{1 - \sin^2\theta}} \qquad \text{etc.}$$

EXERCISE 13

Simplify each of the following.

1 $\dfrac{\sin^3\theta - \cos^3\theta}{\sin\theta - \cos\theta}$

2 $\dfrac{\sec^4\theta - \tan^4\theta}{\sec^2\theta + \tan^2\theta}$

3 $\dfrac{3\sin\theta + \dfrac{1}{\sin\theta}}{\dfrac{1}{2\sin\theta} + \dfrac{1}{6\sin^3\theta}}$

4 $\dfrac{1 + 3\cos\theta + 2\cos^2\theta}{1 + 7\cos\theta + 10\cos^2\theta}$

Reduce each of the following to an expression that involves no function except $\sin\theta$ and $\cos\theta$. Simplify.

5 $\dfrac{\tan\theta + \cot\theta}{\sec\theta\csc\theta}$

6 $\dfrac{\sec\theta}{\cot\theta + \tan\theta}$

7 $\dfrac{\sin\theta - \csc\theta}{\cot\theta} + \dfrac{\cot\theta}{\csc\theta}$

8 $\dfrac{2 + \cot^2\theta}{\csc^2\theta} - 1$

9   Express cot 147° in terms of csc 147°.
10 Express csc 246° in terms of sin 246°.
11 Express tan 278° in terms of sec 278°.
12 Express sin 333° in terms of cos 333°.
13 Express sec 3θ in terms of cos 3θ.

14 Express $\cos 5A$ in terms of $\sin 5A$.

15 Express each of the other trigonometric functions in terms of $\tan \theta$. Why is your expression for $\sin \theta$ not valid when $\theta = 90°$?

16 Express each of the other trigonometric functions in terms of $\cos \theta$.

23 IDENTITIES AND CONDITIONAL EQUATIONS

An identity* is an equation that holds true for all permissible† values of the letters involved.

ILLUSTRATION 1 $x^2 - 9 = (x + 3)(x - 3)$ holds true for all values of x.

ILLUSTRATION 2 $x^2 + xy - 2y^2 = (x + 2y)(x - y)$ holds true for all values of x and y.

ILLUSTRATION 3 $x - \dfrac{x^2 - 7x}{x - 5} = \dfrac{2x}{x - 5}$ holds true for all permissible values of x, that is, for all values of x except $x = 5$.   When $x = 5$, each side of the equation involves a fraction whose denominator is zero.   Such fractions have no meaning, and we say their value does not exist.

ILLUSTRATION 4 $\sin^2 \theta + \cos^2 \theta = 1$ holds true for all values of θ.

ILLUSTRATION 5 $1 + \tan^2 \theta = \sec^2 \theta$ holds true for all permissible values of θ, that is, for all values of θ except $90°$, $270°$, and angles coterminal with them.

ILLUSTRATION 6   The following "trick with numbers" illustrates a simple identity.

Choose any number except 0.
Multiply your number by 5.
To this number add the square of your original number.
Multiply your result by 2.
Divide the number you now have by the original number.

---

* Also called an *identical equation*.
† The *permissible* values of the letters involved are all those values for which each side of the equation has meaning.

Subtract 10.

Divide by your original number.

If you have followed instructions, your result should be 2 regardless of your choice of the original number. To prove this, let x be the original number. Then the numbers that follow are $5x$, $5x + x^2$, $2(5x + x^2)$ or $x(10 + 2x)$, $10 + 2x$, $2x$, and 2. The identity used is

$$\frac{\dfrac{2(5x + x^2)}{x} - 10}{x} = 2$$

It holds for all values of x except 0. Try it for a fraction. For a negative number.

A conditional equation is an equation that *does not* hold true for all permissible values of the letters involved.

ILLUSTRATION 7 $2x - 7 = 3$ holds true for only one value of x, namely $x = 5$.

ILLUSTRATION 8 $x^2 - 8x + 15 = 0$ holds true for only two values of x, namely $x = 3$, and $x = 5$.

ILLUSTRATION 9 $x(x - 7)(x + 4) = 0$ holds true for only three values of x, namely $x = 0$, $x = 7$, and $x = -4$.

ILLUSTRATION 10 $\sin \theta = \cos \theta$ holds true for only two values of θ between $0°$ and $360°$. They are $\theta = 45°$ and $\theta = 225°$.

ILLUSTRATION 11 $\sin \theta = 5 + \cos \theta$ holds true for no value of θ.

The difference between an identity and a conditional equation can easily be seen from the contrasting definitions:

$$\left\{ \begin{array}{l} An\ identity \\ A\ conditional\ equation \end{array} \right\} \quad is\ an\ equation\ that \quad \left\{ \begin{array}{l} holds\ true \\ does\ not\ hold\ true \end{array} \right\}$$

for all permissible values of the letters involved.

An identity *says* that both sides of an equation are equal for all permissible values. The process by which we demonstrate that the two sides are identical is called "proving the identity." A conditional equation *asks,* "For what values of the unknowns is the left side of this equation equal to the right side?" The process by which these values are found is called "solving the equation."

24 TRIGONOMETRIC IDENTITIES   The eight fundamental relations are identities. By using them, we can prove other identities. For one who goes further in mathematics, or in subjects involving mathematics, it is highly important to gain a certain amount of experience in proving identities. For this reason we place considerable emphasis on the following examples and problems.

It is most desirable to prove an identity by *transforming one side* of the equation *to the other side, which should be left unaltered.** The side with which we work is usually the more complicated one. There is no set rule for making these transformations. The following suggestions will, however, indicate the first step in most cases.

1. If one side involves only one function of the angle, express the other side in terms of this function.
2. If one side is factorable, factor it.
3. If one side has only one term in its denominator (and several terms in its numerator), break up the fraction.
4. If one side contains one or more indicated operations (such as squaring an expression, adding fractions, or multiplying two expressions), begin by performing these operations. This is especially helpful if this side involves only sines and cosines.
5. When working with one side, keep an eye on the other side to see which transformation will most easily reduce it to the other side. It is frequently helpful to multiply the numerator and denominator of a fraction by the same expression. If possible, avoid introducing radicals.
6. When in doubt, express the more complicated side in terms of sines and cosines and then simplify.

At each step, look for some combination that can be replaced by a simpler expression.

---

* The instructor may wish to permit the student to reduce each side of the equation independently to a common third expression. This method is sometimes desirable when both sides are quite complicated.

The following examples illustrate the suggestions.

EXAMPLE 1   Prove the identity

$$3 \cos^4 \theta + 6 \sin^2 \theta = 3 + 3 \sin^4 \theta$$

Proof

$$3 \cos^4 \theta + 6 \sin^2 \theta \qquad\qquad\qquad 3 + 3 \sin^4 \theta$$
$$= 3(1 - \sin^2 \theta)^2 + 6 \sin^2 \theta$$
$$= 3 - 6 \sin^2 \theta + 3 \sin^4 \theta + 6 \sin^2 \theta$$
$$= 3 + 3 \sin^4 \theta$$

EXAMPLE 2   Prove the identity

$$\sec^2 \theta + \tan^2 \theta = \sec^4 \theta - \tan^4 \theta$$

Proof

$$\sec^2 \theta + \tan^2 \theta \qquad \sec^4 \theta - \tan^4 \theta$$
$$= (\sec^2 \theta + \tan^2 \theta)(\sec^2 \theta - \tan^2 \theta)$$
$$= (\sec^2 \theta + \tan^2 \theta) \cdot 1$$
$$= (\sec^2 \theta + \tan^2 \theta)$$

EXAMPLE 3   Prove the identity

$$\frac{\sin \theta + \cot \theta}{\cos \theta} = \tan \theta + \csc \theta$$

Proof

$$\frac{\sin \theta + \cot \theta}{\cos \theta} \qquad \tan \theta + \csc \theta$$

$$= \frac{\sin \theta}{\cos \theta} + \frac{\cot \theta}{\cos \theta}$$

$$= \tan \theta + \frac{\dfrac{\cos \theta}{\sin \theta}}{\cos \theta}$$

$$= \tan \theta + \frac{1}{\sin \theta}$$

$$= \tan \theta + \csc \theta$$

EXAMPLE 4   Prove the identity

$$\frac{\sin \theta}{1 + \cos \theta} + \frac{1 + \cos \theta}{\sin \theta} = 2 \csc \theta$$

Proof The left side indicates the addition of two fractions that involve only sines and cosines of θ. We begin by adding these fractions.

$$\frac{\sin\theta}{1+\cos\theta}+\frac{1+\cos\theta}{\sin\theta} \qquad\qquad 2\csc\theta$$

$$=\frac{\sin^2\theta+1+2\cos\theta+\cos^2\theta}{(1+\cos\theta)\sin\theta}$$

$$=\frac{2+2\cos\theta}{(1+\cos\theta)\sin\theta}$$

$$=\frac{2(1+\cos\theta)}{(1+\cos\theta)\sin\theta}$$

$$=\frac{2}{\sin\theta}$$

$$=2\csc\theta$$

EXAMPLE 5   Prove the identity $\dfrac{\cot\theta}{\csc\theta-1}=\dfrac{\csc\theta+1}{\cot\theta}$.

Proof   Since the numerator of the right side is $\csc\theta+1$, let us multiply top and bottom of the left side by $\csc\theta+1$ to get

$$\frac{\cot\theta\,(\csc\theta+1)}{(\csc\theta-1)(\csc\theta+1)}$$

$$=\frac{\cot\theta\,(\csc\theta+1)}{\csc^2\theta-1}$$

$$=\frac{\cot\theta\,(\csc\theta+1)}{\cot^2\theta}$$

$$=\frac{\csc\theta+1}{\cot\theta} \qquad\text{which is the right side}$$

EXAMPLE 6   Prove the identity

$$\frac{\cot A+\csc B}{\tan B+\tan A\sec B}=\cot A\cot B$$

Proof
$$\frac{\cot A+\csc B}{\tan B+\tan A\sec B} \qquad\qquad \cot A\cot B$$

$$=\frac{\dfrac{\cos A}{\sin A}+\dfrac{1}{\sin B}}{\dfrac{\sin B}{\cos B}+\dfrac{\sin A}{\cos A\cos B}}$$

$$= \frac{\dfrac{\sin B \cos A + \sin A}{\sin A \sin B}}{\dfrac{\sin B \cos A + \sin A}{\cos A \cos B}}$$

$$= \frac{\cos A \cos B}{\sin A \sin B}$$

$$= \cot A \cot B$$

The form illustrated in the preceding examples, in which a vertical line separates the two sides of the identity, is a convenient form to follow. It emphasizes the fact that, in trying to prove that the equation holds for all permissible values of the letters involved, we must *not* work with it as if it were an ordinary conditional equation.* The key point, however, is that each step in the transformation from the expression on one side of the identity to the expression on the other side must be clearly justified as an *identity transformation*.

EXERCISE 14

Prove each of the following identities by reducing one side to the other.

1   $\dfrac{2 - \dfrac{x-5}{x-4}}{1 - \dfrac{x-10}{x^2-16}} = \dfrac{x+4}{x+2}$

2   $\dfrac{\dfrac{x}{x+3} - \dfrac{4}{x+5}}{6 - \dfrac{5x+14}{x+3}} = \dfrac{x-3}{x+5}$

---

* For example, to prove that $\cos \theta = \sin \theta / \tan \theta$, *it would not be proper to multiply both sides by* $\tan \theta$ to get $\cos \theta \tan \theta = \sin \theta$ and then replace $\tan \theta$ with $\sin \theta / \cos \theta$ to get $(\cos \theta) \sin \theta / \cos \theta = \sin \theta$ or $\sin \theta = \sin \theta$. This does *not* prove that the given equation is an identity. It merely demonstrates that IF $\cos \theta = \sin \theta / \tan \theta$, THEN (it is necessary that) $\sin \theta = \sin \theta$. We could use this same improper procedure to prove that $3 = 5$ because multiplying both sides by 0 gives us $3(0) = 5(0)$, which is a true statement. Another incorrect procedure is to square both sides of an equation and then claim that the given equation is an identity because the squares of its two sides are identically equal. With this sort of reasoning, we could claim that $-4 = 4$ because $(-4)^2 = 4^2$. If $a^2 = b^2$, it does not necessarily follow that $a = b$.

3 $\dfrac{\dfrac{7}{5} - \dfrac{x-3}{x+1}}{\dfrac{2}{5} - \dfrac{x-1}{3x+3}} = 6$

4 $\dfrac{\dfrac{x^2+16x}{x^2-64} - 1}{3 - \dfrac{2(x+10)}{x+8}} = \dfrac{16}{x-8}$

5 $\sin\theta\cot\theta\sec\theta = 1$

6 $\sin^2 A + \tan^2 B + \cos^2 A = \sec^2 B$

7 $(1 + \cot^2 B)(1 - \cos^2 B) = 1$

8 $\dfrac{\cot\theta}{\csc\theta} = \cos\theta$

9 $\dfrac{\cos\theta}{\sec\theta} + \dfrac{\sin\theta}{\csc\theta} = 1$

10 $\dfrac{\sin A\csc B}{\cos A\sec B} = \tan A\tan B$

11 $\cos^2 A - \sin^2 A = 2\cos^2 A - 1$

12 $(\sin\theta - \cos\theta)^2 + (\sin\theta + \cos\theta)^2 = 2$

13 $\sin^2\theta + \sin^2\theta\cos^2\theta + \cos^4\theta = 1$

14 $\cot^4\theta = \csc^4\theta - 2\csc^2\theta + 1$

15 $\tan^4\theta + \tan^2\theta = \sec^4\theta - \sec^2\theta$

16 $\sec^4\theta + \tan^2\theta = \tan^4\theta + 3\tan^2\theta + 1$

17 $\dfrac{1 - \tan^2\theta}{\cot^2\theta - 1} = \tan^2\theta$

18 $\dfrac{\sin\theta}{1 + \sin\theta + \sin^2\theta} = \dfrac{1}{\csc\theta + 1 + \sin\theta}$

19 $\dfrac{1 + \sec B}{\cos B + 1} = \sec B$

20 $\dfrac{\cos\theta + 7\sec\theta}{\cos\theta + 2\sec\theta} = \dfrac{\cos^2\theta + 7}{\cos^2\theta + 2}$

21 $\dfrac{4\sin\theta + 5\cos\theta\tan\theta}{\cos\theta} = 9\tan\theta$

22 $\dfrac{\cos^2 A + 6 + 7\cot A}{\cos A} = \cos A + 6\sec A + 7\csc A$

23 $\dfrac{\tan B + 3\csc B}{\tan B\sec B} = \cos B + 3\cot^2 B$

24 $\dfrac{\sin\theta + \cos\theta + \tan\theta}{\sin\theta} = 1 + \cot\theta + \sec\theta$

25 $\sec^2\theta - \csc^2\theta = \tan^2\theta - \cot^2\theta$

26 $\dfrac{\sin A \cos B + \cos A \sin B}{\cos A \cos B - \sin A \sin B} = \dfrac{\tan A + \tan B}{1 - \tan A \tan B}$

27 $\dfrac{\cos A + \sec B}{\sec A + \cos B} = \dfrac{\cos A}{\cos B}$

28 $(1 + \csc \theta)(1 - \sin \theta) = \csc \theta - \sin \theta$

29 $\dfrac{\cot^2 A}{\csc A + 1} + 1 - \csc A = 0$

30 $(\sec \theta - \tan \theta)^2 = \dfrac{1 - \sin \theta}{1 + \sin \theta}$

31 $\dfrac{1}{\tan \theta + \cot \theta} = \sin \theta \cos \theta$

32 $\sin \theta \sec \theta + \theta \csc^2 \theta - \tan \theta - \theta = \theta \cot^2 \theta$

In Probs. 33 to 36, either prove or disprove the statement. (Prove that the equation is, or is not, an identity.)

33 $(\sec \theta - \tan \theta)(1 + \sin \theta) = \cos \theta$

34 $\dfrac{1 - \cot \theta}{\sec^2 \theta - 2} = \dfrac{1}{1 + \tan \theta}$

35 $(a \sin \theta - b)(b \sin \theta + a) = (a^2 - b^2) \sin \theta - ab \cos^2 \theta$

36 $(a \sin \theta + b \cos \theta)^2 + (b \sin \theta - a \cos \theta)^2 = a^2 + b^2$

37 $\dfrac{\cot \theta}{\csc \theta - \sin \theta} = \csc \theta$

38 $\dfrac{\csc \theta - 1}{\cot \theta} - \dfrac{\cot \theta}{\csc \theta + 1} = 0$

39 $\dfrac{1}{\csc \theta - \cot \theta} = \dfrac{1}{1 - \cos \theta}$

40 $\dfrac{1}{1 + \cos \theta} + \dfrac{1}{1 - \cos \theta} = 2$

In Probs. 41 to 44, assume the angle is in Q I.

41 $\sqrt{\dfrac{\sec \theta - \tan \theta}{\sec \theta + \tan \theta}} = \sec \theta - \tan \theta$

Hint: Rationalize the left side by multiplying under the radical by $\dfrac{\sec \theta - \tan \theta}{\sec \theta - \tan \theta}$.

42 $\sqrt{\dfrac{1 - \cos \theta}{1 + \cos \theta}} = \csc \theta - \cot \theta$

43 $\sqrt{\dfrac{1 + \sin\theta}{1 - \sin\theta}} = \dfrac{\cos\theta}{1 - \sin\theta}$

44 $\sqrt{\dfrac{\csc\theta + 1}{\csc\theta - 1}} = \sec\theta + \tan\theta$

45 State two nonpermissible values of θ for which identity 21 does not hold true.

46 State four nonpermissible values of θ for which identity 22 does not hold true.

47 State three nonpermissible values of B for which identity 19 does not hold true.

48 State two nonpermissible values of θ for which identity 8 does not hold true.

49 Verify identity 13 for $\sin\theta = \frac{1}{3}$ with θ in Q II.

50 Verify identity 42 for $\cos\theta = \frac{4}{5}$ with θ in Q I.

51 Verify identity 27 for $\sec A = 6$, $\sec B = -5$.

52 Verify identity 12 for $\theta = 120°$.

Prove each of the following identities.

53 $\dfrac{\sin^2\theta}{1 + 5\cos\theta + 4\cos^2\theta} = \dfrac{1 - \cos\theta}{1 + 4\cos\theta}$

54 $\dfrac{\tan A + \tan B}{1 - \tan A \tan B} = \dfrac{\cot A + \cot B}{\cot A \cot B - 1}$

55 $(1 + \sin\theta + \cos\theta)^2 = 2(1 + \sin\theta)(1 + \cos\theta)$

56 $\dfrac{\tan^2\theta - 3}{\tan^2\theta - 5\sec\theta + 7} = \dfrac{\sec\theta + 2}{\sec\theta - 3}$

57 $\sin^6\theta + \cos^6\theta = 1 - 3\sin^2\theta\cos^2\theta$

58 $(\sin A \cos B - \cos A \sin B)^2 + (\cos A \cos B + \sin A \sin B)^2 = 1$

59 $\dfrac{\sin A + \csc B}{\csc A + \sin B} + \dfrac{\sin A + \sin B}{\csc A + \csc B} = \dfrac{\sin B + \csc B}{\csc A}$

60 $\dfrac{\cos\theta - \sin\theta + 1}{\cos\theta + \sin\theta - 1} = \dfrac{\sin\theta}{1 - \cos\theta}$

61 $\dfrac{3\csc\theta + \cot\theta}{4\csc\theta - \cot\theta} + \dfrac{3\sec\theta - 1}{4\sec\theta + 1} = \dfrac{26 - 2\sin^2\theta}{15 + \sin^2\theta}$

62 $\dfrac{\tan^2 A}{\cos^2 B} - \dfrac{\tan^2 B}{\cos^2 A} = \tan^2 A - \tan^2 B$

63 What values of the constants a, b, c will make the following equation an identity?

$$5 + 6 \sec^2 \theta + 7 \sec^4 \theta = a + b \tan^2 \theta + c \tan^4 \theta$$

64 Show that $\frac{1}{4} \sin^2 2x - \frac{1}{6} \sin^4 2x$ is identically equal to $\frac{1}{12} \cos^2 2x - \frac{1}{6} \cos^4 2x + C$, where C is a constant. Determine the value of C.

RELATED
ANGLES

25 RELATED ANGLES The fact that the table of trigonometric functions deals with only acute angles should have implied that functions of larger angles are expressible in terms of functions of acute angles. Such is really the case. In order to determine the functions of angles larger than 90°, we introduce the concept of the related angle.

> **The related angle of a given angle θ is the positive acute angle between the x-axis and the terminal side of θ.**

Hence, *to find the related angle of θ**

---

* Since coterminal angles have the same trigonometric functions, we shall consider only those angles between 0° and 360°.

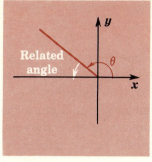 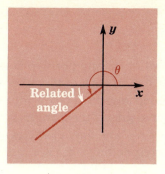 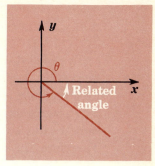

FIGURE 28

(a) *If θ is in* Q II, *subtract θ from* 180°.

(b) *If θ is in* Q III, *subtract* 180° *from θ*.

(c) *If θ is in* Q IV, *subtract θ from* 360°.

It is to be noted that, in finding the related angle, we always work to or from 180° or 360°, never 90° or 270°. Thus

The related angle of 160° is 20°.
The related angle of 260° is 80°.
The related angle of 310° is 50°.
The related angle of 500° is the related angle of
(500° − 360°) = 140°, which is 40°.

EXERCISE 15

Find the related angle of each of the following angles.

1	350°	2	110°
3	212°	4	301°
5	97° 45′	6	207° 25′
7	649°	8	886°
9	−475°	10	−42°
11	−188°	12	−93°

26   REDUCTION TO FUNCTIONS OF AN ACUTE ANGLE   Two numbers are said to be numerically equal if they are equal, except perhaps for sign. If a and b are numerically equal,* then $a = \pm b$.

* They have the same *absolute value.*

RELATED-ANGLE THEOREM   **Any trigonometric function of an angle is** *numerically* **equal to the same function of its related angle.**

To prove this, let θ_1 be any positive acute angle, and let θ_2, θ_3, θ_4 be positive angles, one in each of the other three quadrants, such that their common related angle is θ_1 (see Figure 29).   Choose points P_1, P_2, P_3, P_4 on the terminal sides of these angles so that all four points have the same radius vector r.   The four triangles of reference are congruent. Therefore, the corresponding sides are numerically equal.   Hence

$$x_2 = -x_1 \qquad x_3 = -x_1 \qquad x_4 = x_1$$

and

$$y_2 = y_1 \qquad y_3 = -y_1 \qquad y_4 = -y_1$$

Then

$$\sin \theta_2 = \frac{y_2}{r} = \frac{y_1}{r} = \sin \theta_1$$

$$\sin \theta_3 = \frac{y_3}{r} = \frac{-y_1}{r} = -\frac{y_1}{r} = -\sin \theta_1$$

$$\sin \theta_4 = \frac{y_4}{r} = \frac{-y_1}{r} = -\frac{y_1}{r} = -\sin \theta_1$$

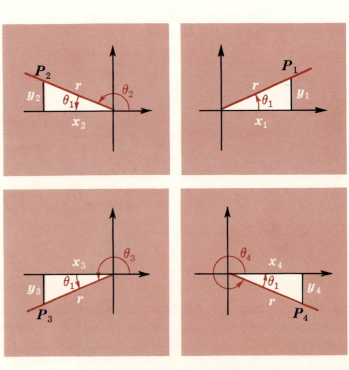

FIGURE 29

and
$$\cos \theta_2 = \frac{x_2}{r} = \frac{-x_1}{r} = -\frac{x_1}{r} = -\cos \theta_1$$

$$\cos \theta_3 = \frac{x_3}{r} = \frac{-x_1}{r} = -\frac{x_1}{r} = -\cos \theta_1$$

$$\cos \theta_4 = \frac{x_4}{r} = \frac{x_1}{r} = \cos \theta_1$$

Similarly, each of the other functions of θ_2, θ_3, θ_4 is *numerically* equal to the same function of θ_1, the common related angle. The proper sign, $+$ or $-$, is determined by the quadrant in which the given angle lies.

EXAMPLE 1   Use the related-angle theorem to find sin 120°, cos 120° without tables.

Solution   The related angle of 120° is 180° $-$ 120° $=$ 60°. Since 120° is in Q II, its sine is positive and its cosine is negative. Hence

$$\sin 120° = +\sin 60° = \frac{\sqrt{3}}{2}$$

$$\cos 120° = -\cos 60° = -\tfrac{1}{2}$$

The remaining functions can be found by using the fundamental identities:

$$\tan 120° = \frac{\sin 120°}{\cos 120°} = \frac{\sqrt{3}/2}{-\tfrac{1}{2}} = -\sqrt{3}$$

$$\cot 120° = \frac{1}{\tan 120°} = -\frac{1}{\sqrt{3}} = \frac{-\sqrt{3}}{3} \qquad \text{etc.}$$

EXAMPLE 2   Use the related-angle theorem to find sin 255° and tan 255°.

Solution   The related angle of 255° is 255° $-$ 180° $=$ 75°. Hence

$$\sin 255° = -\sin 75° = -0.9659$$
$$\tan 255° = +\tan 75° = 3.732$$

EXERCISE 16

Use the related-angle theorem to find the sine and cosine of each of the following angles without the use of tables.

1	300°	2	315°	3	225°
4	240°	5	135°	6	210°

7	150°	**8**	330°	**9**	570°
10	840°	**11**	660°	**12**	495°

Use the related-angle theorem and tables to find the values of the following functions.

13	sin 132°	**14**	cot 235°	**15**	sin 290°
16	tan 111°	**17**	cos 246° 30′	**18**	tan 333° 20′
19	cos 168° 10′	**20**	sin 196° 40′	**21**	cot 357° 12′
22	cos 96° 47′	**23**	tan 251° 34′	**24**	cos 348° 6′

Prove or disprove the following statements without using tables.

25 $1 + \tan^2 10° = \sec^2 170°$

26 $\sin^2 265° + \cos^2 85° = 1$

27 $\csc 306° = \csc 54°$

28 $\cot 123° = -\cot 33°$

29 $\cos 260° < \cos 250°$

30 $\sin 170° > \sin 160°$

31 $\dfrac{\cos 223°}{\cos 133°} = \tan 47°$

32 $\dfrac{\cos 800°}{\sin 280°} = -\cot 80°$

Name one angle in each of the other three quadrants whose trigonometric functions are numerically equal to those of the given angle.

33	55°	**34**	163°	**35**	222°	**36**	321°

27 TRIGONOMETRIC FUNCTIONS OF $(-\theta)$ Let θ be any angle. Then $(-\theta)$ indicates the same amount of rotation but in the *opposite* direction.   Place both angles in standard position on the same coordinate system.   Choose any point $P(x,y)$ on the terminal side of θ.   Drop a perpendicular from P to the x-axis and extend it until it strikes the terminal side of $(-\theta)$ at $P'(x,y')$.   The triangles of refer-

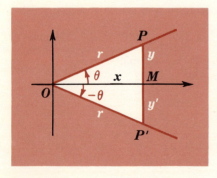
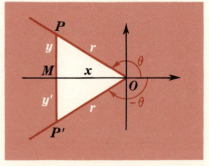

FIGURE 30

ence, OPM and $OP'M$, are congruent. Why? Hence $OP = OP' = r$. But y and y' are only numerically equal. Since $y' = -y$, we conclude that

$$\sin(-\theta) = \frac{y'}{r} = \frac{-y}{r} = -\frac{y}{r} = -\sin\theta$$

$$\cos(-\theta) = \frac{x}{r} = \cos\theta$$

$$\tan(-\theta) = \frac{y'}{x} = \frac{-y}{x} = -\frac{y}{x} = -\tan\theta$$

Similarly,

$$\cot(-\theta) = -\cot\theta$$
$$\sec(-\theta) = \sec\theta$$
$$\csc(-\theta) = -\csc\theta$$

The student should draw figures for θ in the other quadrants and also for θ a negative angle. For all possible positions of θ the following is true.

THEOREM   **For any angle θ,**

$$\sin(-\theta) = -\sin\theta$$
$$\cos(-\theta) = \cos\theta$$
$$\tan(-\theta) = -\tan\theta$$

The other three functions behave as do their reciprocals.

EXAMPLE   Compute $\sin(-225°)$, $\cos(-225°)$, $\tan(-225°)$.

Solution   Using the preceding theorem and the related-angle theorem, we obtain

$$\sin(-225°) = -\sin 225° = -(-\sin 45°) = \frac{\sqrt{2}}{2}$$

$$\cos(-225°) = \cos 225° = -\cos 45° = -\frac{\sqrt{2}}{2}$$

$$\tan(-225°) = -\tan 225° = -\tan 45° = -1$$

A function $f(\theta)$ is said to be an *even* function if, for all permissible values of θ, $f(-\theta) = f(\theta)$.

Examples of even functions* of θ are: $7\theta^2$, $\cos\theta$, $5\theta^6 - \theta^2$, and 1.

> **A function $f(\theta)$ is called an *odd* function if, for all permissible values of θ, $f(-\theta) = -f(\theta)$.**

Examples of odd functions of θ are: $8\theta^3$, $\sin\theta$, $\tan\theta$, and $-\theta^5$. Some functions, such as $\theta^5 + \theta^4$, are neither even nor odd.

EXERCISE 17

Use the theorem in Section 27 to compute the sine, cosine, and tangent of each of the following angles.

1 $-30°$	**2** $-45°$	**3** $-60°$	**4** $-90°$

Prove or disprove the following statements without using tables.

5 $\sin(-999°)\csc 999° = -1$ **6** $\sec 147° \cos(-147°) = -1$

7 $\dfrac{\sin(-\theta)}{\cos\theta} = \tan(-\theta)$ **8** $\dfrac{\sin(-\theta)}{\cos(-\theta)} = \tan\theta$

9 $1 - \tan^2(-\theta) = \sec^2\theta$ **10** $\sin^2(-\theta) + \cos^2\theta = 1$

11 $\cos(-218°) = \cos 38°$ **12** $\tan(-161°) = \tan 19°$

13 $\cos(-110°)$ is a positive number.

14 $\tan(-92°)$ is a positive number.

15 $\sin(-254°)$ is a negative number.

16 Draw a figure and prove that, if θ is a positive angle in Q II, then **(a)** $\sin(-\theta) = -\sin\theta$, **(b)** $\cos(-\theta) = \cos\theta$.

17 Draw a figure and prove that, if θ is a negative angle in Q I, then **(a)** $\sin(-\theta) = -\sin\theta$, **(b)** $\cos(-\theta) = \cos\theta$.

18 Identify each of the following functions as an even function, an odd function, or neither.

(a) $\cos(-4\theta)$	**(b)** $\tan 7\theta$	**(c)** $8\theta^7 + \theta^4$	**(d)** $3\theta^6 + 1$
(e) $\sin^3\theta$	**(f)** $\sin^4\theta$	**(g)** $2\theta^5 - 9\theta$	**(h)** $\sin\theta + \cos\theta$

*See footnote, p. 209.

5
RADIAN MEASURE

28 THE RADIAN   Thus far we have employed the *degree* as the unit of measure for angles.   It may be thought of as $\frac{1}{360}$ of the angular magnitude about a point.   For many practical purposes, the degree is a convenient unit, but most of the applications of the trigonometric functions which require higher mathematics, particularly calculus, are simplified if another unit, the radian, is used. For this reason, students planning to study calculus should make a special effort to learn radian measure thoroughly and to use it, so that it will be familiar to them when they need it.

> A radian is an angle which, if its vertex is placed at the center of a circle, subtends an arc equal in length to the radius of the circle.

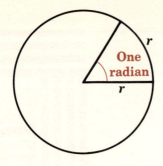

FIGURE 31

FIGURE 32

29 RADIANS AND DEGREES According to the definition of a radian, the number of radians in a circle is equal to the number of times the radius can be laid off along the circumference (Figure 32). Since $c = 2\pi r$, the number of radians in a circle is 2π. But the number of degrees in a circle is 360°. Hence

$$2\pi \text{ radians} = 360°$$

or   **π radians $= 180°$**   (1)

Hence $1 \text{ radian} = \dfrac{180°}{\pi} \doteq \dfrac{180°}{3.14159}$ (2)

$$\doteq 57.2958° \doteq 57° \ 17.75'$$

The symbol $\doteq$ means "is approximately equal to." (It is worth observing that 1 radian is a little less than 60°.)

Also, $1° = \dfrac{\pi}{180} \text{ radians} \doteq 0.017453 \text{ radian}$ (3)

$$1' = \dfrac{1°}{60} \doteq \dfrac{0.017453}{60} \text{ radian} \doteq 0.00029 \text{ radian}$$ (4)

It is better not to try to learn Equations (2), (3), (4). Memorize (1) and derive your results from it. In expressing the more common angles in terms of radians, we usually leave the result in terms of π. Thus $90° = \pi/2$. *When no unit of measure is indicated, it is understood that an angle is expressed in radians.* Thus if we read $\theta = \pi/6$, we understand that $\theta = \pi/6$ radians $= 30°$.

The student should become quite familiar with the radian measure of $30°$, $45°$, $60°$, $90°$, and the angles that are related to them, namely $120°$, $135°$, $150°$, $180°$, $210°$, $225°$, $240°$, $270°$, $300°$, $315°$, $330°$.

EXAMPLE 1   Express $\dfrac{7\pi}{9}$ in terms of degrees.

Solution   Since $\pi = 180°$

$\frac{7}{9}\pi = \frac{7}{9}(180°) = 140°$

EXAMPLE 2   Express $8° \, 25'$ in terms of radians.

Solution 1   Converting $8° \, 25'$ to degrees, we get

$$8\tfrac{25}{60}° = 8\tfrac{5}{12}° = \frac{101°}{12}$$

Since $180° = \pi$

$$1° = \frac{\pi}{180}$$

$$\frac{101°}{12} = \frac{101}{12} \cdot \frac{\pi}{180} = \frac{101\pi}{2160}$$

This is the accurate result. If π is replaced by 3.1416, we get

$$8° \, 25' = 0.1469$$

correct to four decimal places.

Solution 2   Use Table 1, page 244 in the tables, and interpolate:

$$8° \, 20' = 0.1454$$
$$8° \, 25' =$$
$$8° \, 30' = 0.1484$$

Hence $8° \, 25' = 0.1469$

EXAMPLE 3   Express 2 radians in terms of degrees.

Solution 1   Since π radians $= 180°$

$$1 \text{ radian} = \frac{180°}{\pi}$$

$$2 \text{ radians} = \frac{360°}{\pi}$$

This exact result may be approximated by the decimal form $\frac{360°}{3.1416} = 114.591° = 114° \ 35.5'$, which is correct to tenths of minutes.

Solution 2   Since   1 radian $= 57° \ 17.75'$
$$2 \text{ radians} = 2(57° \ 17.75')$$
$$= 114° \ 35.5'$$

EXAMPLE 4   Evaluate $\csc \dfrac{7\pi}{6}$.

Solution $\csc \dfrac{7\pi}{6} = \csc \dfrac{7(\overset{30°}{\cancel{180°}})}{\cancel{6}}$

$$= \csc 210° = \frac{1}{\sin 210°} = \frac{1}{-\sin 30°}$$

$$= \frac{1}{-\frac{1}{2}} = -2$$

EXERCISE 18

Express in degrees.   (The given angle is understood to be in radians.)

1 $\dfrac{7\pi}{9}$ 2 $\dfrac{3\pi}{5}$ 3 $\dfrac{9\pi}{10}$ 4 $\dfrac{\pi}{3}$

5 $\dfrac{11\pi}{6}$ 6 $\dfrac{9\pi}{4}$ 7 $\dfrac{5\pi}{3}$ 8 $\dfrac{7\pi}{20}$

9 $-\dfrac{3\pi}{2}$ 10 $-\dfrac{7\pi}{18}$ 11 10π 12 $\dfrac{9\pi}{2}$

Convert from radians to degrees and minutes; write the result to the nearest minute.

13 3.7   14 4   15 0.6   16 1.05

Express in radians, leaving the result in terms of π.

17 45°   18 150°   19 135°   20 240°
21 120°   22 270°   23 210°   24 585°
25 600°   26 10°   27 $-22\frac{1}{2}°$ 28 $-15°$
29 320°   30 700°   31 800°   32 100°

Convert to radian measure, obtaining the result to four decimal places.

33 22° 11'   **34** 81° 44'   **35** 57° 18'   **36** 40° 23'

Evaluate without using tables.

37 $\cos \dfrac{7\pi}{4}$

38 $\sin \dfrac{2\pi}{3}$

39 $\sin \dfrac{11\pi}{6}$

40 $\cos \dfrac{3\pi}{4}$

41 $\sin \dfrac{5\pi}{3}$

42 $\cos \dfrac{19\pi}{6}$

43 $\cos \dfrac{5\pi}{4}$

44 $\sin \dfrac{\pi}{3}$

45 $\sec \left(-\dfrac{\pi}{6}\right)$

46 $\csc \left(-\dfrac{\pi}{4}\right)$

47 $\tan \dfrac{4\pi}{3}$

48 $\tan \left(-\dfrac{5\pi}{6}\right)$

49 $\cot \pi$

50 $\tan 3\pi$

51 $\csc \dfrac{3\pi}{2}$

52 $\sec \dfrac{\pi}{2}$

Use Table 1 to evaluate the following. (The angle is understood to be in radians.)

53 $\sin 0.1134$ **54** $\cos 1.2770$

55 $\tan 0.4567$ **56** $\sin 1.0007$

57 Through how many radians does the hour hand of a clock rotate in 7 hr.? In 5 days?

58 Express in radians the acute angle made by the hands of a clock at 3:30. At 8:00.

59 One angle of a triangle is $\pi/15$. Another angle is 42°. Express the third angle in radians.

60 Through how many radians does the minute hand of a clock rotate in 55 min.? In 4 hr.?

30 LENGTH OF A CIRCULAR ARC   In a circle of radius r, let an arc of length s be subtended by a central angle θ. Then

$$s = r\theta \qquad \text{where } \theta \text{ is in radians}$$

To prove this, recall that in any circle the arc subtended by a central angle is proportional to this angle. Hence if 1 radian subtends an arc equal to the radius, then θ radians subtend an arc equal to θ times the radius. It is to be remembered that *the central angle must be measured in radians.*

In the equation $s = r\theta$, if two of the quantities s, r, θ are known, the third can be found.

EXAMPLE A circle has a radius of 100 in. (*a*) How long is the arc subtended by a central angle of 72°? (*b*) How large is the central angle that subtends an arc of 30 in.?

Solution* (*a*) Since

$$180° = \pi \text{ radians},$$

$$1° = \frac{\pi}{180} \text{ radians}$$

$$72° = 72 \cdot \frac{\pi}{180} \text{ radians} = \frac{2\pi}{5} \text{ radians}$$

Using

$$s = r\theta, \text{ where } \theta \text{ is in radians,}$$

$$s = 100 \cdot \frac{2\pi}{5} = \mathbf{40\pi \text{ in.}} \text{ (or } \textit{126 in.}\text{)}$$

(*b*) Using

$$s = r\theta,$$

$$30 = 100\theta$$

$$\theta = \frac{30}{100} = \mathbf{\frac{3}{10} \text{ radian}}$$

$$\left(\text{or } \frac{3}{10} \cdot \frac{180°}{\pi} = \frac{54°}{\pi} \doteq \textit{17° 10'} \right)$$

The bold-faced results are exact; the italicized answers are merely three-figure approximations.

31 TRIGONOMETRIC FUNCTIONS OF NUMBERS The equation
$s = r\theta$, θ in radians, implies that $\theta = s/r$, which says, in effect, that the *radian measure* of an angle is the *real number* which is the ratio of the

---

* In all discussions and problems in this chapter, all figures are to be considered as exact. They are not approximations. For the sake of uniformity write all approximate results with three-figure accuracy. It may be convenient to use $1/\pi = 0.3183$.

length of arc subtended by the angle to the length of the radius of a circle, when the vertex of the angle is at the center of the circle. For emphasis, we restate the fact that the radian measure of an angle (the ratio of two lengths) is simply a real number. Since any real number may be interpreted, if we so desire, as the radian measure of some angle, we can define the trigonometric functions as functions of real numbers. Thus, sin 1.245 (the sine of the number 1.245) = sin 1.245 (radians) = 0.9474; and sin $\frac{11\pi}{6} = -\frac{1}{2}$. Hence, for each real number s there corresponds a real number, sin s. We can now deal with the trigonometric functions of real numbers*—with no reference whatsoever to angles or triangles. This concept will be used in some of the succeeding chapters. It is the point of view that is most prevalent in more advanced courses in mathematics.

EXERCISE 19

(*Unless directed to the contrary, round off approximate results to three-figure accuracy.*)

1 On a circle of radius 90 ft., how long is the arc subtended by a central angle of **(a)** $\frac{6}{5}$ radians, **(b)** 23°?

2 Find the radius of a circle on which **(a)** a central angle of 1.6 radians subtends an arc of 60 in., **(b)** a central angle of 58° subtends an arc of 23 in.

3 A railroad curve is to be laid out on a circle. What radius should be used if the track is to change its direction by 7° in a distance of 80 ft.?

4 Find the number of radians in the central angle that subtends an arc of 4 in. on a circle of diameter 3 ft. Express the angle in degrees and minutes.

5 If a locomotive wheel with a diameter of 5 ft. rolls 12 ft., through how many degrees and minutes does it turn?

6 The minute hand of a campus clock is 4 ft. long. How far does its tip travel in 40 min.?

In Probs. 7 to 12, assume that the earth is a sphere of radius 4000 miles. Write results with two-figure accuracy.

7 How far is it from the equator to Tulsa, Okla., latitude 36° N? How far is Tulsa from the north pole? (See Figure 33.)

8 Presque Isle, Maine, is 3300 miles from the equator. Find the latitude of Presque Isle.

---

* The trigonometric functions of real numbers are sometimes called the *circular functions*. See Section 91 of the Appendix.

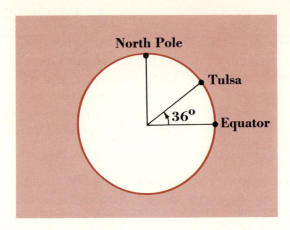

FIGURE 33

9   Cape Town, South Africa (latitude 34° S), is due south of Budapest, Hungary (latitude 47° 30′ N). How far is Cape Town from Budapest?

10   Cincinnati, Ohio, is 350 miles due north of Atlanta, Ga. (latitude 34° N). Find the latitude of Cincinnati.

11   The north pole is 140 miles closer to Athens, Greece, than to Knoxville, Tenn. If the latitude of Knoxville is 36° N, find the latitude of Athens.

12   The latitude of Wellington, New Zealand, is 41° S. The latitude of Flagstaff, Ariz., is 35° N. How many miles is Flagstaff north of Wellington?

In Probs. 13 to 16, the angle being small, we can assume, with little error, that the chord is equal to its subtended arc. Write results with two-figure accuracy.

13   A billboard 10 ft. high subtends an angle of 2° at the observer's eye. How far away is the billboard? (See Figure 34a.)

14   The moon as viewed from the earth subtends an angle of about 31′. If the moon is 240,000 miles from the earth, find the moon's diameter.

15   At a distance of 400 ft., a tree subtends an angle of 4° at the eye of the observer. Find the height of the tree.

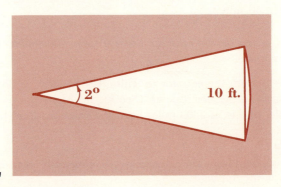

FIGURE 34a

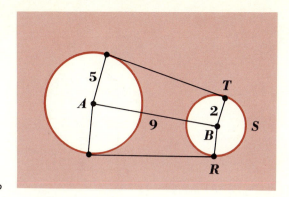

FIGURE 34b

16 A car $4\frac{1}{2}$ ft. high subtends an angle of $1°$ at the eye of an observer. How far is the car from the observer.

17 In Figure 34b, if $AB = 9$, find the length of arc *RST*. (Assume all given distances are exact. Write your result with three-figure accuracy.)

18 An isosceles triangle is inscribed in a circle of radius 100.0 in. Find the angles of the triangle if its base subtends an arc of 162.9 in. *Hint.* An angle inscribed in a circle is measured by one-half its subtended arc.

19 The angle between a chord of a circle and the tangent line passing through one end of the chord is 0.9 radian. The length of the arc subtended by the chord is 18.9 in. Find the radius of the circle.

20 A sector of a circle is the part of the circle bounded by an arc and the radii drawn to its extremities (Figure 35). By geometry, the area of a sector is equal to one-half its arc times the radius of the circle. Show that

area of sector $OAB = \frac{1}{2}r^2\theta$ where θ is in radians

21 Prove that the ratio $\dfrac{\sin\theta}{\theta}$ approaches 1 as θ approaches 0 provided θ

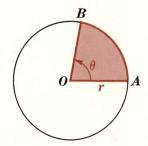

FIGURE 35

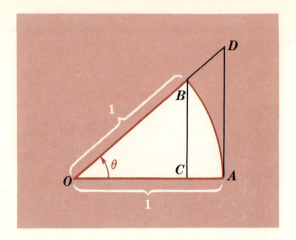

FIGURE 36

is in radians. (See Figure 36.) Hint: area $\triangle OCB <$ area sector $OAB <$ area $\triangle OAD$.

32   LINEAR AND ANGULAR VELOCITY   Consider a point P moving with constant speed on the circumference of a circle with radius r and center O.   If P traverses a distance of s linear units (inches, feet, miles, etc.) in t time units (seconds, minutes, etc.), then $s/t = v$ is called the linear velocity of P.   If the radius OP swings through θ angular units (degrees, radians, etc.), in t time units, then $\theta/t = \omega$ is called the angular velocity of P.   If, further, θ is in radians and ω is in radians per unit of time, then we can divide $s = r\theta$, by t, and get

$$\frac{s}{t} = r \cdot \frac{\theta}{t}$$

or $v = r\omega$

provided ω is in radians per time unit.   This means that the linear velocity of a point on the circumference of a circle is equal to the radius times the angular velocity of the point, in radians per unit of time.

The angular velocity of a rotating body is quite often expressed in revolutions per minute (rpm).   This can be readily converted into radians per minute, by remembering that one revolution represents 2π radians.

EXAMPLE   A flywheel 6 ft. in diameter makes 40 rpm.   (*a*) Find its angular velocity in radians per second.   (*b*) Find the speed of the belt that drives the flywheel.

Solution (*a*) Since 40 rpm represents $40(2\pi) = 80\pi$ radians per min.,

$$\omega = \frac{80\pi}{60} = \frac{4\pi}{3} \text{ radians per sec.}$$

(*b*) The speed of the belt, if it does not slip, is equal to the linear velocity of a point on the rim of the flywheel. Using $v = r\omega$, we have

$$v = 3\left(\frac{4\pi}{3}\right) = 4\pi \text{ ft. per sec.} \doteq 12.6 \text{ ft. per sec.}$$

EXERCISE 20

(Unless directed to the contrary, round off approximate results to three-figure accuracy.)

1　How many revolutions per minute are made by a wheel that has an angular velocity of 80 radians per sec.?

2　A pulley 10 in. in diameter is driven by a belt that moves 3000 ft. per min. How many revolutions per minute are made by the pulley?

3　A car is traveling 30 mph. How many revolutions per minute are made by the wheels, which are 24 in. in diameter?

4　An airplane propeller 5 ft. from tip to tip makes 880 rpm. Find the linear speed of the tip in miles per hour.

5　A rolling wheel travels 15 mph while turning at the rate of 2 revolutions per sec. Find the number of feet in the diameter of the wheel.

6　Find the linear speed, due to the rotation of the earth, of a point on the equator. Use 3960 miles as the radius of the equator. Express the result in miles per hour with three-figure accuracy.

7　Find the linear speed, due to the rotation of the earth, of a building in Tulsa. The latitude of Tulsa is 36° N. Use 3960 miles as the radius of the earth. Express the result in miles per hour with three-figure accuracy. (See Figure 33.)

8　It takes 6 min. to run 100 ft. of movie film at uniform speed through a certain projector. The film is unwound from one reel, which becomes progressively smaller in diameter, and is rewound on another, which becomes larger. Find the number of revolutions per minute of each of the two reels when one roll of film is 3 in. in diameter and the other is 2 in. in diameter.

9　A bicycle is driven by the pedals, which are attached to the sprocket wheel, which has a diameter of 10 in. A chain connects the sprocket to a smaller cogwheel, whose diameter is 3 in. This wheel is fastened to the bicycle's rear wheel, whose diameter is 22 in. Find the speed of the bicycle in miles per hour when the sprocket is making 1 revolution per sec.

10　A cogwheel is driven by a chain that travels 3 ft. per sec. Find in inches the diameter of the wheel if it makes 96 rpm.

GRAPHS OF THE TRIGONOMETRIC FUNCTIONS

33 PERIODIC FUNCTIONS   Since coterminal angles have the same trigonometric functions, we know that any trigonometric function of θ is exactly equal to the same function of $\theta + n{\cdot}360°$, where n is any integer.   For example, $\sin 20° = \sin 380° = \sin 740° = \sin(-340°) =$ etc.   Hence $\sin \theta$ takes on all of its possible values as θ ranges from $0°$ to $360°$; in fact, it takes on all values between -1 and 1 *twice* for $0° \leq \theta < 360°$, once in increasing and once in decreasing, as θ increases (see Section 34).   Then it repeats these values as θ moves from $360°$ to $720°$, from $720°$ to $1080°$, etc.   Inasmuch as

$$\sin \theta = \sin(\theta + 360°) \qquad \text{for all } \theta$$

and since $360°$ is the smallest positive angle for which this is true, we say that the sine is a periodic function of period 360°. In general,

a function $f(\theta)$ is said to be periodic and of period p, provided

$$f(\theta + p) = f(\theta) \qquad \text{for all } \theta$$

where p is the smallest positive constant for which this is true.

The cosine function is periodic with a period of 360° because

$$\cos(\theta + 360°) = \cos\theta \qquad \text{for all } \theta$$

and because this statement is not true if 360° is replaced by any smaller positive angle, such as 180°.

34 VARIATIONS OF THE SINE AND COSINE Knowing that the sine and cosine functions have a period of 360°, we shall confine ourselves to investigating their behavior as the angle increases from 0° to 360°. Consider any angle θ in standard position (Figure 37). On the terminal side of θ, choose the point $P(x, y)$ whose radius vector is 1. As θ varies, make P move so that r is always 1. This will keep P on a circle with center O and radius 1. Recalling that $\sin\theta = y/r = MP/1 = MP$, we see that the length of the *directed* segment MP is equal to $\sin\theta$. Now try to visualize θ as increasing from 0° to 90°.

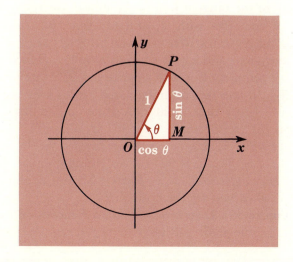

FIGURE 37

Accordingly sin θ (or MP) increases from 0 to 1. As θ increases from 90° to 180°, sin θ decreases from 1 to 0. As θ swings through the third and fourth quadrants, P is below the x-axis. Hence sin θ (or MP) is negative.

Similarly, cos $\theta = x/r = OM$. As θ increases from 0° to 90°, cos θ decreases from 1 to 0. A study of the variations of sin θ and cos θ as θ goes from 0° to 360° reveals the following results:

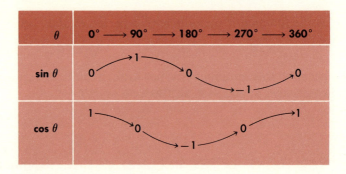

The arrow pointing upward means the function is increasing; downward, decreasing.

35 VARIATION OF THE TANGENT

In discussing tan $\theta = y/x$, we shall keep x numerically equal to 1. For θ in Q I, we have tan $\theta = AP$, the length of the tangent line to the unit circle at A (Figure 38). For $\theta = 0°$, P coincides with A, and tan $0° = 0$. As θ increases toward 90°, P moves upward from A, and tan θ increases. When $\theta = 90°$, P is on the y-axis and $x = 0$ (it cannot be 1). Hence tan $90° = y/0$, which *does not exist*. But as θ approaches 90°, tan θ (or AP) increases rapidly. In fact we can make tan θ just as large as we please by taking θ sufficiently close to 90°. A situation like this is expressed briefly by tan $90° = \infty$ (read "tan 90° is infinite"). Remember, however, that *tan $90° = \infty$ is just an abbreviation for: tan 90° does not exist; but by taking θ sufficiently close to 90° (never letting it equal 90°), we can make tan θ as large numerically as we please.* Memorize this statement and think of it when you see the symbol ∞.

In Q II (Figure 39), keep $x = -1$; then P moves on the tangent line to the unit circle at B; and tan $\theta = y/x = y/-1 = -y = -BP$

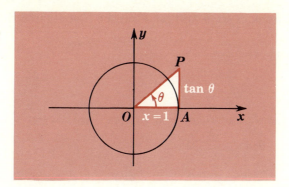

FIGURE 38

(a negative number since BP is positive). As soon as θ leaves $90°$, P starts down on the tangent line. As θ increases from $90°$ to $180°$, BP decreases from very large positive numbers to 0, and $\tan \theta$ increases from very large negative values to 0. Using our symbolic notation, we say that as θ increases from $90°$ to $180°$, $\tan \theta$ increases from $-\infty$ to 0.* By a similar process, the variation of $\tan \theta$ in Q III and Q IV can be investigated.

The variation in $\cot \theta$ can be studied by recalling that $\cot \theta = 1/\tan \theta$. Since $\tan \theta$ is always increasing, $\cot \theta$ is always decreasing. When $\tan \theta$ becomes infinite, $\cot \theta$ approaches 0; when $\tan \theta$ approaches 0, $\cot \theta$ becomes infinite.

* The notation $\tan 90°^- = +\infty$ is frequently used to mean that as θ approaches $90°$ through values *less* than $90°$ (such as $89°$, $89.9°$, $89.99°$, etc.), $\tan \theta$ *increases* without limit. The statement $\tan 90°^+ = -\infty$ is used to indicate that as θ approaches $90°$ through values *greater* than $90°$, $\tan \theta$ *decreases* without limit. Both statements may be incorporated in the single form $\tan 90° = \infty$, which implies that as θ approaches $90°$ (from either side), $\tan \theta$ *increases numerically* without limit.

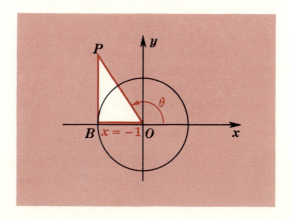

FIGURE 39

The variations in sec θ and csc θ can be investigated through their reciprocals.

In summary we have

θ	Q I $0° \longrightarrow 90°$	Q II $90° \longrightarrow 180°$	Q III $180° \longrightarrow 270°$	Q IV $270° \longrightarrow 360°$
sin θ	0 ↗ 1	1 ↘ 0	0 ↘ −1	−1 ↗ 0
cos θ	1 ↘ 0	0 ↘ −1	−1 ↗ 0	0 ↗ 1
tan θ	0 ↗ ∞	−∞ ↗ 0	0 ↗ ∞	−∞ ↗ 0
cot θ	∞ ↘ 0	0 ↘ −∞	∞ ↘ 0	0 ↘ −∞
sec θ	1 ↗ ∞	−∞ ↗ −1	−1 ↘ −∞	∞ ↘ 1
csc θ	∞ ↘ 1	1 ↗ ∞	−∞ ↗ −1	−1 ↘ −∞

Notice that in Q I the sine, tangent, and secant increase while their cofunctions decrease. Also observe that the sine and cosine range from −1 to 1, the tangent and cotangent take on all values, and the secant and cosecant are always numerically equal to or greater than 1.

36   THE GRAPH OF sin θ

A complete "picture story" of the variation of the sine is presented by its graph. Let us draw a system of coordinate axes and label the horizontal axis as θ and the vertical axis as sin θ. Since the radian is the natural measure of angles, let the θ-axis be laid off in radians. This means that the number 1 on the vertical scale and the number 1 (radian) on the horizontal scale are represented by the same distance. *Hence* 180° *on the θ-axis should be π times as long as 1 unit on the (sin θ)-axis.* Using trigonometric tables and the related-angle theory we form the following table:

θ in degrees	0°	30°	60°	90°	120°	150°	180°	210°	240°	270°	300°	330°	360°
θ in radians	0	$\frac{\pi}{6}$	$\frac{\pi}{3}$	$\frac{\pi}{2}$	$\frac{2\pi}{3}$	$\frac{5\pi}{6}$	π	$\frac{7\pi}{6}$	$\frac{4\pi}{3}$	$\frac{3\pi}{2}$	$\frac{5\pi}{3}$	$\frac{11\pi}{6}$	2π
sin θ	0	.5	.87	1	.87	.5	0	−.5	−.87	−1	−.87	−.5	0

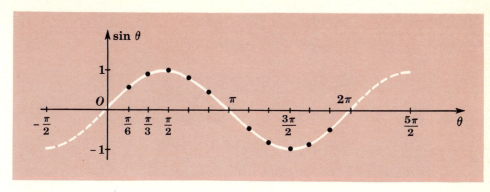

FIGURE 40

After plotting these values on the coordinate axes,* we obtain the curve in Figure 40.

The student should practice drawing the 0° to 360° portion of this curve until he can make a hasty sketch of it from memory. The sine curve† can be used to remember the sine of 0°, 90°, 180°, 270° and also to remember the sign of the sine in the various quadrants. For example, if you encounter sin 270°, make a rapid sketch or form a mental picture of the sine curve and notice that sin 270° = −1. Also, the sine curve enables us to remember that sin θ is positive in Q I and Q II (because in these quadrants the curve is above the horizontal axis) and sin θ is negative in Q III and Q IV (because the curve is below the horizontal axis for   180° < θ < 360°).

Furthermore, the sine curve recalls the related-angle theory. For example, it is obvious from the curve (see Figure 41) that

$$\sin 150° = \sin 30° = \tfrac{1}{2}$$

and $$\sin 225° = \sin 315° = -\frac{\sqrt{2}}{2}$$

Because of its wave form, the sine curve is very important in the study of wave motion in electrical engineering and physics. The maximum distance of the curve from the θ-axis is called the *amplitude* of the curve or wave. The period of the function representing the wave is called the period (or *wave length*) of the curve. The sine curve

---

* A very good approximation to the sine curve can be drawn on quadrille paper. Let 4 quadrille *spaces* represent 1 *unit* vertically and let 1 quadrille space represent 15° horizontally. For θ = 15°, sin θ = 0.26; the corresponding point would be plotted 1 *space* to the right and 4(0.26) = 1.04 *spaces* up from the origin. For θ = 45°, sin θ = 0.71; the corresponding point would be plotted 3 *spaces* to the right and 4(0.71) = 2.88 *spaces* up from the origin.

† Sometimes called the *sinusoid*.

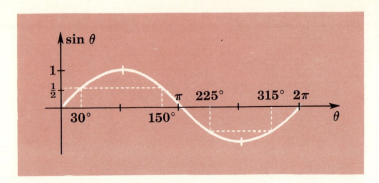

FIGURE 41

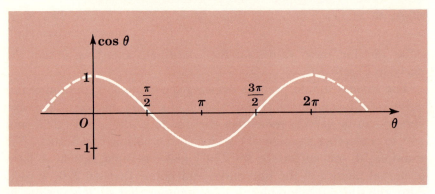

FIGURE 42

has an amplitude of 1 and a period of 2π. The student can see from Figure 41 that $\sin \theta$ does not have a period that is less than $360°$.

37 GRAPHS OF THE OTHER TRIGONOMETRIC FUNCTIONS By using methods exactly like those employed in the preceding article, we can draw the graphs of the other trigonometric functions. For reference purpose we exhibit these graphs (Figures 42–46). The student should draw each of them by preparing a table of values, plotting the points, and then drawing a smooth curve through these points. From these graphs we can see that $\tan \theta$ and $\cot \theta$ have a period of π, whereas $\sec \theta$ and $\csc \theta$ have period 2π.

The student should be able to draw from memory a hasty sketch of $\sin \theta$, $\cos \theta$, $\tan \theta$.

EXAMPLE   Solve the equation $\sin \theta = -\dfrac{\sqrt{3}}{2}$ for all values of θ on the interval $0° \leq \theta < 360°$. (That is, solve for all values of θ that are equal to or greater than $0°$ and less than $360°$.) Use the sine curve to identify the proper quadrants and check the related-angle theory.

Solution   The sine is negative (the sine curve lies below the θ-axis) in Q III and Q IV.

Remember that $\sin 60° = \dfrac{\sqrt{3}}{2}$. Since $\sin \theta = -\dfrac{\sqrt{3}}{2}$, it follows that θ must be equal to those angles in Q III and Q IV that have $60°$ for their related angle (see Figure 47, page 85).

In Q III: $\theta = 180° + 60° = 240°$
In Q IV: $\theta = 360° - 60° = 300°$

Hence if

$$\sin \theta = - \frac{\sqrt{3}}{2}$$

then $\theta = 240°, 300°$

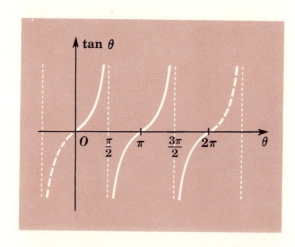

FIGURE 43

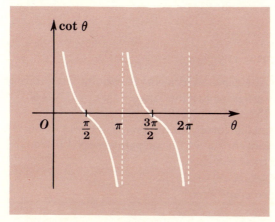

FIGURE 44

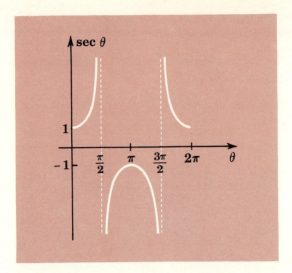

FIGURE 45

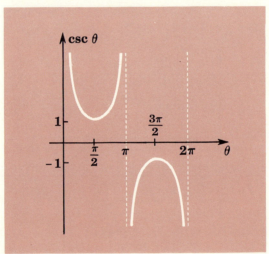

FIGURE 46

1 Sketch the sine curve by drawing a smooth curve through the points obtained by assigning to θ the values $-90°$, $-75°$, $-60°$, $-45°$, $-30°$, $-15°$, $0°$, $15°$, $30°$, ..., $360°$.

2 Sketch the cosine curve. Locate points every $15°$ from $0°$ to $360°$.

3 Sketch the tangent curve. Locate points every $15°$ from $-90°$ to $360°$.

4 Sketch the cotangent curve.

5 Sketch the secant curve.

6 Sketch the cosecant curve.

7 Explain briefly what is meant by the statement "csc $180° = \infty$."
8 Explain briefly what is meant by the statement "sec $90° = \infty$."
9 What is the period of tan θ?
10 Discuss in detail the variation of the cosecant.

Solve the following equations for values of θ on the interval $0° \leq \theta < 360°$. Use the curves to identify the proper quadrants and check the related-angle theory. Do not use tables.

11 $\cos \theta = \dfrac{\sqrt{3}}{2}$ **12** $\sin \theta = \frac{1}{2}$

13 $\sin \theta = 0$ **14** $\cos \theta = -\dfrac{\sqrt{2}}{2}$

15 $\sin \theta = -\dfrac{\sqrt{2}}{2}$ **16** $\cos \theta = -1$

17 $\cos \theta = -\frac{1}{2}$ **18** $\sin \theta = \dfrac{\sqrt{3}}{2}$

19 $\tan \theta = \dfrac{1}{\sqrt{3}}$ **20** $\sec \theta = -\dfrac{2}{\sqrt{3}}$

21 $\csc \theta = \sqrt{2}$ **22** $\cot \theta = -1$

Solve the following equations for values of θ on the interval $0 \leq \theta < 2\pi$ (radians).

23 $\sin \theta = -1$ **24** $\csc \theta = -\sqrt{2}$

25 $\cos \theta = -\dfrac{\sqrt{3}}{2}$ **26** $\cot \theta = \sqrt{3}$

27 $\sec \theta = 2$ **28** $\cos \theta = 0$
29 $\tan \theta = -1$ **30** $\csc \theta = 0$

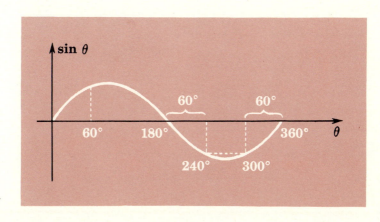

FIGURE 47

Use tables to solve the following equations for values of θ on the interval $0° \leq \theta < 360°$. Obtain results correct to the nearest degree.

31 $\tan \theta = 0.4$ **32** $\cot \theta = -0.5$

33 $\sin \theta = -0.2756$ **34** $\cos \theta = 0.5736$

35 $\cos \theta = -\frac{1}{3}$ **36** $\sin \theta = \frac{1}{8}$

37 $\cot \theta = 1.5$ **38** $\tan \theta = -60$

39 State the maximum and minimum values achieved by each of the following functions of θ:

  (a) $3 + \cos 4\theta$
  (b) $5 - 2 \sin \theta$
  (c) $8 - 3 \cos^2 \theta$

40 State the maximum and minimum values achieved by each of the following functions of θ:

  (a) $10 + 2 \sin 3\theta$
  (b) $11 - 5 \cos \theta$
  (c) $12 + 7 \sin^2 \theta$

41 Correct the following statement: $\sec 270° = \infty$ means (a) $\sec 270°$ does not exist, and (b) by taking θ sufficiently close to $270°$, we can make $\sec 270°$ as large numerically as we please.

In Probs. 42 to 45, either prove or disprove the statement. (Prove that the equation is, or is not, an identity.)

42 $\cos \dfrac{n\pi}{2} = 0$ (n an odd integer)

43 $\sin n\pi = 0$ (n an integer)

44 $\cos n\pi = (-1)^n$ (n an integer)

45 $\sin \dfrac{n\pi}{2} = (-1)^{(n-1)/2}$ (n an odd integer)

FUNCTIONS OF TWO ANGLES

38 FUNCTIONS OF THE SUM OF TWO ANGLES The trigonometric identities in Chap. 3 are relations among the trigonometric functions of one angle. We shall now consider functions of an angle which is the sum of two given angles. It seems reasonable to say that if the functions of 30° and 45° are known, the functions of 75° can be obtained. For instance, is sin 30° + sin 45° = sin 75°? Obviously this is false because $\frac{1}{2} + \frac{\sqrt{2}}{2} = 0.5 + 0.7 = 1.2$ which would make sin 75° greater than 1. This proves that, in general, sin $(A + B) \neq$ sin A + sin B.* Likewise sin $2A$ is not identically equal to 2 sin A

---

* The symbol $\neq$ is read "is not equal to."

because $\sin 60° \neq 2 \sin 30°$. It is, however, possible to express $\sin (A + B)$ in terms of the functions of the separate angles A and B. And it is possible to express $\sin 2A$ in terms of the functions of A. These, and other, formulas will be developed in the following articles.

39   Sin (A + B)   AND   cos (A + B)* The purpose of this article is to prove the following identities:

$$\sin (A + B) = \sin A \cos B + \cos A \sin B \qquad (1)$$
$$\cos (A + B) = \cos A \cos B - \sin A \sin B \qquad (2)$$

We shall first derive a formula for $\cos (A + B)$. On a coordinate system (Figure 48), draw a unit circle with center at the origin. Place the angles A, $A + B$, and $-B$ in standard position. Let the terminal sides of these angles intersect the unit circle at points P_1, P_2, and P_3, respectively; let P_4 designate the point $(1,0)$. If P_1 has coordinates (x_1, y_1), then $\cos A = x_1/1 = x_1$, and $\sin A = y_1/1 = y_1$. Hence the coordinates of P_1 are $(\cos A, \sin A)$†. For the same reason, the coordinates of P_2 and P_3 are as indicated in Figure 48. Angle P_3OP_1 is $(B + A)$ because angle P_3OP_4 is B and angle P_4OP_1 is A. Therefore, triangles P_3OP_1 and P_4OP_2 are congruent. Why? Hence $P_3P_1 = P_4P_2$.

Applying the distance formula (Section 4, p. 4), we get

$$P_3P_1 = \sqrt{[\cos A - \cos (-B)]^2 + [\sin A - \sin (-B)]^2}$$

Squaring and using the identities $\cos (-B) = \cos B$ and $\sin (-B) = -\sin B$, we obtain

$$(P_3P_1)^2 = (\cos A - \cos B)^2 + (\sin A + \sin B)^2$$
$$= \cos^2 A - 2 \cos A \cos B + \cos^2 B$$
$$+ \sin^2 A + 2 \sin A \sin B + \sin^2 B$$
$$= 2 - 2 \cos A \cos B + 2 \sin A \sin B$$

since $\cos^2 \theta + \sin^2 \theta = 1$

Again using the distance formula, we find

$$(P_4P_2)^2 = [\cos (A + B) - 1]^2 + [\sin (A + B) - 0]^2$$
$$= \cos^2 (A + B) - 2 \cos (A + B) + 1$$
$$+ \sin^2 (A + B)$$

---

* The instructor may wish to begin with a simple geometric proof (Section 93).
† This statement is true regardless of the quadrant in which A lies.

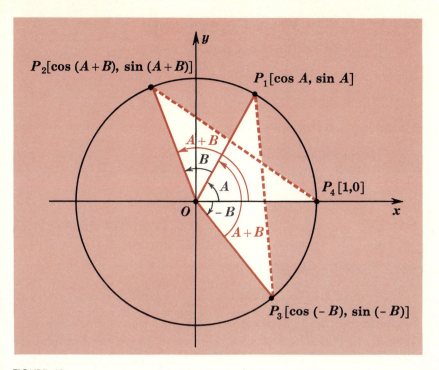

FIGURE 48

Hence $\quad (P_4P_2)^2 = 2 - 2 \cos (A + B)$

since $\quad \cos^2 (A + B) + \sin^2 (A + B) = 1$

Upon equating the expressions for $(P_4P_2)^2$ and $(P_3P_1)^2$, we get

$$2 - 2 \cos (A + B) = 2 - 2 \cos A \cos B + 2 \sin A \sin B$$

which reduces to

$$\cos (A + B) = \cos A \cos B - \sin A \sin B \qquad (2)$$

Note carefully that this proof does not depend upon the quadrants in which A, B, and $A + B$ happen to lie. The formula for $\cos (A + B)$ is valid for all values of A and B, positive or negative.

Before deriving a formula for $\sin (A + B)$, we shall need to obtain general expressions for $\cos (A - C)$, $\cos (90° - C)$, and $\sin (90° - C)$.

Since formula (2) for $\cos (A + B)$ holds for all A and B, we shall rewrite it with B replaced by $-C$. Hence

$$\cos (A - C) = \cos [A + (-C)]$$
$$= \cos A \cos (-C) - \sin A \sin (-C)$$

Since cos $(-C) = \cos C$ and sin $(-C) = -\sin C$, we have

$$\cos (A - C) = \cos A \cos C + \sin A \sin C \tag{7-1}$$

If we replace A with $90°$ in formula (7-1), we find

$$\cos (90° - C) = \cos 90° \cos C + \sin 90° \sin C$$

Since cos $90° = 0$ and sin $90° = 1$, we have

$$\cos (90° - C) = \sin C \tag{7-2}$$

Moreover, since $C = 90° - (90° - C)$,

$$\begin{aligned}
\cos C &= \cos [90° - (90° - C)] \\
&= \cos 90° \cos (90° - C) + \sin 90° \sin (90° - C) \\
&= 0 \cdot \cos (90° - C) + 1 \cdot \sin (90° - C)
\end{aligned}$$

Hence $\cos C = \sin (90° - C)$

or $\sin (90° - C) = \cos C \tag{7-3}$

Formulas (7-2) and (7-3) were established for any acute angle C in Section 11, p. 20. The foregoing arguments prove the formulas for *all* values of C. For example, if $C = 250°$, formula (7-2) says cos $(90° - 250°) = \sin 250°$, which is true because cos $(90° - 250°) = \cos (-160°) = \cos 160° = -\cos 20° = -\sin 70°$, whereas sin $250° = -\sin 70°$, by the related-angle theorem (p. 60).

To derive a formula for sin $(A + B)$, we shall reverse (7-2) and replace C with $(A + B)$ to get

$$\begin{aligned}
\sin (A + B) &= \cos [90° - (A + B)] \\
&= \cos [(90° - A) - B] \\
&= \cos (90° - A) \cos B + \sin (90° - A) \sin B.
\end{aligned}$$

Using formulas (7-2) and (7-3), we may write

$$\sin (A + B) = \sin A \cos B + \cos A \sin B \tag{1}$$

It is impossible to overemphasize the importance of these results. In addition to learning formulas (1) and (2), the student should be able to state them in words. These statements are:

(1) The sine of the sum of two angles is equal to the sine of the first times the cosine of the second, plus the cosine of the first times the sine of the second.

(2) **The cosine of the sum of two angles is equal to the cosine of the first times the cosine of the second, minus the sine of the first times the sine of the second.**

It is equally important for the student to be able to use these formulas backward. For example, when

$$\cos 7\theta \cos 2\theta - \sin 7\theta \sin 2\theta$$

is encountered, it should be recognized as the expansion of $\cos (7\theta + 2\theta)$ or $\cos 9\theta$.

EXAMPLE 1　Compute sin 75° and cos 75° from the functions of 30° and 45°.

Solution

$$\sin 75° = \sin (30° + 45°)$$
$$= \sin 30° \cos 45° + \cos 30° \sin 45°$$
$$= \frac{1}{2} \cdot \frac{\sqrt{2}}{2} + \frac{\sqrt{3}}{2} \cdot \frac{\sqrt{2}}{2}$$
$$= \frac{\sqrt{2}}{4} + \frac{\sqrt{6}}{4} = \frac{\sqrt{2} + \sqrt{6}}{4}$$
$$\doteq \frac{1.414 + 2.449}{4} = \frac{3.863}{4} \doteq 0.966$$

$$\cos 75° = \cos (30° + 45°)$$
$$= \cos 30° \cos 45° - \sin 30° \sin 45°$$
$$= \frac{\sqrt{3}}{2} \cdot \frac{\sqrt{2}}{2} - \frac{1}{2} \cdot \frac{\sqrt{2}}{2}$$
$$= \frac{\sqrt{6} - \sqrt{2}}{4} \doteq 0.259$$

Notice that these decimal approximations agree with the "story" as presented by the sine and cosine curves. Angles near 90° have sines close to 1 and cosines close to 0.

EXAMPLE 2　Given $\sin A = \frac{5}{13}$, with A in Q I, and $\cos B = -\frac{4}{5}$, with B in Q II. Find (*a*) $\sin (A + B)$, (*b*) $\cos (A + B)$, (*c*) the quadrant in which $(A + B)$ lies.

Solution　First find cos A and sin B by drawing triangles of reference (Figure 49). Hence

$$\cos A = \frac{12}{13} \qquad \text{and} \qquad \sin B = \frac{3}{5}$$

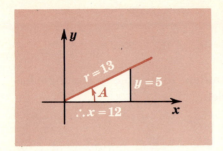 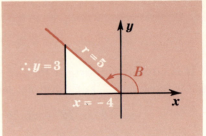

FIGURE 49

(a) $\sin (A + B) = \sin A \cos B + \cos A \sin B$

$$= \left(\frac{5}{13}\right)\left(\frac{-4}{5}\right) + \left(\frac{12}{13}\right)\left(\frac{3}{5}\right)$$

$$= -\tfrac{20}{65} + \tfrac{36}{65} = \tfrac{16}{65}$$

(b) $\cos (A + B) = \cos A \cos B - \sin A \sin B$

$$= \left(\frac{12}{13}\right)\left(\frac{-4}{5}\right) - \left(\frac{5}{13}\right)\left(\frac{3}{5}\right)$$

$$= -\tfrac{48}{65} - \tfrac{15}{65} = -\tfrac{63}{65}$$

(c) The angle $(A + B)$ lies in Q II because $\sin (A + B)$ is positive and $\cos (A + B)$ is negative.

EXERCISE 22

1 Compute $\sin 105°$ from the functions of $60°$ and $45°$.
2 Compute $\cos 285°$ from the functions of $225°$ and $60°$.
3 Compute $\cos 165°$ from the functions of $135°$ and $30°$.
4 Compute $\sin 255°$ from the functions of $210°$ and $45°$.
5 Use tables to show that

$\sin 50° + \sin 35° \neq \sin 85°$
$\cos 70° + \cos 5° \neq \cos 75°$

Simplify by reducing to a single term.

6 $\sin 170° \cos 100° + \cos 170° \sin 100°$

7 $\sin \dfrac{2\theta}{5} \cos \dfrac{3\theta}{5} + \cos \dfrac{2\theta}{5} \sin \dfrac{3\theta}{5}$

8 $\cos 6\theta \cos \theta - \sin 6\theta \sin \theta$

9 $\sin 140° \sin 40° - \cos 140° \cos 40°$

Prove the following identities.

10 $\cos\left(\dfrac{9\pi}{2} + 3\theta\right) = -\sin 3\theta$

11 $\sin\left(\dfrac{5\pi}{2} + 2\theta\right) = \cos 2\theta$

12 $\sin\left(-\dfrac{\pi}{2} + \theta\right) = -\cos\theta$

13 $\sin(30° + \theta) + \cos(60° + \theta) = \cos\theta$

14 $\sin(210° + \theta) - \cos(120° + \theta) = 0$

15 $\sin C \cos(D - C) + \cos C \sin(D - C) = \sin D$

16 $(\sin A \cos B + \cos A \sin B)^2 + (\sin A \sin B - \cos A \cos B)^2 = 1$

17 Express $\cos 55°$ in terms of functions of $22°$ and $33°$.

18 Express $\sin 74°$ in terms of functions of $70°$ and $4°$.

19 Prove $\dfrac{\cos 8\theta}{\sin \theta} - \dfrac{\sin 8\theta}{\cos \theta} = \dfrac{\cos 9\theta}{\sin \theta \cos \theta}.$

20 Prove $\dfrac{\sin 3\theta}{\sin 4\theta} + \dfrac{\cos 3\theta}{\cos 4\theta} = \sin 7\theta \sec 4\theta \csc 4\theta.$

21 Given $\sin A = \frac{7}{10}$ with A in Q I, and $\sin B = \frac{3}{10}$ with B in Q I. Find $\sin(A + B)$. Make a rough check of your result by using tables to approximate the values of A, B, and $(A + B)$ as determined by their sines.

22 Given $\sin(C - D) = \frac{12}{13}$ with $(C - D)$ in Q I, and $\sin D = \frac{15}{17}$ with D in Q I. Find $\sin C$.

23 Given $\cos A = \frac{3}{5}$ with A in Q IV, and $\cos B = -\frac{9}{41}$ with B in Q II. Find (a) $\sin(A + B)$, (b) $\cos(A + B)$, (c) the quadrant in which $(A + B)$ lies.

24 Given $\tan A = \frac{24}{7}$ with A in Q III, and $\cos B = \frac{4}{5}$ with B in Q I. Find (a) $\sin(A + B)$, (b) $\cos(A + B)$, (c) the quadrant in which $(A + B)$ lies.

25 Prove the identity

$$\sin(A + B + C) = \sin A \cos B \cos C + \cos A \sin B \cos C$$
$$+ \cos A \cos B \sin C - \sin A \sin B \sin C$$

Hint: $\sin(A + B + C) = \sin([A + B] + C)$.

26 Prove the identity

$$\cos(A + B + C) = \cos A \cos B \cos C - \sin A \sin B \cos C$$
$$- \sin A \cos B \sin C - \cos A \sin B \sin C$$

27 If A and B are complementary angles, prove that

$$\sin(5A + B) = \cos 4A$$

28 If A and B are complementary angles, prove that

$$\cos(8A + 3B) = \sin 5A$$

In Probs. 29 to 31, either prove or disprove the statement. (Prove that the equation is, or is not, an identity.)

29 $\sin(\theta + n\pi) = (-1)^n \sin\theta$ (n an integer)

30 $\sin\left(\theta + \dfrac{n\pi}{2}\right) = (-1)^{(n-1)/2} \cos\theta$ (n an odd integer)

31 $\cos\left(\theta + \dfrac{n\pi}{2}\right) = (-1)^{(n+1)/2} \sin\theta$ (n an odd integer)

40   Tan (A + B)   Since $\tan\theta = \dfrac{\sin\theta}{\cos\theta}$,

$$\tan(A+B) = \frac{\sin(A+B)}{\cos(A+B)} = \frac{\sin A \cos B + \cos A \sin B}{\cos A \cos B - \sin A \sin B}$$

Dividing top and bottom of this fraction by $\cos A \cos B$, we get

$$\tan(A+B) = \frac{\dfrac{\sin A \cos B}{\cos A \cos B} + \dfrac{\cos A \sin B}{\cos A \cos B}}{\dfrac{\cos A \cos B}{\cos A \cos B} - \dfrac{\sin A \sin B}{\cos A \cos B}}$$

$$= \frac{\dfrac{\sin A}{\cos A} + \dfrac{\sin B}{\cos B}}{1 - \dfrac{\sin A}{\cos A} \cdot \dfrac{\sin B}{\cos B}}$$

or $$\tan(A+B) = \frac{\tan A + \tan B}{1 - \tan A \tan B} \qquad (3)$$

It is important to note that this formula, and all others developed in this chapter, is valid for only those angles for which the functions involved are defined and for which the denominators are not 0.

Stated in words, we have

> **(3) The tangent of the sum of two angles is equal to the sum of their tangents, divided by 1 minus the product of their tangents.**

41   Sin (A − B),   cos (A − B),   AND   tan (A − B)   If A and B are any two angles then,

$$\sin(A - B) = \sin A \cos B - \cos A \sin B \qquad (4)$$

$$\cos(A - B) = \cos A \cos B + \sin A \sin B \qquad (5)$$

$$\tan (A - B) = \frac{\tan A - \tan B}{1 + \tan A \tan B} \qquad (6)$$

Proof Recall that $\sin (-B) = -\sin B$, $\cos (-B) = \cos B$, and $\tan (-B) = -\tan B$. Then

$$\begin{aligned}
\sin (A - B) &= \sin (A + [-B])* \\
&= \sin A \cos [-B] + \cos A \sin [-B] \\
&= \sin A \cos B + (\cos A)(-\sin B) \\
&= \sin A \cos B - \cos A \sin B
\end{aligned}$$

Formulas (5) and (6) are proved by a similar method. The student should state formulas (4), (5), and (6) in words.

In comparing (1), (4), (2), and (5),

$$\begin{aligned}
\sin (A + B) &= \sin A \cos B + \cos A \sin B, \\
\sin (A - B) &= \sin A \cos B - \cos A \sin B, \\
\cos (A + B) &= \cos A \cos B - \sin A \sin B, \\
\cos (A - B) &= \cos A \cos B + \sin A \sin B,
\end{aligned}$$

we notice that "the sines have the same sign, but the cosines have different signs."

There is no need for formulas involving the cotangent, secant, and cosecant of $(A + B)$ and $(A - B)$ because they can be readily expressed in terms of their reciprocals.

42   REDUCTION OF $a \sin \theta + b \cos \theta$ TO $k \sin (\theta + H)$ Any two real nonzero numbers a and b are proportional to two other numbers that represent the cosine and sine, respectively, of some properly chosen angle H. These other two numbers are $a/\sqrt{a^2 + b^2}$ and $b/\sqrt{a^2 + b^2}$. We may write

$$\cos H = \frac{a}{\sqrt{a^2 + b^2}} \quad \text{and} \quad \sin H = \frac{b}{\sqrt{a^2 + b^2}} \qquad (7\text{-}4)$$

as indicated by Figure 50. Then

$$a \sin \theta + b \cos \theta = \sqrt{a^2 + b^2} \left[(\sin \theta) \frac{a}{\sqrt{a^2 + b^2}} + (\cos \theta) \frac{b}{\sqrt{a^2 + b^2}} \right]$$

* This method may be used to convert any "sum formula" to a "difference formula."

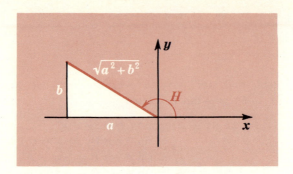

FIGURE 50

$$= \sqrt{a^2 + b^2}\, (\sin \theta \cos H + \cos \theta \sin H)$$

Hence $a \sin \theta + b \cos \theta = \sqrt{a^2 + b^2} \sin (\theta + H)$

where H is an angle that satisfies both equations in (7-4).*

EXAMPLE   Express $4 \sin \theta + 3 \cos \theta$ in the form $k \sin (\theta + H)$, where k and H are constants and $k > 0$.

Solution   Multiply and divide the given expression by $\sqrt{4^2 + 3^2} = 5$ to get

$$5(\tfrac{4}{5} \sin \theta + \tfrac{3}{5} \cos \theta) \quad \text{or} \quad 5[(\sin \theta)\tfrac{4}{5} + (\cos \theta)\tfrac{3}{5}]$$

Put $\tfrac{4}{5} = \cos H$; then $\tfrac{3}{5} = \sin H$. Our expression becomes

$$5(\sin \theta \cos H + \cos \theta \sin H) = 5 \sin (\theta + H)$$

From tables we find that if $\cos H = \tfrac{4}{5}$ and $\sin H > 0$, then $H = 36° 52'$ or 0.6435 radian. Hence

$$4 \sin \theta + 3 \cos \theta = 5 \sin (\theta + 36° 52')$$
$$= 5 \sin (\theta + 0.64) \qquad \text{approx.}$$

EXERCISE 23

1   Compute $\tan 345°$ from $\tan 45°$ and $\tan 300°$.
2   Compute $\tan 75°$ from $\tan 120°$ and $\tan 45°$.
3   Compute $\cos 195°$ from the functions of $225°$ and $30°$.
4   Compute $\sin 15°$ from the functions of $45°$ and $30°$.

---

* Equations (7-4) imply that $\tan H = b/a$. In many cases it will be easiest to obtain H through its tangent, though then we must be careful to choose an angle in the correct quadrant as determined by the signs of a and b.

Simplify by reducing to a single term.

5   $\cos (185° + \theta) \cos (5° + \theta) + \sin (185° + \theta) \sin (5° + \theta)$

6   $\sin 40° \cos 10° - \cos 40° \sin 10°$

7   $\dfrac{\tan 8\theta - \tan 5\theta}{1 + \tan 8\theta \tan 5\theta}$

8   $\dfrac{\tan 20° + \tan 10°}{1 - \tan 20° \tan 10°}$

Prove the following identities.

9   $\tan (2\pi - \theta) = -\tan \theta$

10   $\tan \left(\theta + \dfrac{\pi}{4}\right) = \dfrac{1 + \tan \theta}{1 - \tan \theta}$

11   $\sin \left(\dfrac{3\pi}{2} - \theta\right) = -\cos \theta$

12   $\cos \left(\theta - \dfrac{\pi}{2}\right) = \sin \theta$

13   Given $\sin A = \frac{3}{5}$ with A in Q II, and $\cos B = \frac{8}{17}$ with B in Q I. Find
    (a) $\sin (A - B)$, (b) $\cos (A - B)$, (c) $\tan (A + B)$, (d) $\tan (A - B)$.

14   Given $\cos A = -\frac{7}{25}$ with A in Q III, and $\tan B = -\frac{12}{5}$ with B in Q IV.
    Find (a) $\sin (A - B)$, (b) $\cos (A - B)$, (c) $\tan (A + B)$, (d) $\tan (A - B)$.

15   Given $\sin (C + D) = \frac{5}{6}$ with $(C + D)$ in Q I, and $\sin D = \frac{1}{6}$ with D in
    Q I. Find $\sin C$.

16   Given $\tan A = \frac{9}{7}$ with A in Q I, and $\tan B = \frac{1}{8}$ with B in Q I. Show that
    $A - B = 45°$.

Prove or disprove. (For each of the following equations, show that it is, or is
not, an identity.)

17   $5 \sin 8\theta \cos 2\theta - 5 \cos 8\theta \sin 2\theta = \sin 6\theta$

18   $11 \sin 9\theta \sin 7\theta + 11 \cos 9\theta \cos 7\theta = 11 \cos 2\theta$

19   $\tan 48° = \dfrac{1 + \tan 3°}{1 - \tan 3°}$

20   $\sin C + \sin (D - C) = \sin D$

21   Express $\sin 11°$ in terms of functions of $88°$ and $77°$.

22   Express $\cos 20°$ in terms of functions of $70°$ and $50°$.

23   Express $\tan 36°$ in terms of $\tan 76°$ and $\tan 40°$.

24   Express $\tan 58°$ in terms of $\tan 24°$ and $\tan 34°$.

Prove the following identities.

25   $\sec (A - B) = \dfrac{\sec A \sec B}{1 + \tan A \tan B}$

26 $\dfrac{\sin (A + B)}{\sin (A - B)} = \dfrac{\tan A + \tan B}{\tan A - \tan B}$

27 $\sin (\theta - 60°) + \cos (\theta - 30°) = \sin \theta$

28 $\cos (300° - \theta) + \sin (210° - \theta) = 0$

29 $\cot (A + B) = \dfrac{\cot A \cot B - 1}{\cot A + \cot B}$

30 $\cot (A - B) = \dfrac{\cot A \cot B + 1}{\cot B - \cot A}$

31 $\tan (A + B + C) = \dfrac{\tan A + \tan B + \tan C - \tan A \tan B \tan C}{1 - \tan A \tan B - \tan A \tan C - \tan B \tan C}$

32 $\dfrac{\tan (R - S) + \tan T}{1 - \tan (R - S) \tan T} = \dfrac{\tan R - \tan (S - T)}{1 + \tan R \tan (S - T)}$

33 If A and B are complementary angles, prove that

$\sin 2A = \sin 2B$

34 If A and B are complementary angles, prove that

$\tan (2A + 5B) = \tan 3B$

35 If A and B are supplementary angles, prove that

$\cos (7A - B) = -\cos 8A$

36 If n is an integer, prove that $\cos (n\pi - \theta) = (-1)^n \cos \theta$.

37 Express $8 \sin \theta + 15 \cos \theta$ in the form $k \sin (\theta + H)$.

38 Express $-\sin \theta + \cos \theta$ in the form $k \sin (\theta + H)$.

39 Express $3 \sin \theta - 4 \cos \theta$ in the form $k \sin (\theta + H)$.

40 Express $28 \sin \theta + 96 \cos \theta$ in the form $k \sin (\theta + H)$.

41 Express $10 \sin (\theta + 11\pi/6)$ in the form $a \sin \theta + b \cos \theta$.

42 Express $2 \sin (\theta + 5\pi/4)$ in the form $a \sin \theta + b \cos \theta$.

43 DOUBLE-ANGLE FORMULAS   If A is any angle, then

$$\sin 2A = 2 \sin A \cos A, \tag{7}$$

$$\cos 2A = \cos^2 A - \sin^2 A \tag{8a}$$

$$= 1 - 2 \sin^2 A \tag{8b}$$

$$= 2 \cos^2 A - 1. \tag{8c}$$

Proof $\sin 2A = \sin (A + A)$

$= \sin A \cos A + \cos A \sin A$

$= 2 \sin A \cos A.$

$\cos 2A = \cos (A + A)$

$$= \cos A \cos A - \sin A \sin A$$
$$= \cos^2 A - \sin^2 A$$
$$= 1 - \sin^2 A - \sin^2 A$$
$$= 1 - 2 \sin^2 A, \text{ etc.}$$

Stated in words, formula (7) says

The sine of twice an angle is equal to twice the sine of the angle times the cosine of the angle.

The student should state the three forms of formula (8) in words.
Formulas (7) and (8) are called *double-angle formulas* because the angle on the left side is double the angle on the right side. Formula (7) could have been written

$$\sin B = 2 \sin \frac{B}{2} \cos \frac{B}{2}$$

because B is the double of $B/2$. In other words, formulas (7) and (8) are used to express the sine and cosine of an angle in terms of the functions of an angle that is half as large. To illustrate,

$$\sin 60° = 2 \sin 30° \cos 30° = 2\left(\frac{1}{2}\right)\left(\frac{\sqrt{3}}{2}\right) = \frac{\sqrt{3}}{2}$$
$$\cos 14\theta = 2 \cos^2 7\theta - 1$$
$$\cos \frac{8\theta}{9} = 1 - 2 \sin^2 \frac{4\theta}{9}$$

Furthermore, formula (7) implies that $\sin A \cos A = \frac{1}{2} \sin 2A$.

44   HALF-ANGLE FORMULAS   If θ is any angle, then

$$\sin \frac{\theta}{2} = \pm\sqrt{\frac{1 - \cos \theta}{2}} \tag{9}$$

$$\cos \frac{\theta}{2} = \pm\sqrt{\frac{1 + \cos \theta}{2}} \tag{10}$$

The choice of the sign in front of the radical is determined by the quadrant in which $\theta/2$ lies.

Proof   Formula (8*b*) says

$$\cos 2A = 1 - 2 \sin^2 A$$

Let $A = \theta/2$; then $2A = \theta$, and

$$\cos \theta = 1 - 2 \sin^2 \frac{\theta}{2}$$

Transposing, we obtain

$$2 \sin^2 \frac{\theta}{2} = 1 - \cos \theta$$

$$\sin^2 \frac{\theta}{2} = \frac{1 - \cos \theta}{2}$$

$$\sin \frac{\theta}{2} = \pm \sqrt{\frac{1 - \cos \theta}{2}}$$

By a similar method formula (10) can be derived from formula (8c).

Formulas (9) and (10) are called *half-angle formulas* because the angle on the left side is half the angle on the right side. To illustrate,

$$\sin 30° = \sqrt{\frac{1 - \cos 60°}{2}} = \sqrt{\frac{1 - \frac{1}{2}}{2}} = \sqrt{\frac{\frac{1}{2}}{2}}$$

$$= \sqrt{\frac{1}{4}} = \frac{1}{2}$$

$$\cos 170° = - \sqrt{\frac{1 + \cos 340°}{2}}$$

(Here the minus sign is chosen because $170°$ is in Q II and has a negative cosine.)

$$\sin C = \pm \sqrt{\frac{1 - \cos 2C}{2}}$$

Notice that in the double-angle formulas the large angle is on the left side, while in the half-angle formulas the small angle is on the left side. In fact, the half-angle formulas are merely the double-angle formulas used backward (reading from right to left).

It is desirable to read the left side of formula (9): "the sine of half of θ" rather than to say "the sine of θ over 2" which might be construed as $\frac{\sin \theta}{2}$.

EXAMPLE 1   Prove the identity $\cos 3\theta = 4 \cos^3 \theta - 3 \cos \theta$

Proof $\cos 3\theta$ $4 \cos^3 \theta - 3 \cos \theta$

$= \cos (2\theta + \theta)$

$= \cos 2\theta \cos \theta - \sin 2\theta \sin \theta$

$$= (2 \cos^2 \theta - 1)^* \cos \theta - (2 \sin \theta \cos \theta)$$
$$\cdot \sin \theta$$
$$= 2 \cos^3 \theta - \cos \theta - 2 \sin^2 \theta \cos \theta$$
$$= 2 \cos^3 \theta - \cos \theta - 2(1 - \cos^2 \theta) \cos \theta$$
$$= 2 \cos^3 \theta - \cos \theta - 2 \cos \theta + 2 \cos^3 \theta$$
$$= 4 \cos^3 \theta - 3 \cos \theta$$

This identity is frequently used in proving that it is impossible to trisect a general angle with ruler and compass.

EXAMPLE 2   Express $\cos 20\theta$ in terms of $\sin 5\theta$.

Solution   Since 20θ is the double of 10θ, and 10θ is the double of 5θ, we shall employ double-angle formulas:

$$\cos 20\theta$$
$$= 1 - 2 \sin^2 10\theta$$
$$= 1 - 2(2 \sin 5\theta \cos 5\theta)^2$$
$$= 1 - 8 \sin^2 5\theta \cos^2 5\theta$$
$$= 1 - 8 \sin^2 5\theta(1 - \sin^2 5\theta)$$
$$= 1 - 8 \sin^2 5\theta + 8 \sin^4 5\theta$$

EXAMPLE 3   By using half-angle formulas, reduce $\sin^4 A$ to an expression involving no even exponents.

Solution   Upon squaring both sides of formula (9), we obtain

$$\sin^2 \frac{\theta}{2} = \frac{1 - \cos \theta}{2} = \frac{1}{2}(1 - \cos \theta)$$

Replacing $\theta/2$ with A gives $\sin^2 A = \frac{1}{2}(1 - \cos 2A)$. This equation enables us to change the exponent from 2 to 1, with a doubling of the angle. Hence

$$\sin^4 A = (\sin^2 A)^2 = [\frac{1}{2}(1 - \cos 2A)]^2$$
$$= \frac{1}{4}(1 - 2 \cos 2A + \cos^2 2A)$$

If $\cos^2 2A$ is replaced by $\frac{1}{2}(1 + \cos 4A)$, we get

$$\sin^4 A = \frac{3}{8} - \frac{1}{2} \cos 2A + \frac{1}{8} \cos 4A$$

Such transformations are needed in the integration of even powers of sines and cosines in calculus.

EXERCISE 24

1   Use tables to show that

$2 \sin 35° \neq \sin 70°$
$2 \cos 44° \neq \cos 88°$

* Formula (8c) is used because the right side involves only $\cos \theta$.

2 Use double-angle formulas to compute $\sin 60°$ and $\cos 60°$ from the functions of $30°$.

3 Compute $\sin 22\frac{1}{2}°$ and $\cos 22\frac{1}{2}°$ from the functions of $45°$.

4 Compute $\sin 165°$ and $\cos 165°$ from the functions of $330°$.

Simplify by reducing to a single term involving only one function of an angle.

5 $2\cos^2 8\theta - 1$

6 $9 - 18\sin^2 10°$

7 $\sin 12° \cos 12°$

8 $7\cos^2\dfrac{\theta}{2} - 7\sin^2\dfrac{\theta}{2}$

9 $-\sqrt{\dfrac{1 + \cos 280°}{2}}$

10 $\sqrt{\dfrac{1 - \cos 14\theta}{2}}$

11 $\cos^2\dfrac{\theta}{3} + \sin^2\dfrac{\theta}{3}$

12 $8\sin 5\theta \cos 5\theta$

13 Given $\sin A = \dfrac{24}{25}$ with $90° < A < 180°$. Find $\sin\dfrac{A}{2}$ and $\sin 2A$.

14 Given $\cos A = \dfrac{1}{8}$ with $270° < A < 360°$. Find $\cos\dfrac{A}{2}$ and $\cos 2A$.

15 Given $\cos 6B = -\frac{2}{3}$ with $180° < 6B < 270°$. Find $\cos 3B$ and $\cos 12B$.

16 Given $\sin\dfrac{A}{2} = \dfrac{4}{5}$ with $0° < \dfrac{A}{2} < 90°$. Find $\sin A$ and $\sin\dfrac{A}{4}$.

17 Express $\sin 155°$ in terms of a function of $310°$.

18 Express $\cos 175°$ in terms of a function of $350°$.

19 Express $\sin 280°$ in terms of $\sin 140°$.

20 Express $\cos 66°$ in terms of $\cos 33°$.

21 Express $\cos 4B$ in terms of $\sin 2B$.

22 Express $\sin C$ in terms of functions of $\dfrac{C}{2}$.

23 Express $\sin 3D$ in terms of a function of $6D$.

24 Express $\cos^2 7B$ in terms of $\sin 14B$.

Identify as true or false and give reasons. Do not use tables.

25 $320\sin^5 B \cos^5 B = 10\sin^5 2B$

26 $\cos 6B = 6\cos^2 3B - 1$

27 $\cos^2 2B = \sin^4 B - 2\sin^2 B \cos^2 B + \cos^4 B$

28 $\sin 8\theta \cos 8\theta = \frac{1}{2} \sin 16\theta$

29 $\cos(-2A) = \sin^2 A - \cos^2 A$

30 $\cos^2 \dfrac{\theta}{8} - \sin^2 \dfrac{\theta}{8} = \cos \dfrac{\theta}{4}$

31 $\left(\dfrac{1 + \cos 88°}{2}\right)^{3/2} = \cos^3 44°$

32 $\sqrt{\dfrac{1 - \cos 400°}{2}} = -\sin 20°$

33 $\sqrt{\dfrac{1 + \cos 6\phi}{32}} = \pm\dfrac{1}{4} \cos 12\phi$

34 $\sqrt{50(1 - \cos 2B)} = \pm 10 \sin B$

35 $2 \sin^2 6C = 1 + \cos 12C$

36 $\frac{1}{2} - \sin^2 3B = \frac{1}{2} \cos 6B$

37 Reduce $\sin^4 3A$ to $\frac{3}{8} - \frac{1}{2} \cos 6A + \frac{1}{8} \cos 12A$.

38 Reduce $\cos^4 \theta$ to $\frac{3}{8} + \frac{1}{2} \cos 2\theta + \frac{1}{8} \cos 4\theta$.

39 Reduce $\sin^2 A \cos^2 A$ to $\frac{1}{8}(1 - \cos 4A)$.

Prove each of the following identities by transforming one side to the other.

40 $\cos^4 \theta - \sin^4 \theta = \cos 2\theta$

41 $\sec 2A = \dfrac{\sec^2 A}{2 - \sec^2 A}$

42 $\dfrac{\cos^3 \theta - \sin^3 \theta}{\cos \theta - \sin \theta} = 1 + \dfrac{1}{2} \sin 2\theta$

43 $\dfrac{\sin 2A}{1 + \cos 2A} = \tan A$

44 $\sin^2 2\theta - \cos^2 \theta = \sin^2 \theta - \cos^2 2\theta$

45 $\sec^2 \dfrac{\theta}{2} = \dfrac{2}{1 + \cos \theta}$

46 $\csc^2 \dfrac{\theta}{2} = \dfrac{2 \sec \theta}{\sec \theta - 1}$

47 $\dfrac{\cos 2\theta}{\sin 2\theta} = \dfrac{1}{2}(\cot \theta - \tan \theta)$

48 $\cos 4\theta = 1 - 8 \cos^2 \theta + 8 \cos^4 \theta$

49 $\dfrac{\cot \theta - \tan \theta}{\cot \theta + \tan \theta} = \cos 2\theta$

50 $\sin 4\theta = 4 \sin \theta \cos \theta - 8 \sin^3 \theta \cos \theta$

51 $\tan 2A = \dfrac{2 \tan A}{1 - \tan^2 A}$

52 $\tan \dfrac{\theta}{2} = \dfrac{1 - \cos \theta}{\sin \theta} = \dfrac{\sin \theta}{1 + \cos \theta}$

53 $\dfrac{\cos 2\theta}{1 + \sin 2\theta} = \dfrac{\cos\theta - \sin\theta}{\cos\theta + \sin\theta}$

54 $\tan^2 B = 4\sin^2 B \csc^2 2B - 1$

55 $(4\cos^2 A - 4\cos^4 A)^2 = \sin^4 2A$

56 $\sin 3\theta = 3\sin\theta - 4\sin^3\theta$

57 $\csc 2\theta - \cot 2\theta = \tan\theta$

58 $\sec B = \dfrac{\sec^2\dfrac{B}{2}}{2 - \sec^2\dfrac{B}{2}}$

59 $\dfrac{\tan 9\theta - \tan 3\theta}{\tan 9\theta + \tan 3\theta} = \dfrac{1}{2}\sec 6\theta$

60 $\dfrac{a\cos 2\theta + b\cos\theta + a}{a\sin 2\theta + b\sin\theta} = \cot\theta$

61 $\cos 5A = 16\cos^5 A - 20\cos^3 A + 5\cos A$

62 $\tan^2 A + \cos 2A \tan^2 A + \cos 2A = 1$

63 In right triangle ABC, where $C = 90°$, prove that $\cos\dfrac{A}{2} = \sqrt{\dfrac{b+c}{2c}}$.

64 Prove that the area of right triangle ABC, where $C = 90°$, is $\frac{1}{4}c^2 \sin 2A$.

65 Verify identity 57 for $\theta = 30°, 45°, 60°$.

66 Verify identity 50 for $\theta = 0°, 30°, 45°$.

67 Verify identity 51 for **(a)** $A = 0°$, **(b)** $A = 30°$, **(c)** $A = 90°$.

68 Verify identity 60 for $\theta = 30°, 45°, 90°$.

45   PRODUCT TO SUM FORMULAS; SUM TO PRODUCT FORMULAS

It is sometimes necessary to convert a product of two trigonometric functions into a sum of two functions, and vice versa. For this reason we develop the following formulas. They are not nearly so important as the preceding ten formulas.

When (1) and (4) are added, we get

$$\sin(A + B) + \sin(A - B) = 2\sin A \cos B$$

or

$$\sin A \cos B = \tfrac{1}{2}[\sin(A + B) + \sin(A - B)] \qquad (11)$$

Upon subtracting (1) and (4), we obtain

$$\cos A \sin B = \tfrac{1}{2}[\sin (A + B) - \sin (A - B)] \tag{12}$$

Similarly, by adding and subtracting (2) and (5), we get

$$\cos A \cos B = \tfrac{1}{2}[\cos (A + B) + \cos (A - B)] \tag{13}$$

$$\sin A \sin B = \tfrac{1}{2}[\cos (A - B) - \cos (A + B)] \tag{14}$$

Formulas (11), (12), (13), and (14) are used to convert a product of sines and cosines into a sum or difference* of sines and cosines. They are used in certain problems in integral calculus.

When formula (11) is used backward (from right to left), it converts a sum into a product. Thus

$$\sin (A + B) + \sin (A - B) = 2 \sin A \cos B \tag{7-5}$$

For convenience we shall change notation by making the substitutions

$$A + B = C \qquad \text{and} \qquad A - B = D$$

Adding these two equations and dividing by 2, we then obtain $A = \tfrac{1}{2}(C + D)$. Subtracting and dividing by 2, we get $B = \tfrac{1}{2}(C - D)$. Substituting in (7-5), we obtain

$$\sin C + \sin D = 2 \sin \tfrac{1}{2}(C + D) \cos \tfrac{1}{2}(C - D) \tag{15}$$

Similarly,

$$\sin C - \sin D = 2 \cos \tfrac{1}{2}(C + D) \sin \tfrac{1}{2}(C - D) \tag{16}$$
$$\cos C + \cos D = 2 \cos \tfrac{1}{2}(C + D) \cos \tfrac{1}{2}(C - D) \tag{17}$$
$$\cos C - \cos D = -2 \sin \tfrac{1}{2}(C + D) \sin \tfrac{1}{2}(C - D) \tag{18}$$

Formulas (15), (16), (17), and (18) are used to convert sums and differences into products. Formula (16) is usually employed in the derivation of the formula for the derivative of the sine in differential calculus.

Formula (1) should not be confused with formula (15). The former deals with the sine of the sum of two angles; the latter deals with the sum of the sines of two angles.

EXAMPLE 1   Reduce $\cos 5\theta \cos 3\theta$ to a sum.

Solution   Using formula (13), we get

$$\cos 5\theta \cos 3\theta = \tfrac{1}{2}[\cos (5\theta + 3\theta) + \cos (5\theta - 3\theta)]$$
$$= \tfrac{1}{2} \cos 8\theta + \tfrac{1}{2} \cos 2\theta$$

* The expression "sum or difference" is called the "algebraic sum." The expression $(a - b)$ may be written as $(a + [-b])$, which is the algebraic sum of a and $-b$.

EXAMPLE 2   Prove the identity $\dfrac{\sin 7\theta - \sin 3\theta}{\cos 7\theta + \cos 3\theta} = \tan 2\theta$

Proof

$$\dfrac{\sin 7\theta - \sin 3\theta}{\cos 7\theta + \cos 3\theta} \qquad\qquad \tan 2\theta$$

$$= \dfrac{2 \cos \frac{1}{2}(7\theta + 3\theta) \sin \frac{1}{2}(7\theta - 3\theta)}{2 \cos \frac{1}{2}(7\theta + 3\theta) \cos \frac{1}{2}(7\theta - 3\theta)}$$

$$= \dfrac{\sin 2\theta}{\cos 2\theta}$$

$$= \tan 2\theta$$

EXERCISE 25

Express each of the following as an algebraic sum of sines and cosines.

1 $\sin 208° \cos 67°$
2 $10 \cos 46° \cos 21°$
3 $20 \sin 173° \sin 53°$
4 $14 \sin 9\theta \cos \theta$
5 $2 \cos 3\theta \cos 8\theta$
6 $\sin 5\theta \sin 6\theta$

Express each of the following as a product.

7 $\sin 15\theta + \sin 11\theta$
8 $\sin 42° - \sin 28°$
9 $\cos 190° + \sin 40°$
10 $\cos A - \cos (A + 120°)$

Prove the following identities.

11 $\dfrac{\cos 65° + \cos 25°}{\sin 205° - \sin 65°} = -1$

12 $\dfrac{\cos 7\theta - \cos 3\theta}{\sin 9\theta + \sin \theta} = -2 \sin 2\theta \sec 4\theta$

13 $\dfrac{\sin 11\theta - \sin 5\theta}{\cos 10\theta - \cos 4\theta} = -\dfrac{\cos 8\theta}{\sin 7\theta}$

14 $\sin 295° + \cos 85° = \cos 145°$
15 $\sin 10\theta \cos 4\theta - \sin 5\theta \cos 9\theta = \sin 5\theta \cos \theta$
16 $\cos 14\theta \cos 10\theta - \sin 9\theta \sin 5\theta = \cos 19\theta \cos 5\theta$
17 $\cos^2 C - \cos^2 D = \sin (D + C) \sin (D - C)$
18 $\cos \theta + \cos 2\theta + \cos 3\theta = (2 \cos \theta + 1) \cos 2\theta$

19 $\dfrac{\cos 7A + \sin 5A - \cos 3A}{\sin 7A - \cos 5A - \sin 3A} = -\tan 5A$

20 $\sin{(A + B)} - \sin A = \cos\left(A + \dfrac{B}{2}\right)\dfrac{\sin\dfrac{B}{2}}{\frac{1}{2}}$

21 If $A + B + C = 180°$, prove

$$\sin A + \sin B + \sin C = 4\cos\frac{A}{2}\cos\frac{B}{2}\cos\frac{C}{2}$$

Hint: $\sin C = 2\sin\dfrac{C}{2}\cos\dfrac{C}{2}, \ \sin\dfrac{A + B}{2} = \cos\dfrac{C}{2}$

22 Explain how formula (12) can be derived from formula (11).

23 Verify formula (14) for $B = A$.

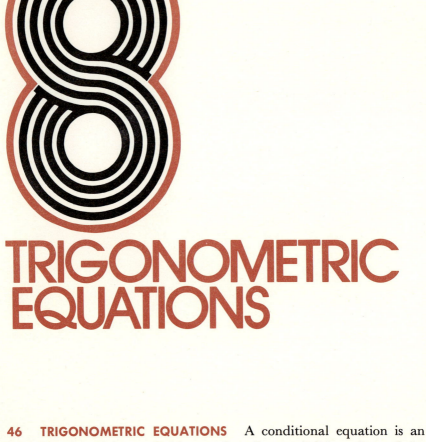

8
TRIGONOMETRIC EQUATIONS

46 TRIGONOMETRIC EQUATIONS   A conditional equation is an equation that does *not* hold true for all permissible values of the letters involved.   (The student should review Section 23 before proceeding.)   If a conditional equation involves trigonometric functions, it is called a trigonometric equation, in contrast to a trigonometric identity.   *A solution of a trigonometric equation is a value of the angle that satisfies the equation.*   For example, $\theta = 90°$ is a solution of the equation $\sin \theta = 1 + \cos \theta$.   Any angle coterminal with 90° is also a solution.   Thus $\theta = 450°$, 810°, etc., are also solutions.   In this book, unless stated to the contrary, *to solve a trigonometric equation shall mean to find all positive (or zero) solutions less than 360°*, that is, all θ on the interval $0° \leq \theta < 360°$, or $0 \leq \theta < 2\pi$.

47 SOLVING A TRIGONOMETRIC EQUATION The process of finding the solutions of a trigonometric equation involves algebraic as
well as trigonometric methods. There is no general rule, but the
following suggestions will take care of most cases.

(A) *If only one function of a single angle is involved, solve algebraically for
the values of the function.* Then determine the corresponding angles.

EXAMPLE 1 Solve for θ: $4 \cos^2 \theta = 3$

Solution $\cos^2 \theta = \frac{3}{4}$

Hence

$$\cos \theta = \frac{\sqrt{3}}{2} \qquad \Bigg| \qquad \cos \theta = -\frac{\sqrt{3}}{2}$$

$$\theta = 30°, 330° \qquad \Bigg| \qquad \theta = 150°, 210°$$

Therefore (arranging the solutions in order of size)

$$\theta = 30°, 150°, 210°, 330°$$

EXAMPLE 2 Solve for θ: $4 \sin^2 \theta = 3 \sin \theta$

Solution Transpose $3 \sin \theta$ to make the right side 0; then factor:

$$\sin \theta (4 \sin \theta - 3) = 0$$

$$\sin \theta = 0 \qquad \Bigg| \qquad 4 \sin \theta - 3 = 0$$
$$\theta = 0°, 180° \qquad \Bigg| \qquad \sin \theta = \tfrac{3}{4} = 0.75$$
$$\Bigg| \qquad \theta = 48° \ 35', 131° \ 25'$$

Hence

$$\theta = 0°, 48° \ 35', 131° \ 25', 180°$$

The student is warned to guard against dividing both sides
of this equation by the variable factor $\sin \theta$. Had this been
done, the solutions 0° and 180° would have been lost.

EXAMPLE 3 Solve for θ: $\sin^2 \theta - 5 \sin \theta - 3 = 0$

Solution The left side is not factorable. The roots of the quadratic
equation $ax^2 + bx + c = 0$ are

$$x = \frac{-b \pm \sqrt{b^2 - 4ac}}{2a}$$

In this case $a = 1$, $b = -5$, $c = -3$, and $x = \sin \theta$. The
formula gives us

$$\sin \theta = \frac{5 \pm \sqrt{25 + 12}}{2} = \frac{5 \pm 6.0828}{2}$$

$\sin \theta = 5.5414$	$\sin \theta = -0.5414$
Impossible	(Related angle is 32° 47′)
	θ is in Q III, IV
	$\theta = 212° 47′, 327° 13′$

Hence

$$\theta = 212° 47′, 327° 13′$$

(B) *If one side of the equation is zero and the other side is factorable, set each such factor equal to zero* and solve the resulting equations.

EXAMPLE 4   Solve for θ: $\cos 2\theta \csc \theta - 2 \cos 2\theta = 0$

Solution   Factor the left side:

$$\cos 2\theta \, (\csc \theta - 2) = 0$$

$\cos 2\theta = 0$	$\csc \theta - 2 = 0$
$2\theta = 90°, 270°, 450°, 630°$	$\csc \theta = 2$
$\theta = 45°, 135°, 225°, 315°$	$\sin \theta = \frac{1}{2}$
	$\theta = 30°, 150°$

Hence

$$\theta = 30°, 45°, 135°, 150°, 225°, 315°$$

In order to solve $\cos 2\theta = 0$ for all values of θ on the interval $0° \le \theta < 360°$, it is necessary to find all values of 2θ on the interval $0° \le 2\theta < 720°$.

(C) *If several functions of a single angle are involved,* use the fundamental relations to *express everything in terms of a single function.* Then proceed as in (A).

EXAMPLE 5   Solve for θ: $\sin^2 \theta - \cos^2 \theta - \cos \theta = 1$

Solution   Replace $\sin^2 \theta$ with $1 - \cos^2 \theta$; collect terms:

$$2 \cos^2 \theta + \cos \theta = 0$$
$$\cos \theta \, (2 \cos \theta + 1) = 0$$

$\cos \theta = 0$	$2 \cos \theta + 1 = 0$
$\theta = 90°, 270°$	$\cos \theta = -\frac{1}{2}$
	(Related angle is 60°)
	θ is in Q II, III
	$\theta = 120°, 240°$

Hence

$$\theta = 90°,\ 120°,\ 240°,\ 270°$$

EXAMPLE 6   Solve for θ: $\sec \theta = \tan \theta - 1$

Solution   Replace $\sec \theta$ with $\pm \sqrt{1 + \tan^2 \theta}$; square both sides:

$$1 + \tan^2 \theta = \tan^2 \theta - 2 \tan \theta + 1$$
$$2 \tan \theta = 0 \qquad \tan \theta = 0$$

Hence

$$\theta = 0°,\ 180°$$

Inasmuch as we squared the equation, we may have introduced some extraneous roots. Consequently, we must check these values in the original equation.

Check for $\theta = 0°$:

$$\sec 0° = \tan 0° - 1$$
$$1 = -1 \qquad \text{False}$$

Check for $\theta = 180°$:

$$\sec 180° = \tan 180° - 1$$
$$-1 = -1 \qquad \text{True}$$

Therefore $\theta = 0°$ is an *extraneous* root and $\theta = 180°$ is a true root. The only solution of the equation is $\theta = 180°$.

(*D*)   *If several angles are involved*, use the fundamental identities to *express everything in terms of a single angle.* Then proceed as in (*C*).

EXAMPLE 7   Solve for θ: $\cos 2\theta = 3 \sin \theta + 2$

Solution   This equation involves two angles. It is not convenient to replace $\sin \theta$ with a function of 2θ because this would introduce the radical

$$\pm \sqrt{\frac{1 - \cos 2\theta}{2}}$$

It is better to replace $\cos 2\theta$ with one of its three forms. Since the right side involves only $\sin \theta$, we choose the form $\cos 2\theta = 1 - 2 \sin^2 \theta$ in order to reduce everything immediately to the same function of a single angle. This gives us

$$1 - 2 \sin^2 \theta = 3 \sin \theta + 2$$

or $$2 \sin^2 \theta + 3 \sin \theta + 1 = 0$$

Factor:

$$(2 \sin \theta + 1)(\sin \theta + 1) = 0$$

$$\begin{array}{c|c} \sin \theta = -\tfrac{1}{2} & \sin \theta = -1 \\ \theta = 210°, 330° & \theta = 270° \end{array}$$

Hence*

$$\theta = 210°, 270°, 330°$$

EXERCISE 26

Solve the following equations for θ on the interval $0° \le \theta < 360°$.

1 $\cos(\theta - 20°) = -\dfrac{\sqrt{2}}{2}$

2 $\sin(\theta + 40°) = \dfrac{\sqrt{3}}{2}$

3 $3 \tan^2 \theta = 1$

4 $\tan^2(\theta + 20°) = 1$

5 $4 \sin^2 \theta = 3$

6 $\sec^2 \theta = 2$

7 $(2 \sin \theta + \sqrt{3})(2 \cos \theta + 1) = 0$

8 $(\sqrt{3} \csc \theta - 2)(\sqrt{2} \cos \theta + 1) = 0$

9 $(\sqrt{3} \sec \theta + 2)(2 \sin \theta - \sqrt{3}) = 0$

10 $(2 \cos \theta - \sqrt{3})(\csc \theta + \sqrt{2}) = 0$

11 $3 \cot^3 \theta = \cot \theta$

12 $2 \sin^4 \theta = \sin^2 \theta$

13 $2 \cos^2 \theta + \cos \theta = 1$

14 $\sin \theta - 2 \sin \theta \cos \theta + 1 - 2 \cos \theta = 0$

Solve the following equations for θ on the interval $0 \le \theta < 2\pi$.

15 $\cos \theta (\csc 2\theta + 1) = 0$

16 $2 \sin \theta \cos \theta - \sqrt{3} \sin \theta = 0$

17 $\sin \theta + \sqrt{3} \cos \theta = 0$

18 $10 \sin^2 \theta + 9 \cos \theta = 3$

19 $\sin \theta + 4 \csc \theta = 3$

20 $4 \cos^2 \theta - 12 \sin \theta - 9 = 0$

21 $12 \tan^2 \theta - \sec \theta + 11 = 0$

22 $\sec^2 \theta + \tan \theta = 1$

* *All* the solutions of this equation could be written as $210° + k \cdot 360°$, $270° + k \cdot 360°$, $330° + k \cdot 360°$, where k is an integer. Another form is $270° + k \cdot 360°$, $\left[k + \dfrac{(-1)^{k+1}}{6} \right] \cdot 180°$, where k is an integer.

23 $\sin 2\theta + \sqrt{2} \sin \theta = 0$

24 $\cos 2\theta + 11 \sin \theta = 6$

25 $\sin \theta + \cos 2\theta = 0$

26 $\sin 4\theta = 2 \cos 2\theta$

27 $\cos 6\theta = \cos^2 3\theta$

28 $\sin \theta + \sqrt{3} \sin \dfrac{\theta}{2} = 0$

29 $\sin 4\theta \sec 2\theta = 2 \cos \theta$

30 $\cos 2\theta = 2 \sin^2 \theta + 2 \sin \theta + 1$

Solve the following equations for θ on the interval $0° \leq \theta < 360°$.

31 $\dfrac{\tan 2\theta + \tan \theta}{1 - \tan 2\theta \tan \theta} = \sqrt{3}$

32 $9 \sin^4 \theta = \cos^4 \theta$

33 $\sec^2 \theta + \csc^2 \theta = \sec^2 \theta \csc^2 \theta$

34 $\tan (\theta + 10°) = \cot (\theta + 30°)$

35 $7 \sin^2 \theta + 9 \sin \theta + 2 = 0$

36 $5 \cos^2 \theta - 8 \cos \theta + 1 = 0$

37 $6 \sin^2 \theta - \cos \theta - 5 = 0$

38 $\cot^2 \theta + \csc \theta + 4 = 0$

39 $2 \tan^2 \theta - 9 \sec \theta - 3 = 0$

40 $5 \tan^2 \theta = 8 \tan \theta$

41 $8 \csc \theta \cot \theta - 27 \sec \theta \tan \theta = 0$

42 $2 + 8 \sin \theta = \csc \theta$

43 $1 + \cos \theta = -\sqrt{3} \sin \theta$

44 $2 + \sin \theta = \cos \theta$

45 $\sec \theta = \tan \theta + \sqrt{3}$

46 $\cos \theta + 1 = \sin \theta$

47 $3 \cot \theta \csc \theta = 2$

48 $1 + \tan \theta = -\sqrt{2} \sec \theta$

49 $4 \sin \theta + 3 \cos \theta = 2$

Hint: Express the left side in the form $k \sin (\theta + H)$.

50 $8 \sin \theta + 15 \cos \theta = 8.5$

51 $5 \sin \theta + 12 \cos \theta = 6.5$

52 $\sin \theta + \sqrt{3} \cos \theta = 1$

53 $\sin 3\theta = \cos (\theta + 50°)$

Hint: Transpose one term to the other side. Replace $\cos (\theta + 50°)$ with $\sin (40° - \theta)$. Express the difference as a product. As a partial check, sketch on the same system of coordinates the graphs of $y_1 = \sin 3\theta$ and $y_2 = \cos (\theta + 50°)$; then approximate the value of θ at each point of intersection.

54 $\cos 4\theta = \cos (2\theta - 60°)$

55 $\sin 3\theta + \sin \theta = \cos \theta$

Hint: Express the left side as a product.

56 $\cos 4\theta + \cos 2\theta = \cos 3\theta$

57 $\cos \theta + \tan \theta - \sec \theta = 0$

58 $\cot \theta \csc \theta = -\sqrt{2}$

59 $3\sqrt{2} \sin \dfrac{\theta}{2} - \sqrt{1 - \cos \theta} = \sqrt{6}$

60 $\sin \dfrac{\theta}{2} = 1 - \cos \theta$

61 $\sin 2\theta + \sin \theta + 4 \cos \theta + 2 = 0$

62 $\sin^2 \theta - 2 \cos^2 \theta + \sin \theta \cos \theta = 0$

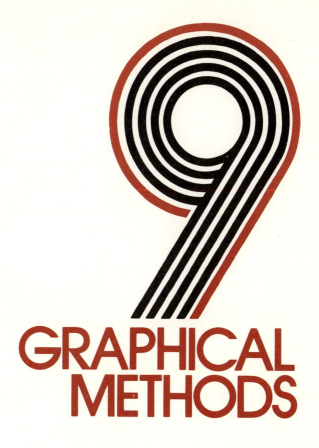

GRAPHICAL METHODS

48   THE GRAPH OF $y = a \sin bx$ The graph of $y = \sin x$ (Figure 40) is a wave that starts at 0, goes up to 1, drops back to 0, then to -1, and finally climbs back up to 0. (We are discussing the changes that take place in y as x varies from 0 to 2π.) The amplitude of the wave (i.e., the maximum distance from the x-axis) is 1. The period, or wave length, which is the same as the period* of the function $\sin x$, is 2π. The graph of $y = 3 \sin x$ (Figure 51) is another wave, with amplitude 3 and period 2π. For any value of x, the magnitude of y in $y = 3 \sin x$ is 3 times the value of y in $y = \sin x$. The coefficient 3 merely *stretches* the *height* of the sine curve by the multiple 3 without affecting its period.

* See Section 33, p. 76.

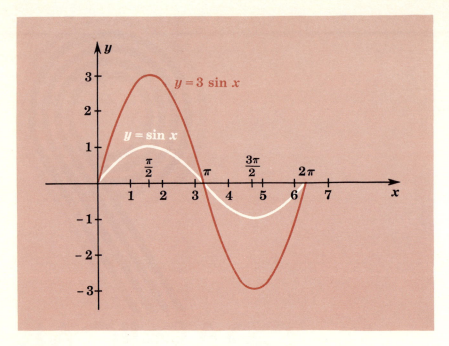

FIGURE 51

The graph of $y = \sin 2x$ is another sine wave* (Figure 52) with amplitude 1 but with period π. As x varies from 0 to $\pi/4$, $2x$ changes from 0 to $\pi/2$, and $\sin 2x$ increases from 0 to 1. The coefficient 2 *compresses* the curve *horizontally* by the multiple $\frac{1}{2}$. The period of the function $\sin 2x$ is half the period of $\sin x$. (Things happen twice as fast.)

To generalize and combine our two previous observations, we may say that

the graph of $y = a \sin bx$ is a sine wave

with amplitude a and period $\dfrac{2\pi}{b}$

A similar statement can be made for $y = a \cos bx$. Figure 53 shows the graph of $y = 3 \sin 2x$.

49   THE GRAPH OF $y = a \sin (bx + c)$ Consider the equation
$y = \sin (x + 30°)$. When $x = 0°$, $y = \sin 30° = \frac{1}{2}$. If $x = 60°$, $y = \sin 90° = 1$. As x varies from $0°$ to $360°$, y takes on the same

---

* We reserve the term "sine *curve*" for the graph of $y = \sin x$.

values assumed by y in the equation $y = \sin x$ *in the same order but starting at a different place.* The graph of $y = \sin (x + 30°)$ may be obtained (Figure 54) by shifting the graph of $y = \sin x$ to the *left* 30°, or $\pi/6$ units. If the reader does not follow the argument, he should plot as many points as necessary using $y = \sin (x + 30°)$. Suggested values to assign to x are 0°, 60°, 150°, 240°, 330°, and others if needed.

The graph of $y = \sin (x - \pi/2)$ is the sine curve shifted $\pi/2$ units to the *right* (Figure 54). Another approach is to use the formula for

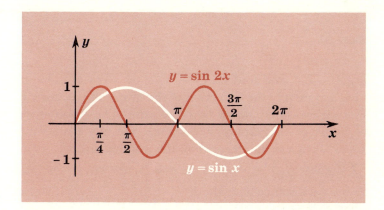

FIGURE 52

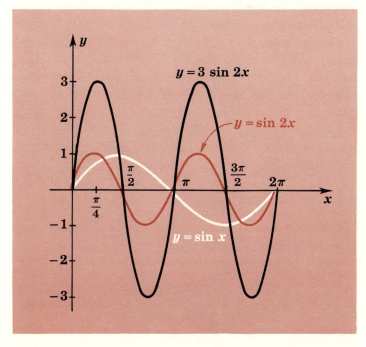

FIGURE 53

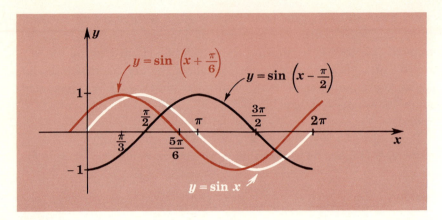

FIGURE 54

sin $(A - B)$* and find sin $(x - \pi/2) = -\cos x$. The graph of $y = -\cos x$ is merely the reflection of the graph of $y = \cos x$ in the x-axis. (Place a two-sided mirror on the x-axis. The graph of $y = -\cos x$ is the image of the graph of $y = \cos x$.) For any given value of x, the y of $y = -\cos x$ is the negative of the y of $y = \cos x$.

The graph of $y = \sin (x + k)$ may be obtained by shifting the graph of $y = \sin x$ a distance $|k|$ to the $\left\{ \begin{array}{l} \textit{left} \ \text{if } k \text{ is } \textit{positive} \\ \textit{right} \text{ if } k \text{ is } \textit{negative} \end{array} \right\}$. If you are in doubt, check with one point; set $x = 0$. The quantity $|k|$ is called the *phase displacement.*

Let us now consider the equation $y = 2.5 \sin (2x + \pi/3)$, which is equivalent† to $y = 2.5 \sin 2(x + \pi/6)$. First, sketch the graph of $y = \sin x$ (Figure 55). Second, *compress* the curve *horizontally* by the multiple $\frac{1}{2}$ to get the graph of $y = \sin 2x$ (Figures 52 and 55). Third, *shift* this curve $\pi/6$ units to the *left* to obtain the graph of $y = \sin 2(x + \pi/6)$. Finally, *stretch* this curve vertically to 2.5 of its original height to get the graph of $y = 2.5 \sin 2(x + \pi/6)$ (Figure 55). Again the reader is encouraged to check the result by assigning a few values to x in the original equation, computing the corresponding values of y, and then plotting the points. Some of the strategic points are $(0, 2.2)$, $(\pi/12, 2.5)$,‡ $(\pi/3, 0)$, $(7\pi/12, -2.5)$,‡ $(5\pi/6, 0)$. For the

---

* Formula (4), p. 94.

† Two equations are equivalent if every solution of one equation is a solution of the other, and vice versa.

‡ The maximum value of y is achieved when sin $(2x + \pi/3) = 1$; $2x + \pi/3 = \pi/2$; $x = \pi/12$. The minimum value of y occurs when $2x + \pi/3 = 3\pi/2$; $x = 7\pi/12$.

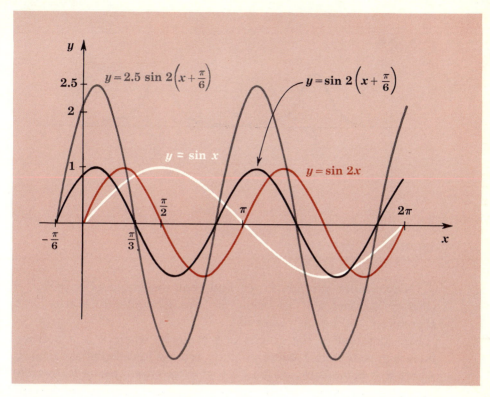

FIGURE 55

function $2.5 \sin (2x + \pi/3)$, the amplitude is 2.5, the period is $2\pi/2 = \pi$, and the phase displacement is $\pi/3 \div 2 = \pi/6$.

The function $a \sin (bx + c)$ has amplitude a, period $2\pi/b$, and phase displacement $|c/b|$. The graph of $y = a \sin (bx + c)$ may be obtained from the graph of $y = a \sin bx$ by shifting it horizontally $|c/b|$ units: to the *left* if c/b is *positive*, to the *right* if c/b is *negative*. A similar statement holds true for $a \cos (bx + c)$.

50   THE GRAPH OF $y = \sin^n x$ The graph of $y = \sin^3 x$ (Figure 56) may be obtained from that for $y = \sin x$ by "cubing all the y's." If $x = 0$, $\sin x = 0$ and $\sin^3 x = 0$, but for $x = \pi/6$, $\sin x = \frac{1}{2}$ whereas $\sin^3 x = \frac{1}{8}$.

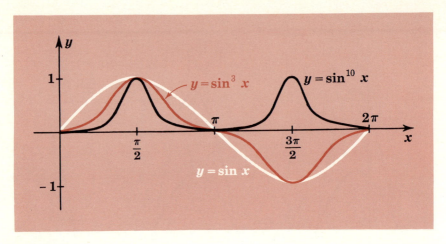

FIGURE 56

Using similar procedure, we find that the graph of $y = \sin^{10} x$ (Figure 56) does not fall below the x-axis because 10 is even and therefore $y \geq 0$ for all values of x.

Consideration of the graphs of $y = \sin^n x$, where n is a positive integer greater than 1, more properly belongs in differential calculus, where it is demonstrated that each of these curves passes *horizontally* through the points $(0,0)$, $(\pi/2,1)$, $(\pi,0)$, etc. Moreover, in integral calculus we learn that the area under one arch of the sine curve (i.e., the first-quadrant area beneath the curve $y = \sin x$ from $x = 0$ to $x = \pi$) is 2 square units. This seems reasonable because the circumscribing rectangle has altitude 1 and base π; its area is π square units. Furthermore, from calculus, the area under one arch of $y = \sin^3 x$ is $1\frac{1}{3}$ square units, and the area under one arch of $y = \sin^{10} x$ is $63\pi/256$ square units.

51 SKETCHING CURVES BY COMPOSITION OF y-COORDINATES

If the graphs of $y = f(x)$ and $y = g(x)$ are drawn to the same scale on the same set of axes, the graph of

$$y = f(x) + g(x)$$

can be sketched by the process of *composition of y-coordinates*. For any value of x, we can determine the y of the equation $y = f(x) + g(x)$ by finding graphically the *algebraic* sum of the y's of the two equations $y = f(x)$ and $y = g(x)$. After a suitable number of points have been located by "adding the heights of the given curves," we connect them with a smooth curve to get the required graph.

EXAMPLE 1   Graph the equation $y = x + \sin x$.

Solution   First draw the graphs of $y = x$ and $y = \sin x$ on the same axes (x being measured in radians). Place a straightedge parallel to the y-axis at M_1. Use compasses to add the segments M_1S_1 and M_1R_1. The sum is M_1P_1, thus locating P_1. To get point P_2, add the negative segment M_2S_2 to the positive segment M_2R_2; their *algebraic* sum is M_2P_2 (see Figure 57).

EXAMPLE 2   Graph the equation $y = \sin x + \sin 2x$.

Solution   After graphing the equations $y = \sin x$ and $y = \sin 2x$, we use the process of composition of y-coordinates (Figure 58).

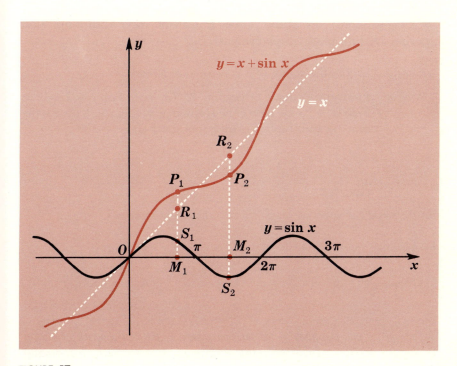

FIGURE 57

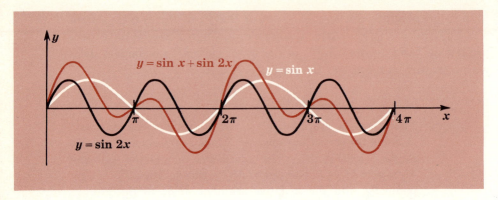

FIGURE 58

52   THE GRAPH OF $y = a \sin x + b \cos x$ One plan of attack in drawing the graph of $y = a \sin x + b \cos x$ is first to sketch (on the same coordinate system) the graphs of $y = a \sin x$ and $y = b \cos x$ and then to use the method of composition of y-coordinates.   A much shorter way, however, is to reduce $y = a \sin x + b \cos x$ to the form $y = k \sin (x + H)$, the graph of which is a sine wave with amplitude k, period 2π, and phase displacement $|H|$.

EXAMPLE   Sketch the graph of $y = 4 \sin x + 3 \cos x$.

Solution   From the example of Section 42, we have $4 \sin x + 3 \cos x = 5 \sin (x + 0.64)$.   Hence the graph of $y = 4 \sin x + 3 \cos x$ is the same as the graph (Figure 59) of $y = 5 \sin (x + 0.64)$.

EXERCISE 27

Find the period and amplitude of each of the following functions of x.

1   $4 \sin 6x$ 2 $5 \sin 3x$ 3 $\sin \dfrac{\pi x}{4}$ 4 $7 \sin \dfrac{\pi x}{3}$

5   $10 \tan \dfrac{3x}{8}$ 6 $\cos \dfrac{2x}{7}$ 7 $\frac{1}{2} \cos 8x$ 8 $\tan 4x$

Find the period and amplitude of the function defined by the first equation in each of the following problems.   Give the position of its graph (number of units left or right) relative to the graph of the second equation.

9   $y = \sin (3x - 6)$ $y = \sin 3x$

10 $y = 5 \sin\left(\dfrac{\pi x}{4} + \pi\right)$ $y = 5 \sin \dfrac{\pi x}{4}$

11 $y = 7 \sin\left(4x + \dfrac{\pi}{3}\right)$ $y = 7 \sin 4x$

12 $y = 9 \sin\left(2x - \dfrac{\pi}{5}\right)$ $y = 9 \sin 2x$

Sketch graphs of the following equations on the specified intervals.

13 $y = 5 \sin 4x$ $x = 0$ to $x = \pi$
14 $y = 1.5 \cos 2x$ $x = 0$ to $x = \pi$

15 $y = \cos \dfrac{\pi x}{3}$ $x = 0$ to $x = 6$

16 $y = 4 \sin 3x$ $x = 0$ to $x = \pi$

17 $y = 3 \cos\left(x - \dfrac{\pi}{2}\right)$ $x = 0$ to $x = 2\pi$

18 $y = \sin\left(x - \dfrac{\pi}{6}\right)$ $x = 0$ to $x = 2\pi$

19 $y = 1.5 \sin\left(x + \dfrac{\pi}{4}\right)$ $x = 0$ to $x = 2\pi$

20 $y = \cos\left(x + \dfrac{\pi}{3}\right)$ $x = 0$ to $x = 2\pi$

21 $y = \sin\left(\dfrac{2x}{3} + 60°\right)$ $x = 0°$ to $x = 540°$

22 $y = 4 \sin(3x + 30°)$ $x = 0°$ to $x = 180°$

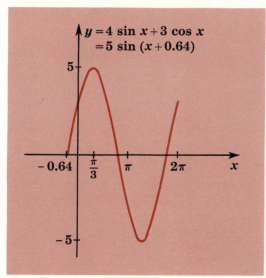

FIGURE 59

23 $y = \sin(2x - 40°)$ $x = 0°$ to $x = 180°$

24 $y = 2\sin\left(\dfrac{x}{2} - 80°\right)$ $x = 0°$ to $x = 720°$

25 $y = \tan\left(x - \dfrac{\pi}{4}\right)$ $x = 0$ to $x = 2\pi$

26 $y = \cos^2 x$ $x = 0$ to $x = 2\pi$
27 $y = \sin^5 x$ $x = 0$ to $x = 2\pi$
28 $y = \tan 4x$ $x = 0$ to $x = \pi$

Use composition of y-coordinates in sketching graphs of the following equations.

29 $y = 2 + \sin x$ 30 $y = x + \cos x$

31 $y = \dfrac{x}{2} + \sin x$ 32 $y = \sin x + \sin 3x$

33 $y = \cos x + \cos 2x$ 34 $y = \sin 3x + \cos x$
35 $y = \frac{1}{2} + \cos 2x$

Graph the following equations after changing them to the form $y = k\sin(x + H)$.

36 $y = \sin x + \cos x$ 37 $y = \sin x + \sqrt{3}\cos x$
38 $y = 3\sin x - \cos x$ 39 $y = \sin x - 2\cos x$

40 (a) Graph $y = \sin^2 x$.
    (b) Graph $y = \frac{1}{2} - \frac{1}{2}\cos 2x$.
    (c) Why are the graphs identical?

LOGARITHMS

53 **THE USES OF LOGARITHMS** Logarithms are used to shorten the labor involved in computing products and quotients, raising to a power, and extracting roots. For example, the operation of multiplying 234.56 by 9876.5 can be reduced to the operation of adding the logarithms of these numbers, namely, 2.37025 and 3.99460. Since the solution of triangles (Chaps. 11 and 12) involves such computations, we shall first consider the theory of logarithms.

54 **SOME LAWS OF EXPONENTS** A logarithm, as we shall see later, is an exponent. Accordingly we shall first review the following laws of exponents.

(1) $a^m \cdot a^n = a^{m+n}$ *Example:* $2^3 \cdot 2^4 = 2^7$

(2) $(a^m)^n = a^{mn}$ *Example:* $(2^3)^4 = 2^{12}$

(3) $\dfrac{a^m}{a^n} = a^{m-n}$ *Example:* $\dfrac{2^8}{2^2} = 2^6$

(4) $a^{m/n} = (\sqrt[n]{a})^m = \sqrt[n]{a^m}$ *Example:* $8^{4/3} = (\sqrt[3]{8})^4 = 2^4 = 16$

(5) $a^0 = 1$ *Example:* $(\frac{2}{3})^0 = 1$

(6) $a^{-n} = \dfrac{1}{a^n}$ *Example:* $3^{-2} = \dfrac{1}{3^2} = \dfrac{1}{9}$

Although these laws are true for all values of m and n and for positive and negative values of a, we shall have occasion to use them for just positive values of a.

55   DEFINITION OF LOGARITHM

The logarithm of a number to a given base is the exponent which must be placed on the base to produce the number.*

Thus the logarithm of 9 to the base 3 (written $\log_3 9$) is 2 because $3^2 = 9$.

ILLUSTRATIONS

$\log_2 8 = 3$	because	$2^3 = 8$
$\log_7 1 = 0$	because	$7^0 = 1$
$\log_3 \frac{1}{3} = -1$	because	$3^{-1} = \frac{1}{3}$
$\log_2 \dfrac{1}{16} = -4$	because	$2^{-4} = \dfrac{1}{2^4} = \dfrac{1}{16}$
$\log_{25} 5 = \frac{1}{2}$	because	$(25)^{1/2} = \sqrt{25} = 5$
$\log_8 4 = \frac{2}{3}$	because	$8^{2/3} = (\sqrt[3]{8})^2 = 2^2 = 4$
$\log_6 6 = 1$	because	$6^1 = 6$

The definition of logarithm implies that

---

* This exponent need not be rational. For example, $\log_{10} 2$ is an irrational number which, rounded off to five decimal places, becomes 0.30103. If the exponent $\dfrac{30{,}103}{100{,}000}$ is placed on 10, the result is 2 *approximately.*

if $\qquad \log_b N = x$

then $\qquad b^x = N$

These two equations, the former logarithmic and the latter exponential, are equivalent. They say the same thing in two different ways. We shall assume in further discussions that N is a positive number and that b is a positive number different from 1. (Explain the necessity for such restrictions.) But x may be any real number: positive, negative, or zero.

Since, by the definition of logarithm, $\log_b N$ is the exponent which must be placed on b to produce N, it follows that if $\log_b N$ is applied as an exponent to the base b, then the result must be N:

$$b^{\log_b N} = N$$

EXERCISE 28

Find the value of each of the following logarithms.

1 $\log_7 49$

2 $\log_5 \frac{1}{5}$

3 $\log_6 6$

4 $\log_3 81$

5 $\log_8 1$

6 $\log_{36} 6$

7 $\log_{1/9} 9$

8 $\log_{10} \frac{1}{1000}$

9 $\log_{64} 4$

10 $\log_{3/2} 1$

11 $\log_8 \sqrt{2}$

12 $\log_{27} 9$

13 $\log_{100} \frac{1}{1000}$

14 $\log_8 \frac{1}{4}$

15 $\log_{2/5} \frac{125}{8}$

16 $\log_{32} \frac{1}{2}$

Find the unknown, N, b, or x in the following.

17 $\log_b \frac{1}{81} = -4$

18 $10^{\log_{10} 2} = x$

19 $\log_{32} N = -\frac{4}{5}$

20 $\log_b 3^8 = 8$

21 $\log_7 7^{365} = x$

22 $\log_6 N = -2$

23 $\log_b 128 = 7$

24 $\log_{16} \frac{1}{8} = x$

25 $\log_{10} N = 1$

26 $\log_b 6 = 3$

27 $e^{\log_e 5} = x$

28 $\log_{12} N = 0$

Express as a logarithmic equation.

29 $625^{3/4} = 125$

30 $p^q = r$

31 $7^0 = 1$

32 $1024^{1/5} = 4$

Express as an exponential equation.

33 $\log_{1/9} 3 = -\frac{1}{2}$

34 $\log_{243} 81 = \frac{4}{5}$

35 $\log_{216} 36 = \frac{2}{3}$

36 $\log_b c = d$

Identify as true or false and give reasons.

37 $\log_2 16^m = 4m$

38 $\log_{100} \dfrac{\sqrt[3]{100}}{10} = -\dfrac{1}{6}$

39 $\log_{25} 5^n = \dfrac{n}{2}$

40 $\log_{16} \sqrt{2} = \frac{1}{8}$

56 PROPERTIES OF LOGARITHMS As a consequence of the definition of logarithm, we have three properties or laws of logarithms. They are used in computations involving logarithms.

PROPERTY 1 **The logarithm of a product is equal to the sum of the logarithms of the factors; i.e.,**

$$\log MN = \log M + \log N*$$

Proof Let $x = \log_a M$ and $y = \log_a N$

Express in exponential form: $M = a^x$ and $N = a^y$

Multiply the equations: $MN = a^x a^y = a^{x+y}$

Change to logarithmic form: $\log_a MN = x + y$
$\log_a MN = \log_a M + \log_a N$

The proof is similar for a product of more than two factors.

ILLUSTRATIONS $\log 35 = \log 5 \cdot 7 = \log 5 + \log 7$
$\log 30 = \log 2 \cdot 3 \cdot 5 = \log 2 + \log 3 + \log 5$

PROPERTY 2 **The logarithm of a fraction is equal to the logarithm of the numerator minus the logarithm of the denominator; i.e.,**

$$\log \dfrac{M}{N} = \log M - \log N$$

Proof Let $x = \log_a M$ and $y = \log_a N$

Express in exponential form: $M = a^x$ and $N = a^y$

Divide the equations: $\dfrac{M}{N} = \dfrac{a^x}{a^y} = a^{x-y}$

Change to logarithmic form: $\log_a \dfrac{M}{N} = x - y.$

$$\log_a \dfrac{M}{N} = \log_a M - \log_a N.$$

* The base is the same for all the logarithms.

ILLUSTRATIONS $\qquad \log \frac{2}{3} = \log 2 - \log 3$

$$\log \frac{6}{35} = \log \frac{2 \cdot 3}{5 \cdot 7} = \log 2 + \log 3 - (\log 5 + \log 7)$$

PROPERTY 3 **The logarithm of the kth power of a number is equal to k times the logarithm of the number; i.e.,**

$$\log N^k = k \log N.$$

Proof Let $\qquad x = \log_a N$

Express in exponential form: $\qquad N = a^x$

Raise to the kth power: $\qquad N^k = (a^x)^k = a^{kx}$

Change to logarithmic form: $\qquad \log_a N^k = kx$

$$\log_a N^k = k \log_a N$$

ILLUSTRATIONS $\qquad \log 8 = \log 2^3 = 3 \log 2$

$\qquad\qquad\qquad \log \sqrt{3} = \log 3^{1/2} = \frac{1}{2} \log 3$

$$\log \frac{125}{49} = \log \frac{5^3}{7^2} = \log 5^3 - \log 7^2 = 3 \log 5 - 2 \log 7$$

NOTE. Since $\sqrt[r]{M} = M^{\frac{1}{r}}$, $\log \sqrt[r]{M} = \frac{1}{r} \log M$.

EXERCISE 29

Given $\log_{10} 2 = 0.30$, $\log_{10} 3 = 0.48$, $\log_{10} 7 = 0.85$, *find the following logarithms.* (Remember $\log_{10} 10 = 1$.)

1	$\log_{10} \frac{2}{3}$	2	$\log_{10} 6$
3	$\log_{10} 210$	4	$\log_{10} \frac{7}{2}$
5	$\log_{10} 64$	6	$\log_{10} \sqrt{2}$
7	$\log_{10} \sqrt[6]{3}$	8	$\log_{10} 81$
9	$\log_{10} \sqrt{5}$	10	$\log_{10} (\frac{3}{7})^{10}$
11	$\log_{10} 2(3^9)$	12	$\log_{10} \sqrt[5]{\frac{2}{7}}$
13	$\log_{10} \frac{21}{16}$	14	$\log_{10} 5 \sqrt[3]{14}$
15	$\log_{10} \frac{\sqrt{2}}{49}$	16	$\log_{10} 30(7^6)$

Express as a single logarithm. (Assume all logarithms have the same base.)

17 $\log (a + 1) + \log (a - 1)$

18 $\log (a - b) + \log b$

19 $\log \pi + \frac{1}{2}(\log l - \log g)$

20　$7 \log a + \frac{1}{4} \log (b + c^2)$

21　$5 \log x - \frac{1}{3} \log y - \frac{2}{7} \log z^7$

22　$\frac{1}{2}[\log (s - a) + \log (s - b) + \log (s - c) - \log s]$

23　$4 \log a - \frac{5}{6} \log b + \frac{2}{3} \log c$

24　$\frac{3}{8} \log x - 7 \log y - \frac{1}{4} \log z$

Identify as true or false and give reasons. (In each equation the base is assumed to be the same for all the logarithms.)

25　$\log \dfrac{1}{a} = -\log a$　　　　　　　**26**　$(\log b)^2 = 2 \log b$

27　$\sqrt{x} = \frac{1}{2} \log x$　　　　　　　　**28**　$\dfrac{a}{b} = \log a - \log b$

29　$y^5 = 5 \log y$　　　　　　　　　**30**　$\log (ab)^5 = (\log a + \log b)^5$

31　$\log a = \dfrac{1}{p} \log a^p$　　　　　　**32**　$7^{4 \log_7 x + \log_7 y} = x^4 y$

33　$10^{3 + \log_{10} 7} = 7000$

34　$\dfrac{1}{2} \log x - \dfrac{1}{3} \log y - \dfrac{1}{6} \log z = \dfrac{1}{6} \log \dfrac{x^3}{y^2 z}$

35　$\log \dfrac{x}{yz} = \dfrac{\log x}{\log y + \log z}$

36　$2 \log (\log x) = \log (\log x)^2$

37　$\log (x^5 - y^6) = 5 \log x - 6 \log y$

38　$8^{(\log_8 a)/3} = \sqrt[3]{a}$

39　$5^{\log_5 x - 3 \log_5 y} = \dfrac{x}{y^3}$

40　$\log (\csc \theta + \cot \theta) = -\log (\csc \theta - \cot \theta)$

41　$\log \dfrac{u - a}{\sqrt{u^2 - a^2}} = \dfrac{1}{2} \log \dfrac{u - a}{u + a} \qquad u > a > 0$

57　SYSTEMS OF LOGARITHMS　There are only two important systems of logarithms in use today. The *natural*, or *Napierian*, system employs the base e, where e is approximately 2.71828. This system is encountered in calculus and higher mathematics. The *common*, or *Briggs*, system employs the base 10. This system is most convenient for computation because our number system uses the base 10. Henceforth, in this text, when the base of a logarithm is not specified, we are to understand that the base is 10. Thus $\log N$ means $\log_{10} N$. And, unless stated to the contrary, the word *logarithm* shall mean common logarithm.

58   CHARACTERISTIC AND MANTISSA   We know that

$$\log 1000 = 3 \quad \text{because} \quad 10^3 = 1000$$
$$\log 100 = 2 \quad \text{because} \quad 10^2 = 100$$
$$\log 10 = 1 \quad \text{because} \quad 10^1 = 10$$
$$\log 1 = 0 \quad \text{because} \quad 10^0 = 1$$
$$\log 0.1 = -1 \quad \text{because} \quad 10^{-1} = 0.1$$
$$\log 0.01 = -2 \quad \text{because} \quad 10^{-2} = 0.01$$
$$\log 0.001 = -3 \quad \text{because} \quad 10^{-3} = 0.001$$

It seems reasonable to assume that, as a number increases, its logarithm increases.*   Consequently, any number lying between 10 and 100 must have a logarithm between 1 and 2.   This logarithm can be written, then, in the form 1 plus a positive decimal.†   For example, $\log 45.7 = 1 + .65992 = 1.65992$.   Also, any number between 0.001 and 0.01 must have a logarithm between -3 and -2.   This logarithm can be written in the form -3 plus a positive decimal.   For example, $\log 0.006 = -3 + .77815$, which can be written $\log 0.006 = 7.77815 - 10$ because $-3 = 7 - 10$.

As a matter of convenience, *every logarithm is usually written as the sum of an integer* (positive, negative, or zero) *plus a positive decimal.*†   The integer is called the **characteristic** of the logarithm; the positive decimal is called the **mantissa** of the logarithm.   This is illustrated in the following table.

Logarithm	Characteristic	Mantissa
4.56789	4	.56789
0.23456	0	.23456
3.00000	3	.00000
8.77665 − 10 or −2 + .77665	8 − 10 or −2	.77665
7.11111 − 10 or −2.88889	7 − 10 or −3	.11111

59   METHOD OF DETERMINING CHARACTERISTICS

If a number has a decimal point immediately to the right of its first nonzero digit, then the decimal point is said to be in *standard position.*

* This is proved in more advanced texts.
† Positive decimal here means a number n such that $0 \leq n < 1$.

For example, the decimal point is in standard position in each of the following: 6.507, 4.17, 3.2, and 8.   Consequently, *if a number N has its decimal point in standard position,* then *N* is between 1 and 10, and log *N* is between 0 and 1; therefore, *the characteristic of log N is 0.*

THEOREM 1   **Whenever a number is multiplied by 10, its logarithm is increased by 1.**

Proof   Let log *N* be the logarithm of any number *N*.   Then

$$\log 10N = \log 10 + \log N \qquad \text{(Property 1)}$$
$$\log 10N = 1 + \log N$$

It is therefore apparent that when a number is multiplied by 10 (i.e., if the decimal point is moved one place to the right), the characteristic of its logarithm is increased by 1, but the mantissa is unaltered.

ILLUSTRATION   If log 2.345 = 0.37014, then log 23.45 = 1.37014.

By repeating this process, we see that if a number is *multiplied* by 10^k (i.e., if the decimal point is moved *k* places to the right), the characteristic of its logarithm is *increased* by *k*.   It also follows that if a number is *divided* by 10^k (i.e., if the decimal point is moved *k* places to the left), the characteristic of its logarithm is *decreased* by *k*.

ILLUSTRATION   If   log 2.345 = 0.37014, then

$$\log 234.5 = 2.37014$$
$$\log 23450 = 4.37014$$
$$\log 0.2345 = 9.37014 - 10$$
$$\log 0.002345 = 7.37014 - 10$$

We may sum up our discussion in the following

THEOREM 2   **If the decimal point in a number is *k* places to the $\left\{ \begin{array}{l} \text{right} \\ \text{left} \end{array} \right\}$ of standard position, then the characteristic of the logarithm of the number is $\left\{ \begin{array}{l} k \\ -k \end{array} \right\}$.**

ILLUSTRATION   The characteristic of log 8765 is 3 because the decimal point is understood to be after the 5, which is 3 places to the right of standard position (i.e., after the 8).
          The characteristic of log 0.08765 is −2 or 8 − 10 because the decimal point is 2 places to the left of standard position.

Number	Characteristic of Logarithm
456780	5
456.78	2
4.5678	0
0.45678	9 − 10 or −1
0.045678	8 − 10 or −2
0.00045678	6 − 10 or −4

An alternate method used in finding characteristics is:

1. For a number N that is greater than 1, the characteristic is one less than the number of digits to the left of the decimal point in N.
2. For a number N that is less than 1, the characteristic is negative and is numerically equal to one more than the number of zeros between the decimal point and the first nonzero digit in N.

THEOREM 3 **The mantissa of the logarithm of a number N is independent of the position of the decimal point in N.**

This means that two numbers differing only in the position of the decimal point have logarithms with the same mantissa. The proof of this theorem is embodied in that of Theorem 1 and the discussion that follows it.

ILLUSTRATION The following numbers have logarithms with the same mantissa:

$$0.04689, \qquad 4.689, \qquad 46.89, \qquad 46890.$$

Theorem 2 serves two purposes:

1. If we look at the position of the decimal point in a *number,* we can determine the characteristic of its *logarithm.*
2. If we look at the characteristic of the *logarithm* of a number, we can determine the position of the decimal point in the *number.*

EXAMPLE 1 Given $\log 1.616 = 0.20844$. Find:

(a) $\log 0.01616$
(b) N if $\log N = 5.20844$
(c) N if $\log N = 6.20844 - 10$

Solution (a)   log 0.01616 = 8.20844 − 10

Since the decimal point in 0.01616 is **2** places to the *left* of standard position, the characteristic is −**2** or 8 − 10. The mantissa is the same as that for log 1.616.

(b) If   log N = 5.20844
             N = 161600

Since log N and log 1.616 have the same mantissa, N is obtainable from 1.616 by moving the decimal point **5** places to the right (from standard position).

(c) If   log N = 6.20844 − 10
             N = 0.0001616

Since the characteristic of log N is −4, we obtain N from 1.616 by moving the decimal point 4 places to the left.

NOTE.   Theorems 1, 2, 3 of this article are valid only if the base of the logarithms is 10. For any other base, the process of finding the characteristic would not be so simple.

EXERCISE 30

Given log 3.456 = 0.53857 *and* log 7.703 = 0.88666, *find the value of the following.*

1	log 34,560	2	log 0.3456
3	log 0.07703	4	log 770.3
5	log 0.007703	6	log 7703
7	log 3,456,000	8	log 0.000 3456
9	log 34.56	10	log 0.000 03456
11	log 0.7703	12	log 770,300

Given log 6.607 = 0.82000 *and* log 12.34 = 1.09132, *find N for each of the following.*

13	log N = 8.09132 − 10	14	log N = 0.09132
15	log N = 7.82000	16	log N = 9.82000 − 10
17	log N = 2.82000	18	log N = 7.82000 − 10
19	log N = 6.09132 − 10	20	log N = 4.09132
21	log N = 9.09132 − 20	22	log N = 5.09132
23	log N = 1.82000	24	log N = 5.82000 − 10
25	log N = −0.18000	26	log N = −3.18000
27	log N = −2.90868	28	log N = −1.90868

Hint: For Probs. 25 to 28, first write log N in a form where the decimal part is positive. For example, −0.18000 = 9.82000 − 10.

60   A FIVE-PLACE TABLE OF MANTISSAS   In Table 2 (pages 250–267 in the tables) there are listed, to five decimal places, the mantissas (with the decimal points omitted) of the logarithms of all whole numbers from 1 to 9999.   Since the mantissa is independent of the decimal point in the number, Table 2 can be used to find the mantissa of the logarithm of any four-figure number.   The problems we shall need to consider are:

     1. Given a number N, to find log N.
     2. Given log N, to find N.

61   GIVEN N, TO FIND log N   The procedure of finding the logarithm of a given number is illustrated by the following examples.

EXAMPLE 1   Find log 0.003467.

Solution   The characteristic of log 0.003467 is $7 - 10$.   To find the mantissa, look for 346 in the left-hand column headed "N" on page 254.   In the line beginning with 346, move over to the column headed by 7.   Here we find 995, which represents the last three figures of the five-place mantissa.   The first two figures, 53, are found in the column headed by 0.   Hence the mantissa of log 3467 (and log 0.003467) is .53995.   Therefore,

$$\log 0.003467 = 7.53995 - 10$$

EXAMPLE 2   Find log 346.8.

Solution   The required characteristic is 2.   To find the mantissa, go down the N column to 346 and then over to the 8 column and find 008.   The asterisk (*) indicates that we are to prefix the 54 of the following line instead of using the 53 as in the preceding example.   Hence the mantissa of log 3468 is .54008.   Therefore,

$$\log 346.8 = 2.54008$$

62   GIVEN log N, TO FIND N   The procedure of finding a number whose logarithm is given is illustrated by the following examples.

EXAMPLE 1   Given log N = 4.38686, find N.

Solution   Look for the mantissa .38686 in the body part of Table 2.   It appears on page 252 in the 243 line and the 7 column.   Hence N is 2437 with the decimal point placed in accordance with a characteristic of 4.   Therefore,

if   log N = 4.38686
              N = 24370

EXAMPLE 2   Given log N = 8.09202 − 10, find N.

Solution   The mantissa .09202 appears in the 123 line and in the 6 column.   Hence N is 1236 with the decimal point moved 2 places to the left of standard position (because the characteristic is 8 − 10 or −2).   Therefore,

if   log N = 8.09202 − 10
              N = 0.01236

EXERCISE 31

Use a five-place table to find the value of each of the following.

1	log 5.921	2	log 77.22
3	log 0.000 07210	4	log 0.002261
5	log 446,600	6	log 0.09126
7	log 37,180	8	log 5678
9	log 0.000 8917	10	log 1,007,000
11	log 3.412	12	log 0.000 006166

Use a five-place table to determine the value of N in each of the following.

13	log N = 3.04060	14	log N = 4.73496 − 10
15	log N = 5.39235	16	log N = 8.46030 − 10
17	log N = 5.80366 − 10	18	log N = 4.43632
19	log N = 7.15045 − 10	20	log N = 0.64197
21	log N = 6.79239	22	log N = 6.95012 − 10
23	log N = 2.52776	24	log N = 1.08350

63   INTERPOLATION   We have already seen (Section 60) that the logarithm of a four-figure number can be found in Table 2.   The logarithm of a five-figure number can also be found from the table by interpolation (Section 17).   It is to be remembered that, since the mantissas of the logarithms of most integers are unending decimals, a five-place table merely rounds off these mantissas to five-figure accuracy. For this reason a five-place table will not give satisfactory results when we are dealing with six-figure numbers. Consequently, whenever we encounter a six-figure number, we shall round it off to five-figure accuracy before using Table 2.

EXAMPLE 1 Find log 24357.

Solution The characteristic is 4. The mantissa lies between the mantissas for 24350 and 24360.

$$
\begin{array}{ll}
\text{mantissa of log } 24350 & = .38650 \\
\text{mantissa of log } 24357 & = \\
\text{mantissa of log } 24360 & = .38668
\end{array}
$$

with brackets indicating 7, 10, and 18.

Using the same procedure as in Section 17, we take $\frac{7}{10}$ of the tabular difference 18 to get our difference $12.6 \rightarrow 13$ and add this to the top mantissa. Hence, mantissa of log 24357 is .38663. Therefore,

$$\log 24357 = 4.38663$$

Instead of writing this three-line arrangement, we can interpolate mentally by using the tables of proportional parts which are at the right of each page. To find $\frac{7}{10}$ of 18, locate the short column with 18 at the top. Follow down to the line that has 7 at the left and read 12.6 in the 18 column. Thus $\frac{7}{10}(18) = 12.6$.

EXAMPLE 2 Find log 0.17083.

Solution The characteristic is $9 - 10$. From page 251, the mantissa of log 17080 is .23249. The tabular difference is 25. Using the proportional parts table, we find $\frac{3}{10}(25) = 7.5$, which must be rounded off to 7 to make the final result even.* Hence

$$\log 0.17083 = 9.23256 - 10$$

EXAMPLE 3 Given log $N = 7.38591 - 10$, find N.

Solution From page 252,

$$
\begin{array}{ll}
\text{mantissa of log } 24310 & = .38578 \\
\text{mantissa of log } N & = .38591 \\
\text{mantissa of log } 24320 & = .38596
\end{array}
$$

with brackets indicating 10, 13, and 18.

* See footnote on page 26.

Proceeding as in Section 17, we take $\frac{13}{18}$ of 10 and get $7\frac{2}{9} \rightarrow 7$, which when added to 24310 gives 24317. Hence

if $\log N = 7.38591 - 10$
$N = 0.0024317$

Here again we can dispense with the three-line arrangement and save time by using the tables of proportional parts. Obviously the result of taking $\frac{13}{18}$ of 10 is the answer to the question, "How many tenths of 18 will be equal to 13."* In the proportional parts table for 18, look for the number nearest to 13. This number is 12.6 and to its left we find 7. Hence, 7 must be added to 24310.

EXAMPLE 4   Given $\log N = 1.52255$, find N.

Solution   From page 254, mantissa of log 33300 is .52244. The tabular difference is 13 and our difference is $(255 - 244) = 11$. In the proportional parts table for 13, we seek the number nearest to 11. This number is 10.4 and it corresponds to 8. Hence

if $\log N = 1.52255$
$N = 33.308$

It is to be noted that, in going from a number to its logarithm, we work from left to right in the small proportional parts table as well as in the large table (Table 2). In going from a logarithm to the corresponding number, the movement is from right to left in both the small and large tables.

EXERCISE 32

Find the value of each of the following. (Use tables of proportional parts—not the three-line arrangement.)

1	log 0.53015	2	log 376.04
3	log 11.654	4	log 0.000 061472
5	log 4,467,800	6	log 0.0019086
7	log 0.000 83177	8	log 29,517
9	log 0.028067	10	log 79,903
11	log 987,510	12	log 0.89989
13	log 6241.3	14	log 4.3217
15	log 0.021438	16	log 31.924

* If $x = \frac{13}{18} (10)$, then $\frac{x}{10} (18) = 13$.

Find the value of N in each of the following. (Use tables of proportional parts.)

17	$\log N = 1.22624$	**18**	$\log N = 9.52412 - 10$
19	$\log N = 5.39719 - 10$	**20**	$\log N = 2.87914$
21	$\log N = 0.95010$	**22**	$\log N = 3.68021$
23	$\log N = 4.89022$	**24**	$\log N = 7.40063 -_. 10$
25	$\log N = 4.56858$	**26**	$\log N = 8.43148 - 10$
27	$\log N = 9.07945 - 10$	**28**	$\log N = 5.72479$
29	$\log N = 6.77820 - 10$	**30**	$\log N = 1.00747$
31	$\log N = 3.14548$	**32**	$\log N = 4.28246 - 10$

64   LOGARITHMIC COMPUTATION   When logarithms are used to compute products, quotients, and powers of numbers, it is advisable to:

1. Make a complete outline indicating the operations to be performed.
2. Fill in all characteristics.
3. Fill in mantissas.
4. Perform the operations outlined in step 1.

These suggestions are offered in the hope that accuracy, speed, and neatness will result. Every logarithm appearing in the solution should be labeled.

EXAMPLE 1   Use logarithms to compute $\dfrac{(2345)(0.6699)}{9.876}$.

Solution   Let
$$N = \frac{(2345)(0.6699)}{9.876}.$$

Then $\log N = \log 2345 + \log 0.6699 - \log 9.876$

After preparing the outline and filling in the characteristics, we have

$$\log 2345 = 3.$$
$$\log 0.6699 = 9. \qquad -10$$
$$\text{———————— Add}$$
$$\log \text{numerator} =$$
$$\log 9.876 = 0.$$
$$\text{———————— Subtract}$$
$$\log N =$$
$$N =$$

After supplying the mantissas and performing the indicated operations, we get

$$\log 2345 = 3.37014$$
$$\log 0.6699 = 9.82601 - 10$$
$$\text{——————} A$$
$$\log \text{numerator} = 13.19615 - 10$$
$$\log 9.876 = 0.99458$$
$$\text{——————} S$$
$$\log N = 12.20157 - 10$$
$$N = 159.1$$

Notice that no interpolation was performed in finding N from $\log N$. The original numbers are all four-figure numbers; hence the computed result should have no more than four-figure accuracy. Since the mantissa .20157 is best approximated by the tabular mantissa .20167, the best four-figure approximation of N is 159.1.

In all logarithmic computations we are really expressing the original numbers as powers of 10. For example, since $\log 2345 = 3.37014$, it follows that $2345 = 10^{3.37014}$.

Consequently

$$N = \frac{(10^{3.37014})(10^{9.82601-10})}{10^{0.99458}}$$

Applying the laws of exponents,

$$N = 10^{3.37014+(9.82601-10)-0.99458}$$
$$= 10^{2.20157} = 159.1$$

EXAMPLE 2   Use logarithms to compute $N = \dfrac{(1.2346)^3}{(60370)(0.045023)}$

Solution   Take the logarithm of each side:

$$\log N = 3 \log 1.2346 - (\log 60370 + \log 0.045023)$$

After taking the four suggested steps, we get

$$\log 1.2346 = \frac{0.09153}{0.27459}\,3 \qquad \log 60370 = 4.78082$$
$$3 \log 1.2346 = \overline{0.27459} \qquad \log 0.045023 = 8.65344 - 10$$
$$\log \text{num.} = 10.27459 - 10 \qquad \text{——————} A$$
$$\log \text{den.} = 3.43426 \leftarrow \qquad \log \text{den.} = 13.43426 - 10$$
$$\text{——————} S \qquad = 3.43426$$
$$\log N = 6.84033 - 10$$
$$N = 0.000\ 69236$$

It is to be noticed that log num. was changed from 0.27459 to 10.27459 − 10 to avoid subtracting 3.43426 from a smaller number.

EXAMPLE 3   Compute $\sqrt[7]{\dfrac{345.80}{4589}}$.

Solution   Let $N = \sqrt[7]{\dfrac{345.80}{4589}}$.

Then log $N = \tfrac{1}{7}$ (log 345.80 − log 4589).
After taking the four suggested steps, we have

$$
\begin{array}{rl}
\log 345.80 = & 12.53882 - 10 \\
\log 4589 = & 3.66172 \\
\hline
& \text{S} \\
\log \text{radicand} = & 8.87710 - 10 \\
\log \text{radicand} = & 68.87710 - 70 \\
7\, \hline \\
\log N = & 9.83959 - 10 \\
N = & 0.6912
\end{array}
$$

Notice that log radicand was changed from 8.87710 − 10 to 68.87710 − 70 to facilitate the division by 7. Had we divided 8.87710 − 10 by 7, the result, $1.26816 - 1.42857 = -0.16041$, would involve a *negative* decimal that does not appear in our table of *positive* mantissas.

The final result is written with only four-figure accuracy because the least accurate number, 4589, in the original data has only four significant figures.

EXAMPLE 4   Use logarithms to compute $x = \dfrac{(-1.2346)^3}{(-60370)(-0.045023)}$

Solution   The value of x is negative since $\dfrac{(-)^3}{(-)(-)} = \dfrac{-}{+} = -$. Discard all minus signs in x and then use logarithms to compute the value of the corresponding expression in which all numbers are positive. This was done in Example 2 with the result 0.000 69236. Hence

$$x = -0.000\ 69236$$

EXERCISE 33

Use logarithms to compute the following, correct to four-figure accuracy. (In finding N from log N, do not interpolate.)

1 $(1.479)^5(8526)$

2 $(5794)(0.001177)$

3 $(9.045)(0.000\ 1122)(87.21)$

4 $(255.2)(3578)$

5 $\dfrac{(31.09)^4}{772,200}$

6 $\dfrac{\sqrt{450.3}}{0.7806}$

7 $\dfrac{39,560}{7954}$

8 $\dfrac{8502}{0.1776}$

9 $(9.000)^{10}$

10 $(0.9238)^{100}$

11 $[(1239)(0.005678)]^8$

12 $\left(\dfrac{34.21}{6.912}\right)^4$

13 $\dfrac{(8.123)(654.1)}{(246.5)(0.007373)}$

14 $\dfrac{(0.03377)(21,920)}{864.2}$

15 $\dfrac{0.08395}{\sqrt[5]{428.4}}$

16 $\sqrt[4]{(236.7)(9191)}$

17 $\sqrt[3]{(0.002838)(0.05767)}$

18 $\sqrt[9]{0.005161}$

19 $\sqrt[6]{0.008172}$

20 $\sqrt[7]{\dfrac{0.004929}{271.4}}$

21 $\dfrac{-0.004004}{(-2.567)(-99.44)}$

22 $\left[\dfrac{-91.31}{(-0.06278)(-4005)}\right]^7$

23 $(-3.047)^4(0.9456)(-0.06723)$

24 $(-5.270)^3(6.804)$

25 $\sqrt{\log 459.2}$

26 $(\log 0.007211)^3$

27 $\dfrac{\log 7.173}{\log 18,450}$

28 $(\log 8091)(\log 18.03)$

29 $\dfrac{0.09292}{\sqrt[3]{582,400} + 14.23}$

30 $\dfrac{(2.143)^9 - 853.2}{15,780}$

31 $(0.6107)^{-10}$

32 $[(92.83)(0.02299)]^{-6}$

Use logarithms to compute the following, correct to five-figure accuracy.

33 $\dfrac{0.54321}{6122.4}$

34 $(585.41)(0.000\ 62530)(4.4309)$

35 $(0.0076543)^2(246.88)$

36 $\dfrac{(876.22)(9.0909)}{48.960}$

37 $\sqrt[6]{0.0060708}$

38 $\sqrt[3]{0.035790}$

39 $\sqrt[7]{\dfrac{0.000\ 32657}{975,030}}$

40 $\sqrt[5]{(0.000\ 77725)(0.054890)}$

41 $(1.1696)^{-30}$

42 $\left(\dfrac{0.096240}{2.9416}\right)^{-4}$

43 $(2,425,500)^{3/10}$

44 $\dfrac{(9.2345)\sqrt{36.480}}{(714.28)(0.81624)^5}$

Use logarithms to compute the following to as much accuracy as is warranted by the numbers involved.

45 $(0.214)(69.120)(8420)$

46 $(0.000\ 1906)(95.372)$

47 $\dfrac{0.2345}{701.68}$

48 $\dfrac{0.001313}{(0.042510)^2}$

49 $\dfrac{(1.830)^2}{(0.77450)^3}$

50 $(0.234)^5$

51 $(0.59979)^{40}(1954)^2$

52 $(0.692)^7 \sqrt[10]{75,980}$

53 $\sqrt[100]{30,405}$

54 $\dfrac{84,260}{\sqrt[5]{39.3}}$

55 $\sqrt[3]{48.0}\ \sqrt[4]{659,300}$

56 $(58,060)^{5/8}$

57   If a principal P is invested at an interest rate i per year, then the compound amount A to which P accumulates at the end of n years is given by $A = P(1 + i)^n$. Compute A when $P = 9725$, $i = 0.065$, $n = 10$.

58   The radius r of the inscribed circle of a triangle whose sides are a, b, and c is given by

$$r = \sqrt{\dfrac{(s - a)(s - b)(s - c)}{s}}$$

where $s = \frac{1}{2}(a + b + c)$. Compute r if $a = 54.32$, $b = 60.06$, $c = 76.54$.

65   **THE GRAPHS OF** $y = a^x$ **AND** $y = \log_a x$ If $a > 1$, the graph of $y = a^x$ is basically the same as that of $y = 2^x$ (Figure 60), which was obtained by use of the following table.

x	$\cdots$	-3	-2	-1	0	1	2	3	$\cdots$
$y = 2^x$	$\cdots$	$\frac{1}{8}$	$\frac{1}{4}$	$\frac{1}{2}$	1	2	4	8	$\cdots$

If $a > 1$, the graph of $y = \log_a x$ is essentially the same as that of $y = \log_2 x$ (Figure 60), which was drawn by use of the following table.

x	$\cdots$	$\frac{1}{8}$	$\frac{1}{4}$	$\frac{1}{2}$	1	2	4	8	$\cdots$
$y = \log_2 x$	$\cdots$	-3	-2	-1	0	1	2	3	$\cdots$

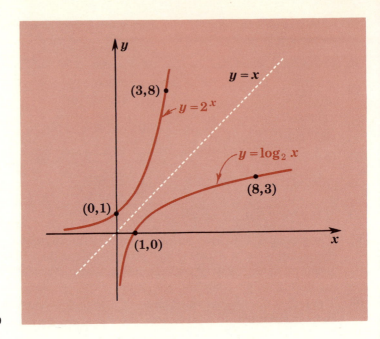

FIGURE 60

Notice that the function 2^x increases as x increases. This function could be called "the doubling function" because whenever x increases by 1, the function doubles its size. $(2^{x+1} = 2 \cdot 2^x.)$

Observe that the function $\log_2 x$ also increases as x increases. Notice that $\log_2 x$ is not defined for $x < 0$. A negative number does not have a (real) logarithm.

The equation $y = \log_2 x$ is equivalent to the equation $x = 2^y$, which is obtainable from $y = 2^x$ by merely interchanging x and y. If (a,b) is a point on the graph of $y = 2^x$, then (b,a) is a point on the graph of $y = \log_2 x$. It can be shown that if a mirror were placed on the line $y = x$, the image of the point (a,b) would be the point (b,a). Hence the graph of $y = \log_2 x$ is the reflection of the graph of $y = 2^x$ in the line $y = x$. As we shall learn in Chap. 13, the functions 2^x and $\log_2 x$ are *inverse functions*.

66   LOGARITHMIC EQUATIONS   A logarithmic equation is an equation that contains the logarithm of some expression involving the unknown. Such an equation can usually be solved by rewriting it in one of the following two forms:

(1) $\log M = \log N,$ which implies $M = N$

or

(2) $\log_b N = w$ which implies $N = b^w$

EXAMPLE 1   Solve for x: $\log x + 3 \log a = \frac{1}{2} \log b.$

   Solution   We shall rewrite the equation in the form $\log M = \log N$ and
              then state that $M = N$. (If two numbers have equal logarithms,
              the numbers must be equal.)

$$\log x + 3 \log a = \tfrac{1}{2} \log b$$
$$\log x + \log a^3 = \log b^{1/2}$$
$$\log a^3 x = \log \sqrt{b}$$
$$a^3 x = \sqrt{b}$$
$$x = \frac{\sqrt{b}}{a^3}$$

   Comment   A common mistake is to say that if $\log A + \log B = \log C$,
             then $A + B = C$. This is incorrect because $\log A + \log B \neq$
             $\log (A + B)$. The proper way to handle this situation is to
             replace $\log A + \log B$ with $\log AB$. Then $\log AB = \log C$;
             $AB = C$.

EXAMPLE 2   Solve for x: $\log_{10} x + \log_{10} a = 3 + 4 \log_{10} b.$

   Solution 1   We shall rewrite the equation in the form $\log_{10} N = w$ and
               then assert that $N = 10^w$ (definition of logarithm). Then

$$\log_{10} x + \log_{10} a - \log_{10} b^4 = 3$$

$$\log_{10} \frac{xa}{b^4} = 3$$

$$\frac{xa}{b^4} = 10^3$$

$$x = \frac{1000 b^4}{a}$$

   This result may be checked by equating the logarithms of the
   two sides of the last equation and then showing that the
   resulting equation is equivalent to the given equation.

   Solution 2   Replace 3 with $\log_{10} 1000$ and proceed as in Example 1.
               Then

$$\log_{10} x + \log_{10} a = \log_{10} 1000 + \log_{10} b^4$$
$$\log_{10} xa = \log_{10} 1000b^4$$
$$xa = 1000b^4$$

$$x = \frac{1000b^4}{a}$$

67   EXPONENTIAL EQUATIONS   An exponential equation is an equation in which the unknown appears in an exponent.   Such an equation can usually be solved by equating the logarithms of the two sides and then finding the roots of the resulting algebraic equation.

EXAMPLE   Solve for x: $(9.55)^x = 0.0345$.

Solution   Take the logarithm of each side:

$$\log (9.55)^x = \log 0.0345$$
$$x \log 9.55 = \log 0.0345$$

$$x = \frac{\log 0.0345}{\log 9.55}$$

$$= \frac{8.53782 - 10}{0.98000}$$

$$= \frac{-1.46218}{0.98} = -1.49$$

In this case it is easier to perform the division without using logs. Had logs been used, we should first have computed the value of the fraction $\dfrac{1.46218}{0.98}$ and then attached a minus sign to the result.

68   CHANGE OF BASE OF LOGARITHMS   In making numerical computations, the most convenient system of logarithms is the *common,* or *Briggs,* system, which employs the base 10.   If we know the logarithm of a number to the base a, we can find the logarithm of that number to the base b by using

(1) $$\log_b N = \frac{\log_a N}{\log_a b} = (\log_b a)(\log_a N)$$

To prove this,

let $\qquad \log_b N = y$
Then $\qquad\quad N = b^y$

Take the logarithm of each side to the base a:

$$\log_a N = \log_a b^y$$
$$= y \log_a b$$
$$\log_a N = \log_b N \log_a b$$

Hence $\qquad \log_b N = \dfrac{\log_a N}{\log_a b}$

If $N = a$,

$$\log_b a = \dfrac{1}{\log_a b}$$

Therefore

$$\log_b N = (\log_b a)(\log_a N).$$

In calculus and higher mathematics, the most suitable system of logarithms is the *natural*, or *Napierian*, system which employs the base e, where e is approximately 2.71828. If $a = 10$ and $b = e$, Equation (1) becomes

$$\log_e N = \dfrac{\log_{10} N}{0.43429} = 2.3026 \log_{10} N.$$

The usual abbreviation for $\log_e N$ is *ln N*.

Hence $\qquad ln\ N = 2.3026 \log_{10} N.$

Thus the natural logarithm of a number may be obtained by multiplying its common logarithm by 2.3026.

EXERCISE 34

Solve the following equations for x.

1 $\log (6 - x) + \log (x + 9) = \log 50$
2 $\log (2 - x) + \log (3 - x) = \log 12$
3 $2 \log (4 - x) = \log (x + 8) + \log 6$
4 $2 \log (1 - x) = \log (x + 11) + \log 2$
5 $\log_{10} (7x + 2) = 1 + \log_{10} (x - 1)$

6 $\log_{10}(x + 11) + 2\log_{10} 5 = 2 + \log_{10} 7$
7 $y = \log_b x + \log_b (x - 6)$
8 $4\log(x - 3) - 5\log a = 6\log b - \log 7$
9 $a^{6x-7} = b^{8x+9}$
10 $a^x = b(c^{x+1})$

Use Table 2 to solve for x.

11 $(2.884)^x = 0.01439$
12 $(50.12)^x = 2500$
13 $(0.1074)^x = 0.003568$
14 $(0.06309)^x = 0.8185$
15 $(0.001995)^x = 0.02378$
16 $(0.3162)^x = 526.4$
17 $(69.23)^x = (76.18)(0.9876)^{x+1}$
18 $(9.265)^x(2.468)^{2x+1} = 67.19$

Find the natural logarithm of each of the following numbers.

19 7438   **20** 469.5   **21** 0.5793   **22** 0.04129

Find the value of each of the following.

23 $\log_3 825.6$ **24** $\log_2 45.98$

Sketch graphs of the following equations.

25 $y = 3^x$ **26** $y = \log_3 x$ **27** $y = (\tfrac{1}{2})^x$ **28** $y = \log_{1/2} x$

RIGHT TRIANGLES

69 LOGARITHMS OF TRIGONOMETRIC FUNCTIONS In Table 3 (pages 271–315 in the tables) there are listed, to five decimal places, the logarithms of the sine, cosine, tangent, and cotangent for acute angles at intervals of one minute. This table is a combination of Tables 1 and 2. We could find the value of log sin 41° 40′ by using Table 1 to find sin 41° 40′ = 0.6648, and then using Table 2 to find log 0.6648 = 9.82269 − 10. It is much easier, however, to use Table 3 and read immediately

$$\log \sin 41° \, 40' = 9.82269 - 10$$

The sine or cosine of any acute angle is less than 1. The same is true of the tangent of an angle between 0° and 45°, or the cotangent of an angle between 45° and 90°. Consequently, the characteristics of the logarithms of such functions are negative. To conserve space in the tables, the "−10" of negative characteristics has been omitted. For the sake of uniformity, the table is so constructed that a "−10" is to be understood with each entry. For example, the table entry for log tan 85° 25′ is 11.09601. But we are to understand that

$$\text{log tan } 85° \; 25′ = 11.09601 - 10 = 1.09601$$

Table 3 can be used to read directly the logarithms of the trigonometric functions of angles measured to the nearest minute. By using interpolation, we can extend this to angles measured to the nearest tenth of a minute. The tabular differences for log sin and log cos appear in the columns marked "d"; those for log tan and log cot appear in the column marked "cd" (common difference). These columns save us the labor of performing the subtraction to obtain the tabular difference.

EXAMPLE 1   Find log tan 73° 14.7′.

Solution   On page 287, we find

$$\text{log tan } 73° \; 14′ = 10.52103 - 10$$

In the "cd" column, we find the tabular difference is 45. Using the table of proportional parts for 45, we find $\frac{7}{10}(45) = 31.5 \rightarrow 31.$* Hence

$$\text{log tan } 73° \; 14.7′ = 0.52134$$

EXAMPLE 2   Find θ if log cos $\theta = 9.86082 - 10$.

Solution   On page 314, we find

$$\text{log cos } 43° \; 27′ = 9.86092 - 10$$

The tabular difference is 12, and our difference is 10. In the proportional parts column headed by 12 we seek the number closest to 10. This number is 9.6, and it corresponds to 8. Hence,

if   log cos $\theta = 9.86082 - 10$
          $\theta = 43° \; 27.8′$

EXERCISE 35

Evaluate the following.

1   log sin 86° 32′   2   log cos 24° 24′

* To make final result *even*.

3 $\log \tan 43° \ 21'$	**4** $\log \cot 69° \ 57'$
5 $\log \cot 38° \ 47.6'$	**6** $\log \tan 57° \ 19.3'$
7 $\log \cos 81° \ 26.2'$	**8** $\log \sin 38° \ 12.8'$
9 $\log \tan 16° \ 49.7'$	**10** $\log \cot 49° \ 29.7'$
11 $\log \sin 74° \ 42.3'$	**12** $\log \cos 12° \ 36.4'$
13 $\log \cos 71° \ 26.9'$	**14** $\log \sin 11° \ 11.1'$
15 $\log \cot 17° \ 7.5'$	**16** $\log \tan 62° \ 0.2'$

Find the value of θ.

17 $\log \cos \theta = 9.96050 - 10$	**18** $\log \sin \theta = 9.97653 - 10$
19 $\log \cot \theta = 9.70152 - 10$	**20** $\log \tan \theta = 9.89411 - 10$
21 $\log \tan \theta = 0.32040$	**22** $\log \cot \theta = 1.23456$
23 $\log \sin \theta = 9.63456 - 10$	**24** $\log \cos \theta = 9.74156 - 10$
25 $\log \cot \theta = 9.16574 - 10$	**26** $\log \tan \theta = 9.21138 - 10$
27 $\log \cos \theta = 9.95852 - 10$	**28** $\log \sin \theta = 9.98120 - 10$
29 $\log \sin \theta = 9.49931 - 10$	**30** $\log \cos \theta = 9.01805 - 10$
31 $\log \tan \theta = 0.01419$	**32** $\log \cot \theta = 0.50246$
33 $\log \cos \theta = 9.48621 - 10$	**34** $\log \sin \theta = 9.12345$
35 $\log \cot \theta = 0.19732$	**36** $\log \tan \theta = 9.99897 - 10$

37 If θ is any angle in Q I, prove that $\log \tan \theta + \log \cot \theta = 0$.

38 If θ_1 and θ_2 are two first-quadrant angles that differ by one minute (or by any amount), show that the tabular difference between $\log \tan \theta_1$ and $\log \tan \theta_2$ is numerically equal to the tabular difference between $\log \cot \theta_1$ and $\log \cot \theta_2$.

70 LOGARITHMIC SOLUTION OF RIGHT TRIANGLES   The solution
of right triangles is discussed in Section 19.*   Since all computations
involved are either multiplication or division, we can usually perform
them with less labor by using logarithms.

In solving a triangle with logarithms the following steps are suggested:

1. Draw the triangle to scale and label numerically the parts that are known.
2. Make a complete outline indicating the operations to be performed.
3. Fill in all logarithms.
4. Perform the operations outlined in step 2.
5. Check the solution.

* The student should review this section before proceeding.

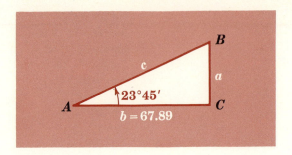

FIGURE 61

It is desirable to find as many as possible of the required parts directly from the given parts.

EXAMPLE 1 Solve the right triangle ABC, for which $A = 23° 45'$ and $b = 67.89$ (see Figure 61).

Solution   (1) $B = 90° - 23° 45'$
                        $= 66° 15'$

(2) To find a, use $\tan 23° 45' = \dfrac{a}{67.89}$.

Hence $a = 67.89 \tan 23° 45'$.
                        $\log 67.89 = \quad 1.83181$
                $\log \tan 23° 45' = \quad 9.64346 - 10$
                ———————————————— A
                        $\log a = 11.47527 - 10$
                        $a = 29.87$

(3) To find c, use $\cos 23° 45' = \dfrac{67.89}{c}$.

Hence $c = \dfrac{67.89}{\cos 23° 45'}$

                $\log 67.89 = 11.83181 - 10$
        $\log \cos 23° 45' = \quad 9.96157 - 10$
                ———————————————— S
                $\log c = \quad 1.87024$
                        $c = 74.17$

(4) A fairly good check* may be obtained by finding c in another way. Use $\sin 23° 45' = \dfrac{a}{c}$. Hence $c = \dfrac{a}{\sin 23° 45'}$.

* The instructor may prefer to use the following check. Since $c^2 = a^2 + b^2$, $a^2 = c^2 - b^2 = (c + b)(c - b)$. Hence $2 \log a = \log (c + b) + \log (c - b)$. Likewise, $2 \log b = \log (c + a) + \log (c - a)$.

$$\log a = 11.47527 - 10$$
$$\log \sin 23° \ 45' = \quad 9.60503 - 10$$
$$\underline{\hspace{3cm}} \text{S}$$
$$\log c = \quad 1.87024$$
$$c = 74.17$$

In this case the two values for log c are exactly the same. Sometimes there is a small discrepancy in the last digit. This is due to the fact that the mantissas in the tables are five-decimal *approximations.*

   This problem illustrates four-figure accuracy in the known parts and the computed parts. No interpolations should be made in finding a and c. Why?

EXAMPLE 2   Solve the right triangle ABC, for which $a = 765.43$ and $c = 898.07$ (see Figure 62).

Solution   (1) To find A, use $\sin A = \dfrac{765.43}{898.07}$.

$$\log 765.43 = 12.88391 - 10$$
$$\log 898.07 = \quad 2.95331$$
$$\underline{\hspace{3cm}} \text{S}$$
$$\log \sin A = \quad 9.93060 - 10$$
$$A = 58° \ 27.9'$$

(2) $B = 90° - 58° \ 27.9' = 31° \ 32.1'$

(3) To find b, use $\cos A = \dfrac{b}{898.07}$. Hence $b = 898.07 \cos A$.

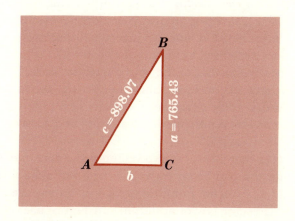

FIGURE 62

$$\log 898.07 = 2.95331$$
$$\log \cos A = 9.71852 - 10$$
$$\overline{\hspace{4cm}} A$$
$$\log b = 12.67183 - 10$$
$$b = 469.71$$

(4) Check by finding the last part, b, in another way: $b = \dfrac{765.43}{\tan A}$.

EXERCISE 36

Make a complete outline of the logarithmic solution of the right triangle in which the following parts are known.

1 A, a	**2** B, a	**3** a, b	**4** b, c

Use logarithms to solve the following right triangles. Check as directed by the instructor.

5 $A = 32° 41', c = 4947$		**6** $A = 61° 17', a = 340.5$
7 $B = 26° 49', a = 67.42$		**8** $B = 51° 38', c = 9.483$
9 $a = 0.06751, b = 0.1942$		**10** $b = 4.179, c = 5.236$
11 $a = 257.0, c = 852.3$		**12** $a = 1492, b = 1776$
13 $B = 69° 4.8', b = 72,893$		**14** $A = 22° 34.8', c = 58,196$
15 $B = 82° 24.0', c = 8.6049$		**16** $A = 33° 46.6', b = 36.912$
17 $b = 7.0234, c = 9.1863$		**18** $a = 57.114, b = 64.128$
19 $a = 75,656, b = 34,081$		**20** $a = 20,304, c = 31,415$

71   VECTORS   A *vector quantity* is a quantity having magnitude and direction. Examples of vector quantities are forces, velocities, accelerations, and displacements. A *vector* is a directed line segment. A vector quantity can be represented by means of a vector if (1) the direction of the vector is the same as that of the vector quantity and (2) the length of the vector represents, to some convenient scale, the magnitude of the vector quantity. For example, a velocity of 30 mph in a northerly direction can be represented by a 3-in. line segment pointed north.

The *resultant* (or vector sum) of two vectors is the diagonal of a parallelogram having the two given vectors as adjacent sides. In physics it is shown that like vector quantities, such as forces, are combined according to this vector law of addition. In Figure 63, vector *OR* is

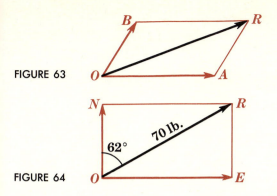

FIGURE 63

FIGURE 64

the resultant of vectors *OA* and *OB*.   If the lengths and directions of *OA* and *OB* are known, then the length and direction of *OR* can be found by solving triangle *OAR*.

Two vectors having a certain resultant are said to be *components* of that resultant.   If the two vectors are at right angles to each other, they are called *rectangular components*.   It is possible to resolve a given vector into components along any two specified directions.   For example (Figure 64), a force of 70 lb. acting in the direction N 62° E can be resolved into an easterly force and a northerly force by solving the right triangle *ORN*.

EXAMPLE 1   An airplane with an *air speed* (speed in still air) of 178 mph is headed due north.   If a west wind of 27.5 mph is blowing, find the direction traveled by the plane and its *ground speed* (actual speed with respect to the ground).

Solution   To find θ (see Figure 65) use $\tan \theta = \dfrac{27.5}{178}$

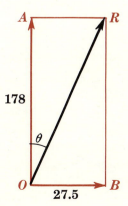

FIGURE 65

Logarithmic computation gives

$$\theta = 8° 47'$$

To find OR, use $\cos \theta = \dfrac{178}{OR}$. Hence $OR = \dfrac{178}{\cos \theta}$. Using logarithms, we find

$$OR = 180.1$$

Inasmuch as the original data indicate only three-figure accuracy, the results should be rounded off to three-figure accuracy. Hence the ground speed of the plane is 180 mph in the direction N 8° 50′ E.

EXAMPLE 2   What is the minimum force required to prevent a 367-lb. barrel from rolling down a plane that makes an angle of 17° 30′ with the horizontal? Find the force of the barrel against the plane.

Solution   The weight or force of 367 lb., OR, acting vertically downward can be resolved into two rectangular components, one parallel to the plane and the other perpendicular to it (see Figure 66). Hence

$$OA = OR \sin \angle ORA* = 367 \sin 17° 30'$$

Computing this product with logarithms, we find $OA = 110.4$

* $\angle ORA = 17° 30'$ because they are acute angles with their sides respectively perpendicular.

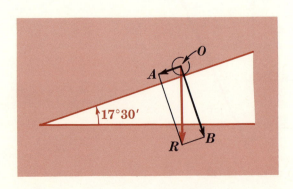

FIGURE 66

lb.   The force required to prevent the barrel from rolling is 111 lb. (to the nearest pound).

The force of the barrel against the plane is

$$OB = OR \cos \angle ROB = 367 \cos 17° 30'$$

Using logarithms, we find $OB = 350$ lb.

EXERCISE 37

1   Angel Falls in Venezuela contains the highest known waterfall. From a point 4732 ft. from, and in the same horizontal plane with, the top of the waterfall, the angle of depression of the foot of the fall is 29° 15′. Find the height of the waterfall.

2   The angle of elevation of the top of New York's World Trade Center from a point on the ground 1093 ft. from its base is 51° 0′. Find the height of the building.

3   From a point on the ground 365 ft. away from the foot of the Washington Monument, the angle of elevation of the top of the monument is 56° 40′. How high is the monument?

4   A dirigible is 1398 ft. directly above one end of the Verrazano-Narrows Bridge in New York. The angle of depression of the other end of the bridge from the dirigible is 18° 10′. How long is the bridge?

5   A car weighing 2995 lb. is parked on a hill that makes an angle of 7° 15′ with the horizontal. **(a)** What force tends to pull it down the hill? **(b)** Find the force exerted by the car against the hill.

6   A cable that can withstand a tension of 4275 lb. is used to pull vehicles up an inclined ramp to the second floor of a storage garage. If the inclination of the ramp is 28° 15′, find the weight of the heaviest truck that can be pulled with the cable.

7   A bus weighing 6075.0 lb. exerts a force of 5835.0 lb. on an inclined ramp. **(a)** What angle does the ramp make with the horizontal? **(b)** What force must be exerted by the brakes to prevent the bus from rolling down the ramp?

8   A barrel weighing 147 lb. rests on an inclined plane. If a force of 38.5 lb. is needed to keep the barrel from rolling down, what angle does the plane make with the horizontal?

9   From a train traveling due north at 3958 ft. per min., a stone is thrown due east at 1125 ft. per min. Find the direction and speed of the stone.

10   The pilot of an airplane wishes to travel due east. His plane has a cruising speed of 105 mph in still air. **(a)** In what direction should he head his plane if a north wind of 22.5 mph is blowing? **(b)** Find the ground speed of the plane.

11   A pilot wants to fly 300 miles due north in 2 hr. If a wind is blowing at 17.5 mph from the east, in what direction should he head his plane

and what air speed must he maintain in order to arrive at his destination on time?

12 A river flows due south. If a motorboat is headed N 75° 0′ E, it will travel due east at 535 ft. per min. Find the speed of the current and the speed of the boat in still water.

13 An airplane with a cruising speed of 185 mph in still air is headed S 10° 0′ W. Find the speed of a west wind that makes the plane travel due south.

14 A river flows due east at 128 ft. per min. By heading his launch in the direction N 25° 0′ W, a fisherman is able to make his boat move due north across the river. How long will it take to cross the river if it is 549 ft. wide?

15 An airplane flying 115 mph is coming down at an angle of 9° 0′ with the horizontal. What is its rate of descent?

16 A ship sails 4.375 miles due north of A and then turns east and sails 2.875 miles. What is the bearing of the ship from A?

17 Two ships leave a port at noon. One ship moves 14.5 mph in the direction N 40° 40′ W. The other ship travels 13.5 mph in the direction S 49° 20′ W. Find the bearing of the first ship from the second one at 1:00 P.M.

18 The Great Pyramid of Gizeh, Egypt, has a square base. Its faces are congruent isosceles triangles that intersect in its four edges. Before vandals removed the outer limestone casing and the top 31.000 ft. of the pyramid, each edge was 719 ft. 2.6 in. long and made an angle of 42° 0.6′ with the horizontal. Find the pyramid's original height and the length of a side of its base.

19 Three forces of 25.25 lb., 35.35 lb., and 45.45 lb. act in the directions N 22° 22′ E, N 33° 33′ E, and N 44° 44′ E, respectively. Find the magnitudes of two forces, one acting due west and the other acting due south, that will counterbalance the three given forces. *Hint:* The force that acts due west must neutralize the easterly components of the three given forces.

20 From the top of a hill 395 ft. above the level of a lake, the angles of depression of two boats are 39° 0′ and 29° 0′, respectively. Find the distance between the boats if one is due east and the other is due north of the hilltop.

Solve* each of the following oblique triangles by dividing it into right triangles. Draw figures.

21 $A = 31° 14′$, $B = 66° 22′$, $b = 7074$.

  Solution Let h represent the length of the perpendicular CD from C to AB. Then use the equations

_____

* May be omitted at the discretion of the instructor.

$$C = 180° - (A + B) \qquad AD = 7074 \cos 31° \, 14'$$

$$h = 7074 \sin 31° \, 14' \qquad DB = \frac{h}{\tan 66° \, 22'}$$

$$c = AD + DB \qquad a = \frac{h}{\sin 66° \, 22'}$$

Computation gives $C = 82° \, 24'$, $AD = 6049$, $\log h = 3.56444$ (the value of h need not be found), $DB = 1605$, $c = 7654$, $a = 4004$. The student is expected to perform the computations.

22 $A = 47° \, 10'$, $B = 59° \, 38'$, $b = 73.38$
23 $A = 55° \, 8'$, $b = 6810$, $c = 4828$

Hint: Let h be the length of the perpendicular CD from C to AB. Use the equations $AD = 6810 \cos 55° \, 8'$, $h = 6810 \sin 55° \, 8'$, $DB = 4828 - AD$, $\tan B = \dfrac{h}{DB}$, $C = 180° - (55° \, 8' + B)$, $a = \dfrac{h}{\sin B}$.

24 $A = 52° \, 36'$, $b = 7.806$, $c = 8.124$
25 $A = 110° \, 31'$, $B = 29° \, 9'$, $a = 33.22$
26 $A = 139° \, 42'$, $b = 1446$, $c = 2468$

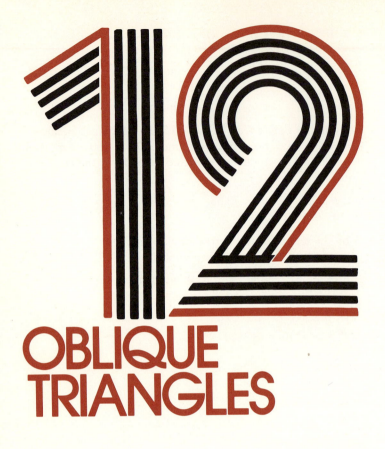

OBLIQUE TRIANGLES

72　INTRODUCTION　A triangle that does not contain a right angle is called an oblique triangle.　Since the ratio of two sides of an *oblique* triangle does not represent a function of an angle of the triangle, additional formulas are needed for solving oblique triangles.　These formulas are the law of sines, the law of cosines, the law of tangents, and the half-angle formulas.

The six parts of a triangle are its three sides a, b, c, and the opposite angles A, B, C, respectively.　If three parts, at least one of which is a side, are given, the remaining parts can be determined.　For convenience, we shall divide the possibilities into the following four cases:

> SAA:　**Given one side and two angles.**
> SSA:　**Given two sides and the angle opposite one**
> **of them.**

SAS: **Given two sides and the included angle.**
SSS: **Given three sides.**

73   THE LAW OF SINES

In any triangle the sides are proportional to the sines of the opposite angles:

$$\frac{a}{\sin A} = \frac{b}{\sin B} = \frac{c}{\sin C}$$

Proof Consider the two cases: all angles acute (Figure 67) and one angle obtuse (Figure 68).   Let h be the perpendicular from C to AB (or AB extended).

In either case

$$\sin B = \frac{h}{a} \qquad \text{hence} \qquad h = a \sin B$$

Also* $\sin A = \dfrac{h}{b}$ hence $h = b \sin A$

Equate the two values of h: $a \sin B = b \sin A$

Divide by $\sin A \sin B$: $\dfrac{a}{\sin A} = \dfrac{b}{\sin B}$

Similarly, $\dfrac{a}{\sin A} = \dfrac{c}{\sin C}$

* In Figure 68, $\sin A = \sin \angle CAD = \dfrac{h}{b}$.

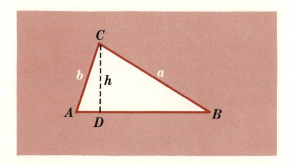

FIGURE 67

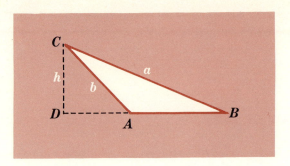

FIGURE 68

Therefore

$$\frac{a}{\sin A} = \frac{b}{\sin B} = \frac{c}{\sin C}$$

The law of sines is equivalent to the following three equations: $\frac{a}{\sin A} = \frac{b}{\sin B}$, $\frac{a}{\sin A} = \frac{c}{\sin C}$, $\frac{b}{\sin B} = \frac{c}{\sin C}$. If the three given parts of the triangle include one side and the opposite angle, then one of these equations will involve three known quantities and one unknown. We can solve for this unknown by using ordinary algebraic methods. Thus, if a, c, C are the known parts, we can find A by solving the equation $\frac{\sin A}{a} = \frac{\sin C}{c}$; $\sin A = \frac{a \sin C}{c}$. The law of sines is well adapted to the use of logarithms because it involves only ratios and products. *Whenever four parts of a triangle are known* (the fourth part may have been obtained from the other three), *we can use the law of sines to find the remaining two parts.* Why?

74 APPLICATIONS OF THE LAW OF SINES: SAA   When one side and two angles of a triangle are known, we can immediately find the third angle from the relation $A + B + C = 180°$. The remaining two sides can then be found by using the law of sines.

Before continuing, the student should review the suggested procedure in Section 70.

EXAMPLE 1   Solve the triangle ABC, given $a = 20$, $A = 30°$, $B = 40°$.

Solution   (See Figure 69.)

(1) $C = 180° - (30° + 40°)$
              $= 110°$

(2) To find b, use

$$b = \frac{a \sin B}{\sin A} = \frac{20 \sin 40°}{\sin 30°} = \frac{20(0.6428)}{0.5000} = 26$$

(3) To find c, use

$$c = \frac{a \sin C}{\sin A} = \frac{20 \sin 110°}{\sin 30°} = \frac{20(0.9397)}{0.5000} = 38$$

The values of b and c are rounded off to two-figure accuracy. The results can be checked by the law of cosines (Section 76) or by finding c with the formula $c = \dfrac{b \sin C}{\sin B}$.

EXAMPLE 2   Solve the triangle ABC, given $b = 1906$, $A = 55°\ 44'$, $C = 81°\ 29'$ (see Figure 70).

Solution   (1) $B = 180° - (55°\ 44' + 81°\ 29')$
                          $= 42°\ 47'$

(2) $a = \dfrac{b \sin A}{\sin B} = \dfrac{1906 \sin 55°\ 44'}{\sin 42°\ 47'}$

An outline of the logarithmic solution is:

$$\begin{aligned}
\log 1906 &= \\
\log \sin 55°\ 44' &= \underline{\hspace{3cm}} \text{A} \\
\log \text{num.} &= \\
\log \sin 42°\ 47' &= \underline{\hspace{3cm}} \text{S} \\
\log a &= \\
a &=
\end{aligned}$$

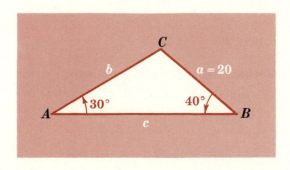

FIGURE 69

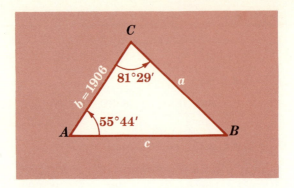

FIGURE 70

The student should fill in the outline and find $a = 2319$.

$$(3) \qquad c = \frac{b \sin C}{\sin B} = \frac{1906 \sin 81° \ 29'}{\sin 42° \ 47'}$$

After preparing the outline and then filling in the numbers, we have:

$$
\begin{aligned}
\log 1906 &= 3.28012 \\
\log \sin 81° \ 29' &= 9.99518 - 10 \\
\hline
\log \text{num.} &= 13.27530 - 10 \quad \text{A} \\
\log \sin 42° \ 47' &= 9.83202 - 10 \\
\hline
\log c &= 3.44328 \quad \text{S} \\
c &= 2775
\end{aligned}
$$

The results may be checked by finding c again using $c = \dfrac{a \sin C}{\sin A}$ or by using the law of tangents (Section 95).*

EXERCISE 38

Solve the following triangles without using logarithms.

1 $b = 300$, $B = 22° \ 20'$, $C = 80° \ 30'$
2 $c = 20$, $A = 100°$, $B = 37°$

Make an outline of the logarithmic solution of the oblique triangle in which the following parts are known.

3 a, A, C 4 b, A, B
5 c, A, B 6 a, A, B

* The instructor may wish to use Mollweide's equations (Probs. 30 and 31 of Exercise 42).

Use logarithms to solve the following triangles.

7 $b = 1945, A = 82° 1', C = 69° 5'$
8 $c = 0.4789, A = 40° 9', B = 108° 23'$
9 $a = 22.95, B = 110° 30', C = 29° 10'$
10 $b = 8.541, A = 71° 9', C = 36° 17'$
11 $c = 435, B = 48° 0', C = 96° 30'$
12 $a = 62.70, A = 23° 40', C = 77° 50'$
13 $b = 7576.7, A = 70° 26.9', B = 50° 34.8'$
14 $c = 51,015, A = 18° 59.2', C = 75° 25.8'$
15 $a = 0.090134, B = 60° 13.5', C = 60° 0.0'$
16 $b = 345.67, B = 86° 12.0', C = 19° 57.0'$

17   A vacationer observes that the angle of elevation of a mountain peak from his cottage is 26° 0'. Leaving the cottage, he walks 2000 ft. *up* a slope of 10° 0' directly toward the mountain and then finds the elevation of the peak to be 31° 0'. How much higher is the mountain peak than the cottage?

18   Boston, Mass., is 150 miles S 30° E from Montpelier, Vt. The bearing of Albany, N.Y., from Montpelier is S 28° W. The bearing of Albany from Boston is N 82° W. How far is Albany from Montpelier? From Boston? *

19   An airplane with a cruising speed of 125 mph in still air is headed N 18° 30' W. An east wind is blowing. Find the speed of the wind if the plane travels in the direction N 25° 30' W.

20   Two angles of a triangle are 58° 46' and 57° 18'. If the largest side is 93.63 ft., find the smallest side.

21   Why does the law of sines fail to solve the *SAS* and *SSS* cases?

22   Show that if $C = 90°$, the law of sines gives the definition of the sine of an acute angle.

75   **THE AMBIGUOUS CASE: SSA**   If two sides and the angle opposite one of them are given, the triangle is not always uniquely determined. With the given parts, we may be able to construct two triangles, or only one triangle, or no triangle at all. Because of the possibility of two triangles, this is usually called the ambiguous case.

To avoid unnecessary confusion, we shall use A to designate the given angle, a to represent the opposite side, and b to indicate the other given side. Construct angle A and lay off b as an adjacent side, thus fixing the vertex C. With C as center and a as radius, strike an

---

* Ignore the curvature of the earth and assume only two-place accuracy. Get angles to the nearest degree. Get distances to the nearest multiple of 10 miles.

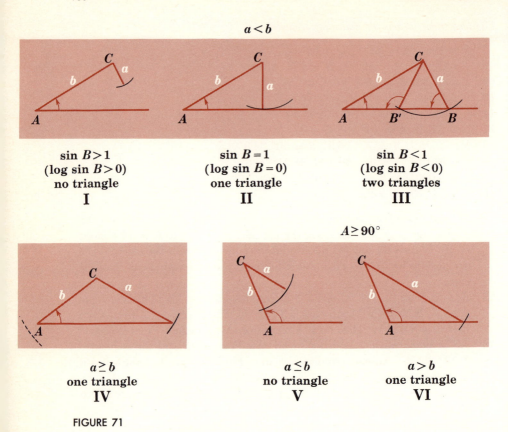

FIGURE 71

arc cutting the other side adjacent to A. Figure 71 illustrates the various possibilities.

The last three cases (diagrams IV, V, and VI) can be quickly identified by merely noting the relative sizes of a and b. If A is acute and $a < b$, it is necessary to begin the solution before we can state how many triangles are possible. To do this, use the law of sines:

$$\frac{\sin B}{b} = \frac{\sin A}{a} \qquad \sin B = \frac{b \sin A}{a}$$

After determining the value of $\sin B$ (or $\log \sin B$ with logarithms), we can definitely classify our problem as one of the various types. It is frequently possible to determine the number of solutions by merely constructing a figure to scale.

ILLUSTRATION 1   Given $a = 100$, $b = 70$, $A = 80°$. This is case IV because $a > b$. There is one triangle.

ILLUSTRATION 2   Given $a = 40$, $b = 42$, $A = 110°$. This is case V because $A > 90°$, and $a < b$. There is no triangle.

ILLUSTRATION 3   Given $a = 80$, $b = 100$, $A = 54°$. Since A is acute and $a < b$, this is case I, II, or III. A carefully constructed figure leaves some doubt as to whether a is long enough to reach the horizontal side of angle A. Using the law of sines, we find

$$\sin B = \frac{b \sin A}{a} = \frac{100(0.8090)}{80} = 1.0112.$$ Since no angle can have a sine greater than 1, B is impossible and there is no triangle. Had logarithms been used, we should have found $\log \sin B = 0.00487$. Since $\log \sin B > 0$, $\sin B > 1$,* B is impossible, and there is no triangle. This is case I.

EXAMPLE   Solve the triangle ABC, given $a = 48.85$, $b = 69.22$, $A = 37° 12'$.

Solution   A scale drawing (Figure 72) indicates that there are two triangles. Using the law of sines, we obtain

$$\frac{\sin B}{b} = \frac{\sin A}{a}$$

$$\sin B = \frac{b \sin A}{a} = \frac{69.22 \sin 37° 12'}{48.85}$$

After preparing the outline and then filling in the numbers, we have

$$
\begin{aligned}
\log 69.22 &= 1.84023 \\
\log \sin 37° 12' &= 9.78147 - 10 \\
&\overline{\qquad\qquad\qquad}\, A \\
\log \text{num.} &= 11.62170 - 10 \\
\log 48.85 &= 1.68886 \\
&\overline{\qquad\qquad\qquad}\, S \\
\log \sin B &= 9.93284 - 10 \\
B &= 58° 57'
\end{aligned}
$$

Continuing with the solution of large triangle ABC (Figure 73), we find

$$C = 180° - (37° 12' + 58° 57')$$
$$= 83° 51'$$

To find c, use

$$\frac{c}{\sin C} = \frac{a}{\sin A} \qquad c = \frac{a \sin C}{\sin A} = \frac{48.85 \sin 83° 51'}{\sin 37° 12'}$$

* Recall that a number greater than 1 has a logarithm greater than 0.

Logarithmic computation gives

$$\begin{aligned}
\log 48.85 &= 1.68886 \\
\log \sin 83°\,51' &= 9.99749 - 10 \\
\hline
\log \text{num.} &= 11.68635 - 10 \\
\log \sin 37°\,12' &= 9.78147 - 10 \\
\hline
\log c &= 1.90488 \\
c &= 80.33
\end{aligned}$$

A

S

Hence the computed parts of triangle ABC are $B = 58°\,57'$, $C = 83°\,51'$, $c = 80.33$.

We shall now solve the small triangle $AB'C'$. Since triangle BCB' (Figure 72) is isosceles, $\angle BB'C = B$. Hence
$$B' = 180° - B = 180° - 58°\,57' = 121°\,3'$$

Also (see Figure 74),
$$C' = 180° - (37°\,12' + 121°\,3') = 21°\,45'$$

Using the law of sines with logarithms, we find $c' = 29.94$.
Hence the computed parts of triangle $AB'C'$ are $B' = 121°\,3'$, $C' = 21°\,45'$, $c' = 29.94$.

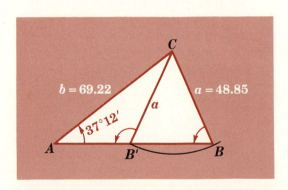

FIGURE 72

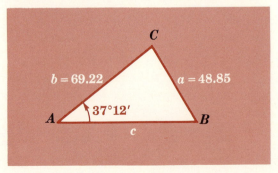

FIGURE 73

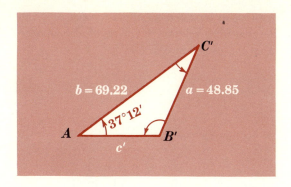

FIGURE 74

Draw a figure and solve all possible triangles.

1 $a = 16.20$, $b = 19.49$, $A = 54° 40'$
2 $a = 0.8142$, $b = 0.6237$, $A = 73° 12'$
3 $a = 9000$, $b = 9911$, $A = 80° 4'$
4 $a = 5061$, $b = 5426$, $A = 58° 44'$
5 $a = 1335$, $b = 6986$, $A = 11° 1'$
6 $a = 2497$, $b = 3129$, $A = 123° 45'$
7 $a = 4.968$, $b = 7.054$, $A = 41° 23'$
8 $a = 27.34$, $b = 22.15$, $A = 58° 19'$
9 $a = 79,925$, $c = 88,665$, $A = 71° 35.5'$
10 $a = 142.43$, $c = 222.77$, $A = 38° 51.6'$
11 $b = 39,170$, $c = 30,405$, $B = 114° 27.0'$
12 $b = 73,101$, $c = 65,402$, $C = 92° 31.5'$

13 Cleveland, Ohio, is 390 miles N 51° E from Evansville, Ind. Lexington is due east of Evansville and is 280 miles from Cleveland. How far is Lexington from Evansville?*

14 A launch capable of a speed of 840 ft. per min. in still water is at the east bank of a river that flows south at 175 ft. per min. In what direction should the launch be headed if it is to land at a pier 1000 ft. northwest of its starting point? How long will the trip take?

15 A television tower is 25 miles N 67° 40′ E from a railroad station. A train moving east at 30 mph on a straight east-west track passes the station at 2 P.M. When will the train be 19 miles from the tower?

* Ignore the curvature of the earth and assume only two-place accuracy. Get angles to the nearest degree. Get distances to the nearest multiple of 10 miles.

16   A body is acted upon by forces of 52.34 lb. and 30.72 lb. The larger force acts due south. Find the direction of the smaller force if the resultant acts in the direction S 32° 0′ W.

76   THE LAW OF COSINES

The square of any side of a triangle is equal to the sum of the squares of the other two sides minus twice their product times the cosine of the included angle:

$$a^2 = b^2 + c^2 - 2bc \cos A$$
$$b^2 = c^2 + a^2 - 2ca \cos B$$
$$c^2 = a^2 + b^2 - 2ab \cos C$$

Proof **1**   If all the angles are acute: Draw CD perpendicular to AB. Then (see Figure 75)

$$a^2 = h^2 + \overline{DB}^2$$
$$= h^2 + (c - AD)^2$$
$$= \underbrace{h^2 + \overline{AD}^2} + c^2 - 2cAD$$
$$= \quad b^2 \quad + c^2 - 2c(b \cos A)$$
$$a^2 = b^2 + c^2 - 2bc \cos A$$

2   If one angle is obtuse: Draw CD perpendicular to AB extended. Notice that if AB is considered a positive directed segment, then AD is negative in Figure 76 while DB is negative in Figure 77. With this understanding, the proofs for

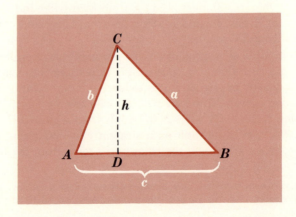

FIGURE 75

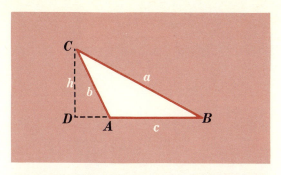

FIGURE 76

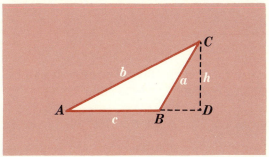

FIGURE 77

Figures 76 and 77 are exactly the same as that for Figure 75. The student should verify this.

This is a general proof because a may be used to represent any side of the given triangle. The other two forms of the law of cosines can be obtained from the first form by the method of *cyclic permutation* in which

a is changed to b, A is changed to B,
b is changed to c, B is changed to C,
c is changed to a, C is changed to A.

If $A = 90°$, $\cos A = 0$, and $a^2 = b^2 + c^2 - 2bc \cos A$ reduces to the form $a^2 = b^2 + c^2$. We conclude that the Pythagorean theorem is a special case of the law of cosines; i.e., the law of cosines is a generalization of the Pythagorean theorem.

77   APPLICATIONS OF THE LAW OF COSINES: SAS AND SSS   The law of cosines is used in many geometric problems, one of which is the solution of triangles. Since three sides and one angle are

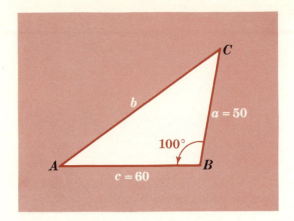

FIGURE 78

involved in every form of the law of cosines, it can be used to solve the *SAS* and *SSS* cases described in Section 72.

EXAMPLE 1   Given $a = 50$, $c = 60$, $B = 100°$, find b (see Figure 78).

Solution   This is the two sides and included angle case, *SAS*.   Hence

$$b^2 = c^2 + a^2 - 2ca \cos B$$
$$= (60)^2 + (50)^2 - 2(60)(50) \cos 100°$$
$$= 3600 + 2500 - 6000(-0.1736)$$
$$= 7141.6$$
$$b = 85$$

This example illustrates two-figure accuracy in the given data and the computed result.   Notice that $\cos 100° = -\cos 80° = -0.1736$.

The square root of 7141.6 may be obtained by use of logarithms or by using Table 4, with interpolation.

EXAMPLE 2   Given $a = 8.00$, $b = 5.00$, $c = 4.00$, find the largest and smallest angles (see Figure 79).

Solution   The largest and smallest angles lie opposite the largest and smallest sides, respectively.
To find A, use

$$a^2 = b^2 + c^2 - 2bc \cos A$$
$$64 = 25 + 16 - 40 \cos A$$
$$40 \cos A = 41 - 64 = -23$$
$$\cos A = -\tfrac{23}{40} = -0.5750$$
(Related angle is 54° 50′)
$$A = 180° - 54° 50' = 125° 10'$$

To find C, use

$$c^2 = a^2 + b^2 - 2ab \cos C$$
$$16 = 64 + 25 - 80 \cos C$$
$$80 \cos C = 73, \qquad \cos C = \tfrac{73}{80} = 0.9125$$
$$C = 24° \; 10'$$

This example illustrates three-figure accuracy.

If the numbers are not easy to square, use logarithms to compute the various *terms* in the equation; then collect the terms and finish with or without using logarithms. For example, if $a = 51.66$, $b = 39.15$, $C = 76° \; 0'$, use the law of cosines to get

$$c^2 = (51.66)^2 + (39.15)^2 - 2(51.66)(39.15) \cos 76° \; 0'$$

Using logarithms to compute the two squares and the product on the right side, we obtain

$$c^2 = 2668.7 + 1532.7 - 978.56$$
$$= 3222.8$$
$$c = 56.77 \qquad \text{(Use logs to extract the square root)}$$

EXERCISE 40

1   $a = 4.0$, $b = 7.0$, $c = 5.0$, find B.
2   $a = 11$, $b = 20$, $c = 30$, find C.
3   $a = 12$, $b = 13$, $c = 11$, find A.
4   $a = 600$, $b = 300$, $c = 700$, find B.
5   $a = 10$, $b = 30$, $C = 56°$, find c.
6   $b = 7.00$, $c = 3.00$, $A = 45° \; 20'$, find a.
7   $a = 20.00$, $c = 10.00$, $B = 155° \; 30'$, find b.
8   $a = 10$, $b = 40$, $C = 151°$, find c.

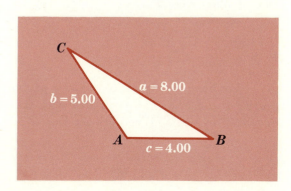

FIGURE 79

9   $b = 5.0$, $c = 2.0$, $A = 133°$, find a.
10 $a = 80$, $c = 50$, $B = 96°$, find b.
11 $a = 3.0$, $b = 5.0$, $c = 7.0$, find A, B, and C.
12 $a = 10.0$, $b = 6.0$, $c = 5.0$, find A, B, and C.

13 The resultant of forces of 20 lb. and 50 lb. is a force of 60 lb. Find the angles the resultant makes with each of the components.
14 Two sides of a parallelogram are 70 and 90, and one diagonal is 30. Find the angles of the parallelogram.
15 Paris is 210 miles S 29° E from London. Amsterdam is 220 miles N 74° E from London. Find the bearing and the distance of Amsterdam from Paris. *
16 In a certain baseball park the distance from home plate to a point A in straightaway center field is 400 ft. Find the distance from A to third base. (Straightaway center field is the extension of the line drawn from home plate through second base. The distance between consecutive bases is exactly 90 ft.)

In the following problems use logarithms to compute squares, products, square roots, and quotients as needed in using the law of cosines.

17 $a = 19.3$, $b = 28.5$, $c = 30.7$, find C.
18 $a = 6.66$, $b = 8.15$, $c = 9.78$, find A.
19 $a = 3.92$, $b = 5.09$, $C = 28° \ 50'$, find c.
20 $b = 75.4$, $c = 68.1$, $A = 39° \ 10'$, find a.

21 Given $a = 73.1$, $c = 94.7$, $B = 19° \ 30'$. Use the law of cosines to find b. Then use the law of sines to find A and C. (Carry four-figure accuracy throughout the computations. Then round off the final results.) *Hint:* Since $c^2 > a^2 + b^2$, C must be obtuse.
22 Given $a = 25.93$, $b = 42.60$, $C = 63° \ 40'$. Use the law of cosines to find c. Then use the law of sines to find A and B. (Carry five-figure accuracy throughout the computations. Then round off the final results.)
23 Discuss the equation $c^2 = a^2 + b^2 - 2ab \cos C$ **(a)** when C approaches 180°, **(b)** when C approaches 0°.

78 SUMMARY   We have seen that, of the four problems listed in Section 72, two (SAA and SSA) can be solved with the law of sines while the other two (SAS and SSS) require the law of cosines. Since

---

* Ignore the curvature of the earth and assume only two-place accuracy. Get angles to the nearest degree. Get distances to the nearest multiple of 10 miles.

the law of cosines is not well adapted to logarithmic computation,* it can occasionally be a matter of some importance to take advantage of alternative formulas—the "law of tangents" and the "half-angle formulas"—which are better suited to computation using logarithms and frequently produce more accurate results. These formulas, together with an outline to assist the student in choosing the best plan of attack, are presented in the Appendix, Sections 94–98. But if extreme accuracy is not of the essence,† all four cases can be treated by using only the law of sines and the law of cosines. The student should recall that the law of sines can be used to complete the solution as soon as four parts are known.

79   THE AREA OF A TRIANGLE

I. The area of a triangle is equal to one-half the product of any two sides times the sine of the included angle:

$$K = \tfrac{1}{2}bc \sin A = \tfrac{1}{2}ca \sin B = \tfrac{1}{2}ab \sin C$$

Proof   Let h be the altitude from C to AB (Figure 80).

---

* Recall that $\log (R + S) \neq \log R + \log S$.

† In earlier days, the study of trigonometry was motivated almost entirely by its use in solving triangles. Today the use of trigonometric functions for other purposes is more important. For some time there has been a definite trend toward emphasizing the analytic aspects of trigonometry at the expense of the computational features.

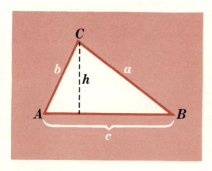
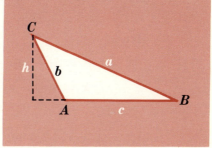

FIGURE 80

$$\text{Area } \triangle ABC = K = \tfrac{1}{2}ch = \tfrac{1}{2}bc \sin A$$

since $h = b \sin A$. The other forms can be obtained by cyclic permutation.

II. The area of triangle ABC is

$$K = \sqrt{s(s - a)(s - b)(s - c)}$$

where $s = \tfrac{1}{2}(a + b + c)$

Proof   Since $K = \tfrac{1}{2}bc \sin A$,

$$K^2 = \frac{1}{4} b^2 c^2 \sin^2 A = \frac{b^2 c^2}{4} (1 - \cos^2 A)$$

$$= \frac{bc}{2} (1 + \cos A) \frac{bc}{2} (1 - \cos A)$$

Using the law of cosines, we get

$$K^2 = \frac{bc}{2} \left(1 + \frac{b^2 + c^2 - a^2}{2bc} \right) \frac{bc}{2} \left(1 - \frac{b^2 + c^2 - a^2}{2bc} \right)$$

$$= \frac{2bc + b^2 + c^2 - a^2}{4} \cdot \frac{2bc - b^2 - c^2 + a^2}{4}$$

$$= \frac{(b + c)^2 - a^2}{4} \cdot \frac{a^2 - (b - c)^2}{4}$$

$$= \frac{b + c + a}{2} \cdot \frac{b + c - a}{2} \cdot \frac{a - b + c}{2} \cdot \frac{a + b - c}{2}$$

Let $s = \dfrac{1}{2} (a + b + c)$; then $\dfrac{b + c - a}{2} = s - a$, etc.

Hence

$$K^2 = s(s - a)(s - b)(s - c)$$

$$K = \sqrt{s(s - a)(s - b)(s - c)}$$

III. The area of triangle ABC is

$$K = \frac{a^2 \sin B \sin C}{2 \sin A} = \frac{b^2 \sin C \sin A}{2 \sin B} = \frac{c^2 \sin A \sin B}{2 \sin C}$$

Proof We start with $K = \frac{1}{2}ab \sin C$ and use the law of sines to replace b with $\dfrac{a \sin B}{\sin A}$. This gives $K = \dfrac{a^2 \sin B \sin C}{2 \sin A}$. This formula should be used when one side and two angles are given.

EXERCISE 41

Find the areas of the following triangles without using logarithms.

1 $b = 6.0$, $c = 1.5$, $A = 43°$
2 $a = 50.0$, $b = 80.0$, $C = 165° \; 30'$
3 $c = 10$, $A = 31°$, $B = 119°$
4 $b = 40$, $B = 80°$, $C = 20°$
5 $a = 13$, $b = 40$, $c = 45$ (Consider the numbers as exact.)
6 $a = 9$, $b = 11$, $c = 14$ (Consider the numbers as exact.)

Use logarithms to find the areas of the following triangles.

7 $a = 17.28$, $c = 22.95$, $B = 69° \; 30'$
8 $b = 4.766$, $c = 7.258$, $A = 33° \; 9'$
9 $c = 54.19$, $B = 127° \; 27'$, $C = 29° \; 57'$
10 $b = 7.654$, $A = 58° \; 0'$, $C = 77° \; 7'$
11 $a = 3.012$, $b = 4.077$, $c = 4.555$
12 $a = 51.23$, $b = 66.55$, $c = 39.78$

13 Use area formula I to derive an expression for the area of an equilateral triangle. Check your result by using (a) formula II, (b) formula III.
14 Find the number of acres in a triangular lot whose sides are 500 ft., 600 ft., and 700 ft. (One acre contains 43,560 sq. ft.)
15 The area of a triangle is 91.68 sq. ft. Two of its sides are 27.95 ft. and 31.47 ft. Find the included angle.
16 Use area formula II to show that the area of an isosceles triangle with sides a, a, b is $\dfrac{b}{4} \sqrt{4a^2 - b^2}$.
17 The area of a triangle is 6; its perimeter is 20. If one side of the triangle is 4, find the other two sides. (Assume all numbers are exact.) *Hint:* $s - c = s - (2s - a - b) = a + b - s$.

EXERCISE 42 Miscellaneous Problems

For each of the eight following problems, (a) draw the triangle, (b) write the formula that should be used in solving for the required part, (c) prepare an outline of the solution, (d) find the required part.

1   $a = 60$, $b = 70$, $c = 90$, find C.
2   $b = 3.0$, $c = 2.0$, $A = 86°$, find a.
3   $a = 634.9$, $A = 37° 47'$, $C = 38° 47'$, find c.
4   $a = 88.66$, $b = 99.77$, $A = 40° 0'$, find B.
5   $a = 5176$, $c = 9345$, $A = 25° 50'$, find C.
6   $c = 9282$, $A = 27° 15'$, $B = 41° 45'$, find a.
7   $a = 20$, $b = 21$, $C = 100°$, find c.
8   $a = 5.0$, $b = 7.0$, $c = 8.0$, find C.

9   A railroad track crosses a highway at an angle of 77°. A locomotive is 40 yd. from the intersection when a car is 50 yd. from the crossing. What is the distance between the two vehicles?

10   Three forces of 30 lb., 70 lb., and 80 lb. act on a body in such a way that they are in equilibrium. Find the angles they make with each other.

11   A patrol boat is due south of a freighter moving along the shore line in the direction S 77° W. If the freighter travels 12 mph and the patrol boat can make 24 mph, in what direction should the patrol boat travel in order to intercept the freighter as quickly as possible?

12   Starting at the water's edge, a person walks 120 ft. directly away from a river and up a slope of 18° 0'. Turning around, he finds that the angle of depression of the water's edge on the opposite bank is 4° 0'. How wide is the river?

13   An airplane heads S 70° W. A 25-mph wind, blowing toward N 40° W, causes the plane to move in the direction S 75° W. Find the air speed and the ground speed of the plane.

14   The Leaning Tower of Pisa measures 180 ft. from its top to its bottom. When the angle of elevation of the sun is 17°, the shadow of the tower (falling on a horizontal plane) is 602 ft. long and appears on the side *toward which* the tower leans. Find the angle the tower makes with the vertical.

15   An airplane heads west with an air speed of 210 mph. A wind speed of 20 mph causes the plane to travel with a ground speed of 200 mph. In what direction is the wind blowing? In what direction does the plane travel?

16   An airplane heads N 25° W with an air speed of 100 mph. Find the speed and the direction of the wind that makes the plane travel north at 110 mph.

17 Two sides of a parallelogram are 50 and 60. One diagonal is 80. Find the angles of the parallelogram.

18 An airplane with a cruising speed of 115 mph in still air is headed N 10° W. A wind is blowing from S 67° W with a speed of 32 mph. Find the direction the plane travels and its ground speed.

19 One diagonal of a parallelogram makes angles of 28° 19′ and 32° 41′ with the sides. If the length of the diagonal is 84.56 in., how long are the sides?

20 The pilot of an airplane wishes to travel due north. His plane has a cruising speed of 160 mph in still air. In what direction should he head his plane if a southwest wind of 40 mph is blowing? Find the ground speed of the plane.

21 The planet Jupiter is 483,000,000 miles from the sun. The earth is 93,000,000 miles from the sun. How far is Jupiter from the earth when the sun is about to rise in the east and Jupiter is 36° above the horizon in the west? (Get result to the nearest multiple of a million miles.)

22 A vertical tree stands on a hillside that makes an angle α with the horizontal. From a point m ft. *down* the hill from the tree, the angle of elevation of the treetop is β. If the tree is y ft. tall, express y in terms of m, α, and β.

23 A ship is anchored 20 miles S 25° E from a lighthouse. Find its distance from the lighthouse after it has moved 15 miles in the direction N 60° E.

24 Baton Rouge, La., is 140 miles S 25° W from Jackson, Miss.; Little Rock, Ark., is 210 miles from Jackson and 300 miles from Baton Rouge. What is the bearing of Little Rock from Baton Rouge? From Jackson? (Little Rock, of course, lies north and west from the Jackson–Baton Rouge line.)

25 Prove that the bisector of an angle of a triangle divides the opposite side into segments that are proportional to the adjacent sides.

26 If a triangle contains an obtuse angle, this angle, of course, is the largest angle in the triangle. As this obtuse angle increases, its sine decreases. Is it possible for the sine of the obtuse angle to be less than the sine of one of the acute angles of the triangle? Explain.

27 A *segment* of a circle is the region bounded by an arc of the circle and the chord that subtends it. Show that the area of the shaded segment of the circle in Figure 81 is equal to $\frac{1}{2}r^2(\theta - \sin\theta)$, where θ is in radians. (See Prob. 20, page 73.)

28 Given A, a, and b, use the law of cosines to show that

$$c = b\cos A \pm \sqrt{a^2 - b^2\sin^2 A}$$

Discuss this equation and interpret the geometric significance if (1) $a < b\sin A$, (2) $a = b\sin A$, (3) $a > b\sin A$, (4) $a = b$. Compare with the possibilities in the *SSA* case.

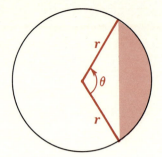

FIGURE 81

29 Prove that in any triangle,

$$a^2 + b^2 + c^2 = 2ab \cos C + 2bc \cos A + 2ac \cos B$$

30* Prove that in any triangle

$$\frac{a + b}{c} = \frac{\cos \frac{1}{2}(A - B)}{\sin \frac{1}{2} C}$$

Hint: Add the equations $\dfrac{a}{c} = \dfrac{\sin A}{\sin C}$ and $\dfrac{b}{c} = \dfrac{\sin B}{\sin C}$. Apply formula (15) of Section 45 to $\sin A + \sin B$. Replace $\sin C$ with $2 \sin \frac{1}{2} C \cos \frac{1}{2} C$. Notice that $\frac{1}{2} C$ is the complement of $\frac{1}{2}(A + B)$.

---

* The formulas in Probs. 30 and 31 are called **Mollweide's equations.** They serve as good checks because each of them involves all six parts of the triangle.

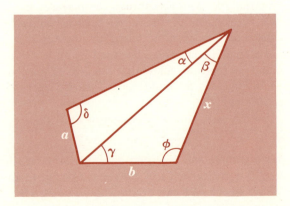

FIGURE 82

31* Prove that in any triangle

$$\frac{a - b}{c} = \frac{\sin \frac{1}{2}(A - B)}{\cos \frac{1}{2}C}$$

Use cyclic permutation and the interchange of pairs of letters to derive five similar formulas.

32 In Figure 82 on page 180, show that

$$x = \csc \alpha \sqrt{a^2 \sin^2 \delta + b^2 \sin^2 \alpha - 2ab \sin \alpha \sin \delta \cos \gamma}$$

---

* The formulas in Probs. 30 and 31 are called **Mollweide's equations.** They serve as good checks because each of them involves all six parts of the triangle.

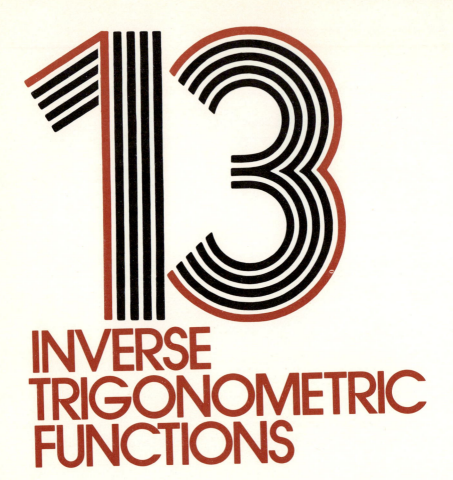

80 INVERSE TRIGONOMETRIC RELATIONS (Before proceeding, the student should make a thorough review of Section 31, page 70.) The equation

$$u = \sin \theta$$

says that u is a number representing the sine of the number* θ. Another interpretation is

θ is a number* whose sine is u.

This statement is usually written in the form

---

* This number may be regarded as the radian measure of an angle.

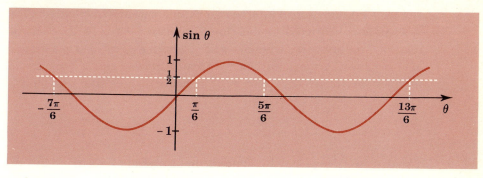

FIGURE 83

$$\theta = \arcsin u$$

With this understanding, we can say

$$\arcsin \frac{1}{2} = \frac{\pi}{6}, \frac{5\pi}{6}, \frac{13\pi}{6}, -\frac{7\pi}{6}, \text{ etc.,}$$

because the sine of each of these numbers is $\frac{1}{2}$. Values of arcsin $\frac{1}{2}$ may be obtained by finding all the points on the sine curve that are $\frac{1}{2}$ unit above the θ-axis (Figure 83).

Similarly, arccos u denotes a number* whose cosine is u, arctan u denotes a number* whose tangent is u, etc. The six inverse trigonometric relations are arcsin u, arccos u, arctan u, arccot u, arcsec u, arccsc u. Thus arccos $0 = \pi/2$, $3\pi/2$, etc., and arctan $1 = \pi/4$, $5\pi/4$, etc.

81 INVERSE TRIGONOMETRIC FUNCTIONS The sine function is defined as the infinite set of ordered pairs $\{(\theta, \sin \theta)$, where θ is a real number$\}$. This trigonometric relation† is a function because to each real number θ there corresponds one and only one number sin θ. Let us now consider the arcsin relation $\{(u, \arcsin u)$, where $-1 \leq u \leq 1\}$. As demonstrated in the preceding section, to each value of u there corresponds more than one value of arcsin u. Therefore, the arcsin relation is not a function.

It will be quite useful to modify the arcsin relation in such a way as to make it a function. A careful inspection of the sine curve (Figure 84) shows that as θ varies from $-\pi/2$ to $\pi/2$, then sin θ moves from

---

* This number may be regarded as (the radian measure of) an angle.
† Recall that a relation is a set of ordered pairs of numbers, whereas a function is a relation $\{(x,y)\}$ such that to each x there corresponds one and only one y.

-1 to $+1$, taking on all intervening values once and only once. Thus, if arcsin u is restricted to the closed interval $[-\pi/2, \pi/2]$, then the relation $\{(u, \text{arcsin } u), \text{ where } -\pi/2 \leq \text{arcsin } u \leq \pi/2\}$ is a function. The notation Arcsin u or Sin$^{-1} u$ (read "inverse sine u") means *the* number (between $-\pi/2$ and $\pi/2$, inclusive) whose sine is u; that is,

$$-\frac{\pi}{2} \leq \text{Sin}^{-1} u \leq \frac{\pi}{2}$$

Thus, $\text{Sin}^{-1} \dfrac{1}{2} = \dfrac{\pi}{6}$. It should be carefully noted that the superscript -1 is *not* to be interpreted as an exponent; that is, $\text{Sin}^{-1} u$ is *not* $(\text{Sin } u)^{-1}$ or $\dfrac{1}{\sin u}$.

The symbols arccos u and arctan u may be modified to refer to functions rather than relations by defining Arccos $u = \text{Cos}^{-1} u$ and Arctan $u = \text{Tan}^{-1} u$ as follows:

$$0 \leq \text{Cos}^{-1} u \leq \pi \qquad -\frac{\pi}{2} < \text{Tan}^{-1} u < \frac{\pi}{2}$$

The following summary shows the domain, range, and defining equation for each of the previously discussed inverse trigonometric functions.

Function	Inverse Function	Defining Equation	Domain	Range
sin	Sin^{-1}	$y = \text{Sin}^{-1} x$	$-1 \leq x \leq 1$	$-\dfrac{\pi}{2} \leq \text{Sin}^{-1} x \leq \dfrac{\pi}{2}$
cos	Cos^{-1}	$y = \text{Cos}^{-1} x$	$-1 \leq x \leq 1$	$0 \leq \text{Cos}^{-1} x \leq \pi$
tan	Tan^{-1}	$y = \text{Tan}^{-1} x$	$-\infty < x < \infty$	$-\dfrac{\pi}{2} < \text{Tan}^{-1} x < \dfrac{\pi}{2}$

Figure 84 restates these definitions in terms of the sine, cosine, and tangent curves in which the unbroken lines indicate the portions of the curves that are to be used.

Hence, by definition, $\text{Sin}^{-1} u$, $\text{Cos}^{-1} u$, and $\text{Tan}^{-1} u$ refer to functions rather than relations. Thus $\text{Cos}^{-1} u$ means *the* number (between 0 and π) whose cosine is u.

ILLUSTRATIONS $\text{Sin}^{-1}\left(-\dfrac{\sqrt{3}}{2}\right) = -\dfrac{\pi}{3}$

$\text{Cos}^{-1}\left(-\dfrac{\sqrt{2}}{2}\right) = \dfrac{3\pi}{4} \qquad \text{Tan}^{-1}(-1) = -\dfrac{\pi}{4}$

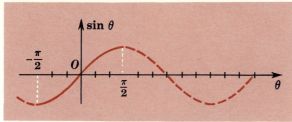

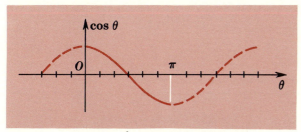

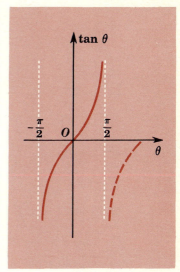

Find Sin^{-1} u between $-\frac{\pi}{2}$ and $\frac{\pi}{2}$.

Find Cos^{-1} u between 0 and π.

**Find Tan^{-1} u
between $-\frac{\pi}{2}$ and $\frac{\pi}{2}$.**

FIGURE 84

In calculus, frequent use is made of the inverse trigonometric functions. For example, the area bounded by the curve $y = 1/(x^2 + 1)$,
the x-axis, and the lines $x = 0$ and $x = 2$ is Tan^{-1} 2 = 1.1071 (square
units).

The functions Cot^{-1} u, Sec^{-1} u, and Csc^{-1} u are of less importance.
They can readily be expressed in terms of Tan^{-1} u, Cos^{-1} u, and
Sin^{-1} u, respectively. For example, Sec^{-1} $u =$ Cos^{-1} $(1/u)$ [the number
whose secant is u is the same number whose cosine is $(1/u)$]. Thus,
Sec^{-1} 2 = Cos^{-1} $\dfrac{1}{2} = \dfrac{\pi}{3}$.

EXERCISE 43

Find the value of each of the following. Do not use tables.

1 Sin^{-1} $(-\frac{1}{2})$

2 Sin^{-1} $\dfrac{\sqrt{3}}{2}$

3 Cos^{-1} $\frac{1}{2}$

4 Cos^{-1} $\left(-\dfrac{\sqrt{3}}{2}\right)$

5 Cos^{-1} $\dfrac{\sqrt{3}}{2}$

6 Cos^{-1} $(-\frac{1}{2})$

7 Sin^{-1} 0

8 Sin^{-1} (-1)

9 $\text{Csc}^{-1}\sqrt{2}$ 10 $\text{Sin}^{-1}\dfrac{\sqrt{2}}{2}$

11 $\text{Cos}^{-1}(-1)$ 12 $\text{Cos}^{-1}\dfrac{\sqrt{2}}{2}$

13 $\text{Cos}^{-1}0$ 14 $\text{Cos}^{-1}1$

15 $\text{Sin}^{-1}\left(-\dfrac{\sqrt{2}}{2}\right)$ 16 $\text{Csc}^{-1}2$

17 $\text{Cot}^{-1}1$ 18 $\text{Sec}^{-1}(-1)$

19 $\text{Sec}^{-1}\dfrac{2}{\sqrt{3}}$ 20 $\text{Tan}^{-1}\left(-\dfrac{1}{\sqrt{3}}\right)$

21 $\text{Tan}^{-1}(-\sqrt{3})$

Use Table 1 to find the value of each of the following.

22 $\text{Sin}^{-1}0.7509$ 23 $\text{Sin}^{-1}(-0.5544)$
24 $\text{Cos}^{-1}0.2447$ 25 $\text{Cos}^{-1}(-0.9914)$
26 $\text{Tan}^{-1}(-0.4770)$ 27 $\text{Tan}^{-1}4.511$

28 Explain why $\text{Cos}^{-1}u$ could not be taken on the interval $-\pi/2$ to $\pi/2$.

82 OPERATIONS INVOLVING INVERSE TRIGONOMETRIC FUNC-TIONS Since every inverse trigonometric function may be thought of as (the radian measure of) an angle, it is frequently convenient to place this angle in standard position and label its triangle of reference in accordance with the inverse function (Example 1). Sometimes it is advisable to replace the inverse functions with angle symbols, such as θ, A, B, and then try to express the problem in terms of ordinary functions.

EXAMPLE 1 Evaluate $\cos\left(\text{Tan}^{-1}\tfrac{2}{3}\right)$.

Solution We are asked to find the cosine of the angle whose tangent is $\tfrac{2}{3}$. Draw a right triangle with legs 2 and 3. The acute angle opposite the side 2 has a tangent of $\tfrac{2}{3}$. It can be labeled $\text{Tan}^{-1}\tfrac{2}{3}$. After finding the hypotenuse is $\sqrt{13}$, we see (Figure 85) that

$$\cos\left(\text{Tan}^{-1}\frac{2}{3}\right) = \frac{3}{\sqrt{13}}$$

EXAMPLE 2 Find the value of

$$\sin\left(\text{Sin}^{-1}u + \text{Cos}^{-1}v\right)$$

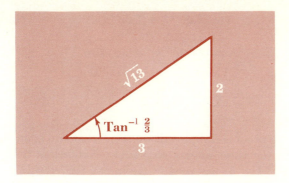

FIGURE 85

Solution Since $\operatorname{Sin}^{-1} u$ and $\operatorname{Cos}^{-1} v$ are angles, we have the sine of the sum of two angles. Let $A = \operatorname{Sin}^{-1} u$ and $B = \operatorname{Cos}^{-1} v$. Using

$$\sin (A + B) = \sin A \cos B + \cos A \sin B$$

we have

$$\begin{aligned} \sin (\operatorname{Sin}^{-1} u + \operatorname{Cos}^{-1} v) &= \sin (\operatorname{Sin}^{-1} u) \cos (\operatorname{Cos}^{-1} v) \\ &\quad + \cos (\operatorname{Sin}^{-1} u) \sin (\operatorname{Cos}^{-1} v) \\ &= uv + \sqrt{1 - u^2} \sqrt{1 - v^2} \end{aligned}$$

The value of $\cos (\operatorname{Sin}^{-1} u)$ is found by use of the triangle in Figure 86. Notice that if u is negative, $\operatorname{Sin}^{-1} u$ is in Q IV (its value in radians lies between $-\pi/2$ and 0) but $\cos (\operatorname{Sin}^{-1} u)$ is still positive. Explain why $\sin (\operatorname{Cos}^{-1} v)$ is always the positive radical $\sqrt{1 - v^2}$.

EXAMPLE 3 Prove that

$$\operatorname{Tan}^{-1} \frac{1}{2} + \operatorname{Tan}^{-1} \frac{1}{3} = \frac{\pi}{4}$$

Solution The left side is the sum of two acute angles, each being less than $\pi/4$. Why? To prove that their sum is $\pi/4$, let us take the tangent of each side of the equation:

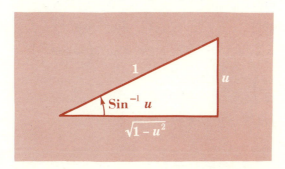

FIGURE 86

$$\tan \left(\mathrm{Tan}^{-1}\tfrac{1}{2} + \mathrm{Tan}^{-1}\tfrac{1}{3}\right) \qquad \bigg| \qquad \tan\frac{\pi}{4}$$

$$= \frac{\tan \left(\mathrm{Tan}^{-1}\tfrac{1}{2}\right) + \tan \left(\mathrm{Tan}^{-1}\tfrac{1}{3}\right)}{1 - \tan \left(\mathrm{Tan}^{-1}\tfrac{1}{2}\right) \tan \left(\mathrm{Tan}^{-1}\tfrac{1}{3}\right)} \qquad \bigg| \qquad = 1$$

$$= \frac{\tfrac{1}{2} + \tfrac{1}{3}}{1 - \tfrac{1}{2} \cdot \tfrac{1}{3}} = \frac{\tfrac{5}{6}}{\tfrac{5}{6}} = 1$$

The proof is complete if we recall that two acute angles having the same tangent are equal. Notice that the formula for tan $(A + B)$ was used in evaluating the left side.

EXERCISE 44

Find the value of the following without using tables.

1 $\tan \left(\mathrm{Cos}^{-1} u\right)$ 2 $\sin \left(\mathrm{Tan}^{-1} u\right)$

3 $\sec \left(\mathrm{Cos}^{-1} u\right)$ 4 $\cot \left(\mathrm{Cot}^{-1} u\right)$

5 $\cos \left[\mathrm{Sin}^{-1}\left(-\tfrac{12}{13}\right)\right]$ 6 $\tan \left[\mathrm{Cos}^{-1}\left(-\tfrac{1}{3}\right)\right]$

7 $\cos \left[\mathrm{Tan}^{-1}\left(-2\right)\right]$ 8 $\sin \left[\mathrm{Cos}^{-1}\left(-\tfrac{24}{25}\right)\right]$

9 $\cos \left(\mathrm{Cos}^{-1} u + \mathrm{Cos}^{-1} v\right)$ 10 $\sin \left(\mathrm{Sin}^{-1} u - \mathrm{Sin}^{-1} v\right)$

11 $\sin \left(\mathrm{Cos}^{-1} u + \mathrm{Sin}^{-1} v\right)$ 12 $\cos \left(\mathrm{Sin}^{-1} u - \mathrm{Cos}^{-1} v\right)$

13 $\sin \left(\dfrac{3\pi}{2} - \mathrm{Sin}^{-1}\dfrac{4}{7}\right)$ 14 $\cos \left(\pi + \mathrm{Cos}^{-1}\tfrac{2}{3}\right)$

15 $\cos \left(\dfrac{\pi}{3} - \mathrm{Sin}^{-1} u\right)$ 16 $\sin \left(\dfrac{\pi}{2} + \mathrm{Sin}^{-1}\dfrac{3}{4}\right)$

17 $\cos \left(2\,\mathrm{Cos}^{-1} u\right)$

Hint: Let $A = \mathrm{Cos}^{-1} u$. We seek cos $2A$.

18 $\sin \left(2\,\mathrm{Sin}^{-1} u\right)$ 19 $\sin \left(2\,\mathrm{Cos}^{-1} 8u\right)$

20 $\cos \left(2\,\mathrm{Sin}^{-1} 3u\right)$ 21 $\sin \left(\mathrm{Sin}^{-1} 2u\right)$

22 $\cos \left(\mathrm{Cos}^{-1} 6u\right)$ 23 $\mathrm{Cos}^{-1}\left[\cos \left(-\dfrac{3\pi}{8}\right)\right]$

24 $\mathrm{Sin}^{-1}\left(\sin \dfrac{9\pi}{10}\right)$ 25 $\mathrm{Tan}^{-1}\left(\tan \dfrac{4\pi}{5}\right)$

26 $\mathrm{Sin}^{-1}\left(\sin 140°\right)$ 27 $\mathrm{Sin}^{-1}\left(\cos^3 \pi\right)$

28 $\mathrm{Tan}^{-1}\sqrt{2 - \sec \pi}$ 29 $\mathrm{Cos}^{-1}\left(-\sin^2 \dfrac{\pi}{4}\right)$

30 $\mathrm{Tan}^{-1}\left(\sin \dfrac{3\pi}{2}\right)$

Assume $u > 0$. Copy the following and fill in the blanks.

31 $\mathrm{Sin}^{-1}\dfrac{u}{\sqrt{u^2 + 1}} = \mathrm{Cos}^{-1}\underline{\hspace{1.5cm}} = \mathrm{Tan}^{-1}\underline{\hspace{1.5cm}}$

32 $\mathrm{Cos}^{-1}\dfrac{\sqrt{u^2-9}}{u} = \mathrm{Sin}^{-1}\underline{\hspace{1cm}} = \mathrm{Tan}^{-1}\underline{\hspace{1cm}},\ u \geq 3$

33 $\mathrm{Tan}^{-1}\dfrac{\sqrt{4-u^2}}{u} = \mathrm{Sin}^{-1}\underline{\hspace{1cm}} = \mathrm{Cos}^{-1}\underline{\hspace{1cm}},\ 0 < u \leq 2$

34 $\mathrm{Sec}^{-1}\dfrac{\sqrt{u^2+4u+13}}{u+2} = \mathrm{Csc}^{-1}\underline{\hspace{1cm}} = \mathrm{Cot}^{-1}\underline{\hspace{1cm}}$

35 $\mathrm{Cos}^{-1}u - \mathrm{Sin}^{-1}u = \mathrm{Tan}^{-1}\underline{\hspace{1cm}},\ 0 < u < 1$

Prove each of the following identities without using tables. (Consider all permissible values—positive, negative, and zero—of the letters involved.)

36 $\sin\left(\dfrac{1}{2}\mathrm{Cos}^{-1}u\right) = \sqrt{\dfrac{1-u}{2}}$

37 $\cos\left(\dfrac{1}{2}\mathrm{Cos}^{-1}8u\right) = \sqrt{\dfrac{1+8u}{2}}$

38 $\cos\left(\dfrac{1}{2}\mathrm{Sin}^{-1}a\right) = \sqrt{\dfrac{1+\sqrt{1-a^2}}{2}}$

39 $\sin\left(\dfrac{1}{2}\mathrm{Sin}^{-1}3u\right) = \pm\sqrt{\dfrac{1-\sqrt{1-9u^2}}{2}}$

40 $\mathrm{Tan}^{-1}8 - \mathrm{Tan}^{-1}3 = \mathrm{Tan}^{-1}\frac{1}{5}$

41 $\mathrm{Tan}^{-1}10 + \mathrm{Tan}^{-1}\dfrac{11}{9} = \dfrac{3\pi}{4}$

42 $\mathrm{Cos}^{-1}\frac{4}{5} + \mathrm{Cos}^{-1}\frac{12}{13} = \mathrm{Cos}^{-1}\frac{33}{65}$

43 $\mathrm{Sin}^{-1}\frac{5}{8} - \mathrm{Cos}^{-1}\frac{19}{20} = \mathrm{Sin}^{-1}\frac{7}{20}$

44 $\frac{1}{2}\mathrm{Cos}^{-1}\frac{7}{25} = \mathrm{Sin}^{-1}\frac{56}{65} - \mathrm{Sin}^{-1}\frac{5}{13}$

45 $2\,\mathrm{Sin}^{-1}\frac{5}{8} + \mathrm{Cos}^{-1}\left(-\frac{7}{32}\right) = \pi$

Identify each statement as true or false and give reasons. (Consider all permissible values—positive, negative, and zero—of the letters involved.)

46 $\mathrm{Sin}^{-1}u + \mathrm{Cos}^{-1}u = \dfrac{\pi}{2}$ **47** $\mathrm{Sin}^{-1}(-u) = -\mathrm{Sin}^{-1}u$

48 $\mathrm{Tan}^{-1}(-u) = -\mathrm{Tan}^{-1}u$ **49** $\mathrm{Cos}^{-1}(-u) = \pi - \mathrm{Cos}^{-1}u$

50 $\mathrm{Tan}^{-1}(-\infty) = -\dfrac{\pi}{2}$ **51** $\sin\left(\mathrm{Cos}^{-1}u\right) = \cos\left(\mathrm{Sin}^{-1}u\right)$

52 $\sin\left(2\,\mathrm{Sin}^{-1}a\right) = 2a\sqrt{1-a^2}$ **53** $\mathrm{Tan}^{-1}u + \mathrm{Tan}^{-1}\dfrac{1}{u} = \dfrac{\pi}{2}$

54 $\sec\left(\mathrm{Sin}^{-1}\dfrac{1}{a}\right) = a$ **55** $\cot\left(\mathrm{Cos}^{-1}u\right) = \pm\dfrac{u}{\sqrt{1-u^2}}$

56 $\sin\left(\mathrm{Cos}^{-1}a\right) = \pm\sqrt{1-a^2}$ **57** $\sin\left(\frac{1}{2}\mathrm{Sin}^{-1}2u\right) = u$

58   Given the equation $\text{Tan}^{-1} u = \dfrac{1}{2} \text{Cos}^{-1} \dfrac{1 - u^2}{1 + u^2}.$

(a) Prove that this equation is, or is not, an identity for $u \geq 0$.
(b) Prove that the equation does not hold for any value of $u < 0$.

Solve for x.

59 $\text{Cos}^{-1} (2 - 3x) = 2 \text{Sin}^{-1} x$ **60** $\text{Sin}^{-1} x = 2 \text{Cos}^{-1} x$

61 $\cot (\text{Cos}^{-1} x) = \dfrac{x}{\sqrt{1 - x^2}}$ **62** $\text{Cos}^{-1} x + \text{Cos}^{-1} (1 - x) = \dfrac{\pi}{2}$

83   INVERSE FUNCTIONS*

Two equations are equivalent if they have the same solution set. That is, every solution of one equation is a solution of the other, and vice versa. Thus $y = 2^x$ and $x = \log_2 y$ are equivalent equations by the definition of logarithm. If f and g are functions and if $y = f(x)$ and $x = g(y)$ are equivalent equations, then f and g are said to be inverses of each other. To find the inverse of the function f, where $f(u) = 2u + 7$, we let $y = 2x + 7$† and then solve for x: $x = \frac{1}{2}(y - 7)$. Hence the inverse of f is g, where $g(u) = \frac{1}{2}(u - 7)$. Likewise, if $y = \sin x$, where $-\pi/2 \leq x \leq \pi/2$, then $x = \text{Sin}^{-1} y$, where $-1 \leq y \leq 1$. Hence, if $-\pi/2 \leq u \leq \pi/2$, then the inverse of the trigonometric function $(u, \sin u)$ is the inverse trigonometric function $(u, \text{Sin}^{-1} u)$. For this reason Sin^{-1} is called an inverse trigonometric function.

If f and g are inverse functions, then it can be shown that $g[f(u)] = u$ and $f[g(u)] = u$. If $f(u) = 2^u$, then $g(u) = \log_2 u$ (See Section 65, page 143). Thus $g[f(u)] = \log_2 (2^u) = u$ and $f[g(u)] = 2^{\log_2 u} = u$. Moreover, if $f(u) = \sin u$, where $-\pi/2 \leq u \leq \pi/2$, then $f[g(u)] = \sin (\text{Sin}^{-1} u) = u$ and $g[f(u)] = \text{Sin}^{-1} (\sin u) = u$.

Furthermore, if the function g‡ is the inverse of the function f, then the graph of $g(u)$ is the reflection of the graph of $f(u)$ in the line that bisects the first and third quadrants. (See Section 65, page 143.) Hence the graph of $y = \text{Sin}^{-1} x$ is the reflection of the graph of $y = \sin x$ (for $-\pi/2 \leq x \leq \pi/2$) in the line $y = x$ (Figure 87).

---

* For a detailed treatment of inverse functions, see the author's *College Algebra,* McGraw-Hill, New York, 1973.

† The quantity $2x + 7$ is a function of x. [This is another way of describing the function consisting of all ordered pairs $(x, 2x + 7)$, for real values of x.] The expression $2u + 7$ is the *same* function of u. In each case the independent variable is doubled and added to 7.

‡ It would be consistent to use f^{-1} to designate the inverse of f. Then $g[f(u)] = f^{-1}f^{1}(u) = f^0(u) = u$, *symbolically.* Of course we are not multiplying (in the narrow sense) f^{-1} by f.

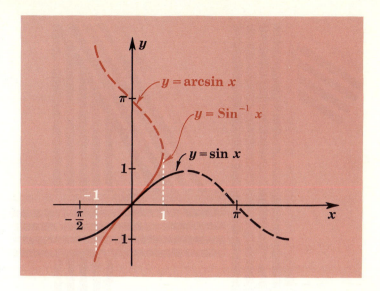

FIGURE 87

EXERCISE 45

In the following problems g designates the inverse of the function f.

1   Given $f(u) = \dfrac{5u - 6}{7}$.

  (a) Find $g(u)$.
  (b) Show that $g[f(u)] = f[g(u)] = u$.

2   Given $f(u) = \dfrac{2u + 3}{9u - 2}$.

  (a) Find $g(u)$.
  (b) Show that $g[f(u)] = f[g(u)] = u$.

3   Given $f(u) = 4 - \sqrt{u + 7}$, where $u \geq -7$.
  (a) Find $g(u)$.
  (b) Show that $g[f(u)] = u$.
  (c) Show that $f[g(u)] = u$ if and only if $u \leq 4$.

4   Given $f(u) = \operatorname{Sin}^{-1} 2^u$.

  (a) Find $g(u)$.
  (b) Show that $g[f(u)] = u$ if $u \leq 0$.

  (c) Show that $f[g(u)] = u$ if $0 < u \leq \dfrac{u}{2}$.

5   Sketch the graph of $y = \operatorname{Cos}^{-1} x$.
6   Sketch the graph of $y = \operatorname{Tan}^{-1} x$.

COMPLEX
NUMBERS

84 COMPLEX NUMBERS It will be recalled that there are quadratic equations with real coefficients that have no real roots. The simplest such equation is $x^2 + 1 = 0$, which obviously has no real solution since the square of any real number—whether positive, negative, or zero—is nonnegative. (An analogous linear equation with positive coefficients is $x + 1 = 0$, which necessitates the postulation of negative numbers for its solution—though the negative numbers probably arose from other considerations.) Another example of a quadratic equation with real coefficients but with no real roots is $x^2 - 4x + 13 = 0$. It was for the solution of such equations that the system of *complex numbers* was first investigated, but this remarkable system has proved to be invaluable in many other connections.

The system of complex numbers is really the system of *ordered pairs* of *real* numbers—a first, then a second—(a,b), in which equality, addition, and multiplication are defined in a certain specified way. Usually, however, complex numbers are written as $a + bi$, and in this form the defining properties* are as follows:

> **Equality:** $a + bi = c + di$ if and only if $a = c$ and $b = d$.
>
> **Addition:** $(a + bi) + (c + di) = (a + c) + (b + d)i$.
>
> **Multiplication:** $(a + bi)(c + di) = (ac - bd) + (ad + bc)i$.

It is easy to show that all the ordinary rules for adding and multiplying real numbers—the associative and commutative laws, etc.— carry over to the system of complex numbers.   For this reason, we speak of the *field* of complex numbers.

There is no need to memorize the definition of addition and multiplication of complex numbers.   Just remember that addition and multiplication are ordinary addition and multiplication, with the single special property that

$$i^2 = -1$$

There is, however, one property of the real numbers that does not carry over to the complex numbers.   *The complex number field is not ordered;* we do not say that one complex number is less than or greater than another.

Complex numbers have been defined as numbers that obey the definitions of equality, addition, and multiplication listed above.   This approach is equivalent to the following definition.

> A *complex number* is a number of the form $a + bi$, where a and b are real numbers and $i = \sqrt{-1}$.
>
> If $b \neq 0$, the complex number $a + bi$ is called an *imaginary number*.
>
> If $a = 0$ and $b \neq 0$, the complex number $a + bi$ is called a *pure imaginary number*.

---

* In terms of the (a, b) notation, which we shall not adopt, these definitions are:
  Equality: $(a, b) = (c, d)$ if and only if $a = c$ and $b = d$
  Addition: $(a, b) + (c, d) = (a + c, b + d)$
  Multiplication: $(a, b)(c, d) = (ac - bd, ad + bc)$

If $b = 0$, the complex number $a + bi$ becomes a, a real number.

Hence we see that the field of complex numbers includes all real numbers and all imaginary numbers.

ILLUSTRATIONS

Imaginary numbers. $3i, 2 + 5i, -7 + 8i, 9 - i, -1 - 6i$
Pure imaginary numbers. $3i, -4i, \sqrt{-49}, -\sqrt{-2}$

Real numbers. $4, \frac{1}{7}, 0, -\frac{2}{9}, 5, -\sqrt{2}, \pi$

All these numbers are complex numbers.

Since $\sqrt{a} \cdot \sqrt{a} = a$ (by the definition of square root), we see that if $i = \sqrt{-1}$, then $i^2 = -1$. Moreover, $i^3 = i^2 \cdot i = -i$, $i^4 = (i^2)^2 = (-1)^2 = 1$, $i^5 = i^4 \cdot i = i$, $i^6 = i^4 \cdot i^2 = -1$, $i^{87} = i^{84} \cdot i^3 = (i^4)^{21} i^3 = 1^{21}(-i) = -i$.

In solving the equation $x^2 - 4x + 13 = 0$, we find the roots to be $x = 2 \pm \sqrt{-9}$. Remembering that $i = \sqrt{-1}$, we have $x = 2 \pm 3i$. Notice that if $2 + 3i$ is substituted for x in the equation $x^2 - 4x + 13 = 0$, we get

$$4 + 12i + 9i^2 - 8 - 12i + 13 = 0$$

If i^2 is replaced by -1, we obtain $17 + 12i - 12i - 17 = 0$. This shows that $2 + 3i$ is a perfectly good root of the equation, provided we understand that i is a number whose square is -1.

The complex numbers $a + bi$ and $a - bi$ are said to be *complex conjugates* of each other. Notice that the roots of $x^2 + 4 = 0$ are $2i$ and $-2i$, which are pure imaginary complex conjugates. The roots of the equation $x^2 - 4x + 13 = 0$ are the conjugate imaginary numbers $2 + 3i$ and $2 - 3i$. It can be shown that if an imaginary number $(a + bi)$ is a root of an equation with real coefficients, then the conjugate imaginary $(a - bi)$ is also a root of this equation.

Since i is a number whose square is -1, the best procedure in handling complex numbers is to perform all operations as if i were an ordinary letter and then replace i^2 with -1. It is to be noted that the quotient of two complex numbers is obtained by multiplying numerator and denominator by the conjugate of the denominator. For example,

$$\frac{7 + 5i}{3 - i} = \frac{(7 + 5i)(3 + i)}{(3 - i)(3 + i)} = \frac{21 + 7i + 15i + 5i^2}{9 - i^2}$$

$$= \frac{16 + 22i}{10} = \frac{8}{5} + \frac{11}{5}i$$

This result can be checked by multiplying $\frac{8}{5} + \frac{11}{5} i$ by $3 - i$. What should the result be?

All complex numbers should first be written in the form $a + bi$. Thus $3 + \sqrt{-49} = 3 + \sqrt{49} \sqrt{-1} = 3 + 7i$. This procedure is suggested to avoid mistakes such as $\sqrt{-5} \cdot \sqrt{-5} = \sqrt{(-5)(-5)} = \sqrt{25} = 5$. This is obviously incorrect because, by the definition of square root, $\sqrt{-5}$ is a number which when multiplied by itself becomes -5. The correct way of handling this is $\sqrt{-5} \cdot \sqrt{-5} = i\sqrt{5} \cdot i\sqrt{5} = 5i^2 = -5$. This result agrees with the definition of square root.

EXERCISE 46

Perform each of the indicated operations and express the result in the form $a + bi$.

1 $(9 + 4i) + (3 - 7i)$

2 $(6 - 8i) + (-1 - 5i)$

3 $(-2 - i) + (4 + i) - (8 - 3i)$

4 $(7 + 3i) - (6 - 2i)$

5 $(7 - 6i)(4 + 3i)$

6 $(2 - i\sqrt{7})(6 + i\sqrt{7})i$

7 $(9 + \sqrt{-5})(1 - \sqrt{-5})$

8 $(3 + 4i)(5 + 8i)$

9 $(4 + 7i)^2$

10 $(5 - 2i)^2$

11 i^{21}

12 i^{22}

13 i^{1975}

14 $i^{39} + i^{40} + i^{41}$

15 $\dfrac{9 + 2i}{6 + i}$

16 $\dfrac{7 - 3i}{2 - 5i}$

17 $\dfrac{8 + i\sqrt{6}}{1 - i\sqrt{6}}$

18 $\dfrac{5 - 8i}{3 + 4i}$

19 $\dfrac{7 + 5i}{i}$

20 $\dfrac{2 - 6i}{-1 + 3i}$

Find the values of the real numbers x and y.

21 $9x - 2yi = 63 + 16i$

22 $(5 - ix)(6 - i) = y + 7i$

23 $(x + i)(1 + yi) = 4 + 46i$

24 $(3 - 2x) + 5yi = x + yi$

25 Find the value of $4x^2 - 12x + 13$ if $x = \frac{3}{2} + i$.

26 Show by substitution that $\dfrac{1 - i\sqrt{3}}{2}$ is a root of the equation $x^2 - x + 1 = 0$.

27 Prove that the sum of two conjugate imaginary numbers is a real number.

28 Prove that the product of two conjugate imaginary numbers is a positive real number.

29 Write a complex number that is not an imaginary number.

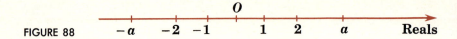

FIGURE 88

85 GRAPHICAL REPRESENTATION OF COMPLEX NUMBERS   Let us
represent (as in the case of the x-axis of a rectangular coordinate
system) the real numbers by points on a horizontal directed line
(Figure 88). Let the vector V represent the directed segment connecting
the origin O to the point corresponding to the real number a. Since
$ai^2 = -a$, it can be said that multiplying a by $i \cdot i$ is geometri-
cally equivalent to rotating V through 180° about O. Consequently,
it is logical to represent the multiplication of a by i as a rotation of V
through 90° about O. Accordingly, the number ai will be represented
as a point a units above O on the *vertical* line through O. We shall
refer to the horizontal axis as the **axis of reals** and the vertical axis as
the **axis of (pure) imaginaries**. This system of axes defines a region
called the **complex plane.** It is to be noted that, while the unit on
the axis of reals is the number 1, the unit on the axis of imaginaries is
the imaginary number i. Hence the complex number $(a + bi)$ is
represented by the point a units from the axis of imaginaries and
b units from the axis of reals. Figure 89 illustrates the graphical
representation of complex numbers in the complex plane.

It is often convenient to think of the complex number $(a + bi)$ as
representing the vector OP (Figure 89).

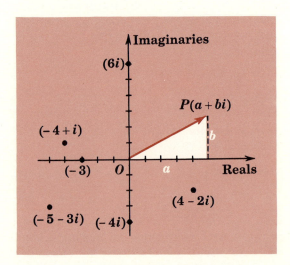

FIGURE 89

86 GRAPHICAL ADDITION OF COMPLEX NUMBERS Since the sum of $(a + bi)$ and $(c + di)$ is $(a + c) + (b + d)i$, we can add the numbers graphically by adding the real parts, a and c, to get the real part of the sum, and adding the imaginary coefficients, b and d, to get the imaginary coefficient. This is illustrated in Figure 90. The result is exactly the same as if we had applied the parallelogram law to the vectors representing the numbers $(a + bi)$ and $(c + di)$. Three complex numbers can be added graphically by first obtaining the sum of two of them and then adding this to the third.

We can subtract $(c + di)$ from $(a + bi)$ graphically by adding $(a + bi)$ to $(-c - di)$.

87 TRIGONOMETRIC FORM OF A COMPLEX NUMBER Let point P in the complex plane represent the complex number $a + bi$. The **absolute value*** of $a + bi$ is the distance r from O to P. It is always considered positive. The **amplitude*** of $a + bi$ is the angle measured from the positive axis of reals to the line OP. From Figure 91, it is obvious that

$$r = \sqrt{a^2 + b^2} \qquad \tan \theta = \frac{b}{a} \tag{1}$$

and
$$a = r \cos \theta \qquad b = r \sin \theta \tag{2}$$

---

* Absolute value is also called *modulus;* amplitude is sometimes called *argument*.

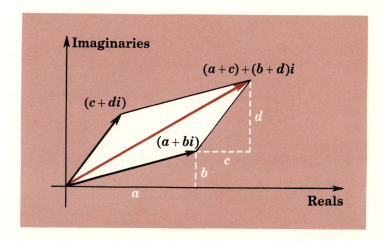

FIGURE 90

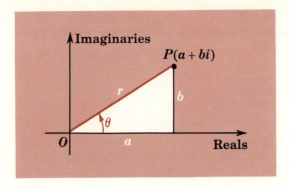

FIGURE 91

These equations hold regardless of the quadrant in which P lies. If the last equation is multiplied by i and added to the preceding one, we get

$$a + bi = r(\cos \theta + i \sin \theta) \tag{3}$$

The expression $r(\cos \theta + i \sin \theta)$ is called the **trigonometric*** **form** of a complex number. The expression $a + bi$ is called the **algebraic form** of a complex number. The trigonometric form is useful in finding powers and roots of complex numbers.

Any complex number in algebraic form can be expressed in trigonometric form by use of Equations (1). After the value of $\tan \theta$ has been obtained, θ can be found by use of a table of trigonometric functions. In general, there are two angles between 0° and 360° having the same tangent. In order to be certain to get the correct angle, we should *always plot the complex number†* in the complex plane. The amplitude of a real number or a pure imaginary number can be obtained by inspection of its location in the complex plane. For example, the amplitude of $-4i$ is 270° (Figure 89).

Any complex number in trigonometric form can be expressed in algebraic form by use of Equations (2).

EXAMPLE   Express each of the following in trigonometric form:

(a) $3 - 3i$ (b) -4

Solution   (a) Plot the number in the complex plane. Equations (1) give us $r = \sqrt{18} = 3\sqrt{2}$, $\tan \theta = -3/3 = -1$. From the

---

* Also called the *polar* form. It is sometimes written in the abbreviated form r cis θ.
† The expression *plot the complex number* is an abbreviation we shall use for the more rigorous statement, *plot the point corresponding to the complex number.*

last equation, θ could be 135° or 315°. From Figure 92 we see that θ must be 315°. Hence

$$3 - 3i = 3\sqrt{2}\,(\cos 315° + i \sin 315°)$$

This result can be checked by replacing cos 315° and sin 315° with $\sqrt{2}/2$ and $-\sqrt{2}/2$, respectively, and then demonstrating that the right side is actually equal to the left side.

(*b*) After plotting the number, Figure 92, we find by inspection that $r = 4$ and $\theta = 180°$. Hence we can see immediately that

$$-4 = 4(\cos 180° + i \sin 180°)$$

It is to be carefully noted that, regardless of the signs of a and b, r is *always positive*, and *the signs in front of* cos θ *and* i sin θ *are always positive*.

EXERCISE 47

Perform the indicated operations graphically and check the results algebraically.

1 $(4 - 3i) + (2 + i)$ 2 $(-1 + 5i) + (6 - 2i)$
3 $(5 - 2i) - (3 - 4i)$ 4 $(6 + i) + (-2 - 5i)$
5 $(-4 + i) - (-1 - 3i)$ 6 $(-2 - 3i) - (5 + 3i)$
7 $(1 + 2i) + (-5 + i) + (3 + 3i)$
8 $(3 + i) + (2 + 4i) - (3 - 2i)$

Plot each of the following complex numbers and then express it in trigonometric form.

9 $6 + 6i$ 10 $\sqrt{2} - i\sqrt{2}$
11 $-5 + 5i$ 12 $-1 - i$
13 $-5i$ 14 -3

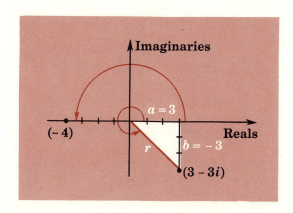

FIGURE 92

15 7	**16** $2i$	
17 $-1 + i\sqrt{3}$	**18** $8 + 15i$	
19 $-4 - 3i$	**20** $2\sqrt{3} - 2i$	

Plot each of the complex numbers and then express it in algebraic form.

21 $10(\cos 250° + i \sin 250°)$

22 $6(\cos 120° + i \sin 120°)$

23 $2(\cos 330° + i \sin 330°)$

24 $3(\cos 40° + i \sin 40°)$

25 On one system of coordinates, plot and label the number $4 - 5i$, its conjugate, and its negative.

26 What is the amplitude **(a)** of a positive real number? **(b)** of a negative real number? **(c)** of bi if $b > 0$? **(d)** of bi if $b < 0$?

27 Show that the negative of $r(\cos \theta + i \sin \theta)$ is

$r[\cos (\theta + 180°) + i \sin (\theta + 180°)]$

28 Show that the conjugate of $r(\cos \theta + i \sin \theta)$ is $r[\cos (-\theta) + i \sin (-\theta)]$.

88 MULTIPLICATION OF COMPLEX NUMBERS IN TRIGONOMETRIC FORM.

THEOREM   The absolute value of the product of two complex numbers is the product of their absolute values; the amplitude of the product is the sum of their amplitudes;

$$r_1(\cos \theta_1 + i \sin \theta_1) \cdot r_2(\cos \theta_2 + i \sin \theta_2)$$
$$= r_1 r_2 [\cos (\theta_1 + \theta_2) + i \sin (\theta_1 + \theta_2)]$$

Proof   Let $r_1(\cos \theta_1 + i \sin \theta_1)$ and $r_2(\cos \theta_2 + i \sin \theta_2)$ be any two complex numbers in trigonometric form. Their product is

$$r_1(\cos \theta_1 + i \sin \theta_1) \cdot r_2(\cos \theta_2 + i \sin \theta_2)$$
$$= r_1 r_2(\cos \theta_1 \cos \theta_2 + i \sin \theta_1 \cos \theta_2 + i \cos \theta_1 \sin \theta_2$$
$$+ i^2 \sin \theta_1 \sin \theta_2)$$
$$= r_1 r_2 [(\cos \theta_1 \cos \theta_2 - \sin \theta_1 \sin \theta_2)$$
$$+ i(\sin \theta_1 \cos \theta_2 + \cos \theta_1 \sin \theta_2)]$$
$$= r_1 r_2 [\cos (\theta_1 + \theta_2) + i \sin (\theta_1 + \theta_2)]$$

ILLUSTRATION $2(\cos 130° + i \sin 130°) \cdot 3(\cos 50° + i \sin 50°)$
$$= 2 \cdot 3 [\cos (130° + 50°) + i \sin (130° + 50°)]$$
$$= 6(\cos 180° + i \sin 180°) = 6(-1 + i \cdot 0) = -6$$

This theorem can be extended to include the product of any number of complex numbers:

$$r_1(\cos\theta_1 + i\sin\theta_1) \cdot r_2(\cos\theta_2 + i\sin\theta_2) \cdots r_n(\cos\theta_n + i\sin\theta_n)$$
$$= r_1 r_2 \cdots r_n[\cos(\theta_1 + \theta_2 + \cdots + \theta_n) + i\sin(\theta_1 + \theta_2 + \cdots + \theta_n)]$$

89 DE MOIVRE'S THEOREM

If n is any real number,

$$[r(\cos\theta + i\sin\theta)]^n = r^n(\cos n\theta + i\sin n\theta)$$

Proof   For n a positive integer (by mathematical induction)

PART 1   *Verification*

For $n = 1$:
$$[r(\cos\theta + i\sin\theta)]^1 = r(\cos\theta + i\sin\theta) \qquad\qquad \text{True}$$

For $n = 2$:
$$[r(\cos\theta + i\sin\theta)]^2 = r^2(\cos 2\theta + i\sin 2\theta) \qquad\qquad \text{True}$$

For $n = 3$:
$$[r(\cos\theta + i\sin\theta)]^3 = r^3(\cos 3\theta + i\sin 3\theta) \qquad\qquad \text{True}$$

PART 2   A *proof* that *if* the theorem is true for $n = k$, then the theorem is true for $n = k + 1$. Let k represent any particular value of n. Assuming that

$$[r(\cos\theta + i\sin\theta)]^k = r^k(\cos k\theta + i\sin k\theta) \qquad\qquad (A)$$

we must prove that

$$[r(\cos\theta + i\sin\theta)]^{k+1}$$
$$= r^{k+1}[\cos(k + 1)\theta + i\sin(k + 1)\theta] \quad (B)$$

An examination of the left sides of the equations suggests that we multiply Equation (A) by $r(\cos\theta + i\sin\theta)$. Doing this, we obtain

$$[r(\cos\theta + i\sin\theta)]^{k+1} = r^k(\cos k\theta + i\sin k\theta) \cdot r(\cos\theta + i\sin\theta)$$

Applying the theorem on page 200, we get

$$[r(\cos\theta + i\sin\theta)]^{k+1} = r^{k+1}[\cos(k\theta + \theta) + i\sin(k\theta + \theta)]$$
$$= r^{k+1}[\cos(k + 1)\theta + i\sin(k + 1)\theta]$$

which is identical with Equation (B). This proves that if the theorem is true for $n = k$, then it must be true for $n = k + 1$.

PART 3   *Conclusion*   The theorem is true for $n = 1, 2, 3$ (Part 1). Since it is true for $n = 3$, it is true for $n = 4$ (Part 2, where $k = 3$ and $k + 1 = 4$). Since it is true for $n = 4$, it is true for $n = 5$, and so on for all positive integers n.*

It can be shown that De Moivre's theorem is true for all real values of n. We shall use it for only two cases: (1) when n is a positive integer and (2) when n is the reciprocal of a positive integer. The proof of the latter case is omitted in this text.

EXAMPLE   Use De Moivre's theorem to find the value of $(-1 + i)^{10}$.

Solution   After plotting $(-1 + i)$ and putting it in trigonometric form, we have

$$-1 + i = \sqrt{2}(\cos 135° + i \sin 135°)$$

Apply De Moivre's theorem:

$$\begin{aligned}(-1 + i)^{10} &= [\sqrt{2}(\cos 135° + i \sin 135°)]^{10} \\ &= (\sqrt{2})^{10}(\cos 10 \cdot 135° + i \sin 10 \cdot 135°) \\ &= 32(\cos 1350° + i \sin 1350°) \\ &= 32(\cos 270° + i \sin 270°) \\ &= -32i\end{aligned}$$

90   ROOTS OF COMPLEX NUMBERS

THEOREM   **The n nth roots of $r(\cos \theta + i \sin \theta)$ are given by the formula**

$$\sqrt[n]{r}\left(\cos \frac{\theta + k \cdot 360°}{n} + i \sin \frac{\theta + k \cdot 360°}{n}\right)$$

where $k = 0, 1, 2, \ldots, n - 1$

* We could, of course, merely cite the axiom of mathematical induction as the authority for stating that De Moivre's theorem is true for all natural numbers n. The author feels, however, that the argument in Part 3 is helpful in establishing the plausibility of the axiom.

Proof Assuming De Moivre's theorem is true when n is the reciprocal of a positive integer, we have

$$\sqrt[n]{r(\cos \theta + i \sin \theta)} = [r(\cos \theta + i \sin \theta)]^{\frac{1}{n}}$$

$$= r^{\frac{1}{n}} \left(\cos \frac{\theta}{n} + i \sin \frac{\theta}{n} \right)$$

Since $\cos \theta$ and $\sin \theta$ are periodic functions (Section 33) with a period of $360°$, we can say that $\cos \theta = \cos (\theta + k \cdot 360°)$ and $\sin \theta = \sin (\theta + k \cdot 360°)$, where k is an integer. Hence

$$\sqrt[n]{r(\cos \theta + i \sin \theta)}$$

$$= \sqrt[n]{r} \left(\cos \frac{\theta + k \cdot 360°}{n} + i \sin \frac{\theta + k \cdot 360°}{n} \right)$$

It is easy to show that the right side of this equation takes on n distinct values when k takes on the values $0, 1, 2, \ldots, n - 1$. But if k takes on a value larger than $(n - 1)$, the result is merely a duplication of one of the n roots already found.

EXAMPLE Find the three cube roots of $-8i$.

Solution After plotting the number and putting it in trigonometric form, we have

$$-8i = 8(\cos 270° + i \sin 270°)$$

Apply the theorem on roots. The three cube roots of $-8i$ are

$$\sqrt[3]{8} \left(\cos \frac{270° + k \cdot 360°}{3} + i \sin \frac{270° + k \cdot 360°}{3} \right)$$

$$= 2[\cos (90° + k \cdot 120°) + i \sin (90° + k \cdot 120°)]$$

Let the three roots be r_1, r_2, r_3. Then

$$
\begin{aligned}
r_1 &= 2(\cos 90° + i \sin 90°) = 2i & (k = 0) \\
r_2 &= 2(\cos 210° + i \sin 210°) = -\sqrt{3} - i & (k = 1) \\
r_3 &= 2(\cos 330° + i \sin 330°) = \sqrt{3} - i & (k = 2)
\end{aligned}
$$

The three roots are equally spaced on a circle with radius 2 and center at the origin (Figure 93). Notice that for $k = 3$, we obtain r_1 again.

EXERCISE 48

Perform the indicated multiplications and then express the results in algebraic form.

1 $7(\cos 130° + i \sin 130°) \cdot 2(\cos 170° + i \sin 170°)$

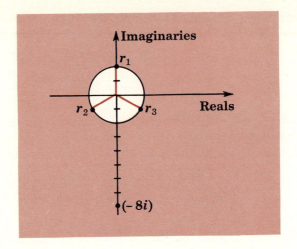

FIGURE 93

2 $6(\cos 40° + i \sin 40°) \cdot (\cos 200° + i \sin 200°)$
3 $5(\cos 85° + i \sin 85°) \cdot 4(\cos 50° + i \sin 50°)$
4 $3(\cos 100° + i \sin 100°) \cdot 8(\cos 290° + i \sin 290°)$

For each of the following products, (a) express the factors in trigonometric form, (b) find their product trigonometrically, (c) check your result by finding the product algebraically.

5 $(-1 + i)(-3 - 3i)$ **6** $(2\sqrt{3} - 2i)(3 + 3i\sqrt{3})$
7 $(4 - 4i\sqrt{3})(-\sqrt{3} + 3i)$ **8** $-4i(-6 - 2i\sqrt{3})$

For each of the following products, (a) express the factors in algebraic form, (b) find their product algebraically, (c) check your result by finding the product trigonometrically.

9 $5(\cos 60° + i \sin 60°) \cdot 6(\cos 330° + i \sin 330°)$
10 $7(\cos 210° + i \sin 210°) \cdot 4(\cos 30° + i \sin 30°)$
11 $(\cos 53° + i \sin 53°) \cdot (\cos 37° + i \sin 37°)$
12 $3(\cos 300° + i \sin 300°) \cdot 2(\cos 120° + i \sin 120°)$

Use De Moivre's theorem to find the value of each of the following. Express results in algebraic form.

13 $[\sqrt{2}(\cos 15° + i \sin 15°)]^{10}$ **14** $[3(\cos 54° + i \sin 54°)]^5$
15 $(1 + i)^{11}$ **16** $(-2 + 2i)^6$
17 $(\sqrt{2} - i\sqrt{2})^7$ **18** $(-1 - i)^{10}$

19 $(-\sqrt{3} + i)^8$ **20** $\left(\dfrac{1}{2} + \dfrac{\sqrt{3}}{2}i\right)^{100}$

21 $(6 + 8i)^4$ **22** $(3 - 3i\sqrt{3})^4$

Find all the indicated roots of the following complex numbers. Express results in algebraic form.

23 The cube roots of $27(\cos 27° + i \sin 27°)$.

24 The fourth roots of $625(\cos 80° + i \sin 80°)$.

25 The square roots of $36i$.

26 The square roots of $-2i$.

27 The square roots of $-2 - 2i\sqrt{3}$.

28 The cube roots of $27i$.

29 The cube roots of 8.

30 The cube roots of $-4\sqrt{2} + 4i\sqrt{2}$.

31 The fourth roots of $16i$.

32 The fifth roots of $16\sqrt{2} + 16i\sqrt{2}$.

33 The fourth roots of $-8 - 8i\sqrt{3}$.

34 The fourth roots of -256.

Find all the roots of the following equations.

35 $x^8 - 256 = 0$. *Hint:* The roots of the equation $x^8 - 256 = 0$ are the eight eighth roots of 256.

36 $x^4 + 10{,}000 = 0$

37 $x^5 = -32$

38 $x^6 = 4096$

39 Prove that

$$\frac{r_1(\cos \theta_1 + i \sin \theta_1)}{r_2(\cos \theta_2 + i \sin \theta_2)} = \frac{r_1}{r_2}[\cos(\theta_1 - \theta_2) + i \sin(\theta_1 - \theta_2)]$$

APPENDIX

91 THE CIRCULAR FUNCTIONS Consider a unit circle (radius 1) with center at O (Figure 94). Let s be any real number. Beginning with the point $A(1,0)$, lay off on the circumference of the circle (counterclockwise if s is positive, clockwise if s is negative) an arc of length s. Let the terminal point of the arc be $P(x,y)$. The circular functions are defined as follows:

$$\sin s = y \qquad \cos s = x \qquad \tan s = \frac{\sin s}{\cos s}$$

$$\cot s = \frac{\cos s}{\sin s} \qquad \sec s = \frac{1}{\cos s} \qquad \csc s = \frac{1}{\sin s}$$

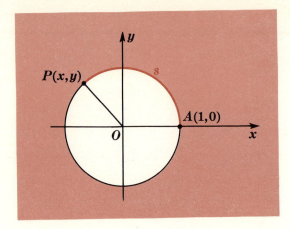

FIGURE 94

Observe that s is a real number that designates the measure of the length of the directed arc AP, the unit being the length OA. The circular functions are defined as functions of real numbers, with no reference whatsoever to angles or triangles.

In order to establish a relationship between the circular functions and the trigonometric functions of an angle expressed in radians, let us observe that the radian measure (Section 31, page 70) of angle AOP is $s/1 = s$. Using the definitions of the trigonometric functions of a general angle (page 8), we find that the sine of angle AOP is $y/1 = y$. Hence sin s (the sine of the real number s) = sin s (radians). Similar statements can be made for the other five trigonometric functions.

92 CIRCULAR AND EXPONENTIAL FUNCTIONS   In calculus it is shown that the exponential function e^θ and the circular functions $\cos \theta$ and $\sin \theta$ can be represented by *convergent infinite series.** These are

$$e^\theta = 1 + \theta + \frac{\theta^2}{2!} + \frac{\theta^3}{3!} + \frac{\theta^4}{4!} + \frac{\theta^5}{5!} + \cdots + \frac{\theta^n}{n!} + \cdots$$

$$\cos \theta = 1 - \frac{\theta^2}{2!} + \frac{\theta^4}{4!} - \frac{\theta^6}{6!} + \cdots + (-1)^n \frac{\theta^{2n}}{(2n)!} + \cdots$$

_____

* Speaking loosely, as we include more and more terms of a series, their sum more closely approaches the expression on the left side. In the case of these three series, this statement is true for all values of θ.

$$\sin \theta = \theta - \frac{\theta^3}{3!} + \frac{\theta^5}{5!} - \frac{\theta^7}{7!} + \cdots + (-1)^n \frac{\theta^{2n+1}}{(2n+1)!} + \cdots$$

It is series such as these* that are used in constructing tables like those appearing at the back of this book. For example, with $\theta = 0.16$, and using only two terms of the cosine and sine series, we have

$$\cos 0.16 \doteq 1 - \frac{(0.16)^2}{2} = 1 - 0.0128 = 0.9872$$

$$\sin 0.16 \doteq 0.16 - \frac{(0.16)^3}{6} \doteq 0.16 - 0.0007 = 0.1593$$

which agree, to four decimal places, with the corresponding entries in Table 1.

Regardless of how these series were derived, we can take them as alternative definitions of functions that we shall designate by e^θ, $\cos \theta$, and $\sin \theta$, respectively—a far cry from triangles.

Except for the alternating signs in the series for $\sin \theta$ and $\cos \theta$, the sum of these series is the same as the series for e^θ. This prompts us to substitute $i\theta$ for θ in the series for e^θ, which yields

$$e^{i\theta} = 1 + (i\theta) + \frac{(i\theta)^2}{2!} + \frac{(i\theta)^3}{3!} + \frac{(i\theta)^4}{4!} + \frac{(i\theta)^5}{5!} + \cdots$$

$$+ \frac{(i\theta)^m}{m!} + \cdots$$

$$= 1 + i\theta - \frac{\theta^2}{2!} - i\frac{\theta^3}{3!} + \frac{\theta^4}{4!} + i\frac{\theta^5}{5!} + \cdots$$

$$+ (-1)^n \frac{\theta^{2n}}{(2n)!} + i(-1)^n \frac{\theta^{2n+1}}{(2n+1)!} + \cdots$$

$$= \left[1 - \frac{\theta^2}{2!} + \frac{\theta^4}{4!} - \cdots + (-1)^n \frac{\theta^{2n}}{(2n)!} + \cdots \right]$$

$$+ i \left[\theta - \frac{\theta^3}{3!} + \frac{\theta^5}{5!} - \cdots + (-1)^n \frac{\theta^{2n+1}}{(2n+1)!} + \cdots \right]$$

$$= \cos \theta + i \sin \theta$$

Thus in strictly nonrigorous fashion, but with heuristic intent, we

---

* Notice that $\cos \theta$, an *even* function, is expressed as a series of even powers of θ. Also, the *odd* function $\sin \theta$ is expressed as a series of odd powers of θ.

have obtained *Euler's formula:*

$$e^{i\theta} = \cos\theta + i\sin\theta$$

An important special case, $\theta = \pi$, yields the unusual relation

$$e^{i\pi} + 1 = 0$$

which involves 0, the real unit 1, the imaginary unit i, and two famous irrational numbers, e and π. Moreover, since $e^{\pi i} = -1$, it follows that $\log_e(-1) = \pi i$.*

Raising both sides of Euler's formula to the power n, we get $(e^{i\theta})^n = (\cos\theta + i\sin\theta)^n$. But $(e^{i\theta})^n = e^{in\theta} = \cos n\theta + i\sin n\theta$. Hence we have an alternative derivation of De Moivre's theorem:

$$(\cos\theta + i\sin\theta)^n = \cos n\theta + i\sin n\theta$$

EXERCISE 49

1 Use the series definition of $\cos\theta$ to prove that $\cos(-\theta) = \cos\theta$.
2 Use the series definition of $\sin\theta$ to prove that $\sin(-\theta) = -\sin\theta$.
3 Use the series for $\sin\theta$ to compute $\sin 0.1$ correct to 8 decimal places.
4 Use Euler's formula to derive the formulas for $\sin(A + B)$ and $\cos(A + B)$. *Hint:* Replace θ with $A + B$; notice that $e^{A+B} = e^A e^B$; apply the definition of equality of complex numbers.
5 Use Euler's formula to prove that

(a) $\cos\theta = \dfrac{e^{i\theta} + e^{-i\theta}}{2}$

(b) $\sin\theta = \dfrac{e^{i\theta} - e^{-i\theta}}{2i}$

6 Use formula (a) of Prob. 5 to prove that

$$\cos 5\theta = 16\cos^5\theta - 20\cos^3\theta + 5\cos\theta.$$

Hint: Express the right side in terms of exponential functions; use the binomial formula to expand; simplify to $\frac{1}{2}(e^{5i\theta} + e^{-5i\theta})$, which is $\cos 5\theta$.

93   A GEOMETRIC PROOF, FOR ACUTE ANGLES, OF THE FORMULAS FOR sin (A + B) AND cos (A + B)

Let A and B be any two positive acute angles whose sum, $A + B$, is also acute. Place A in standard position and place B so its initial side coincides with the

---

* The general expression is $(1 + 2k)\pi i$, where k is an integer. This is in partial explanation of the statement that "negative numbers do not have real logarithms."

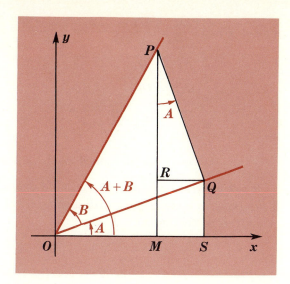

FIGURE 95

terminal side of A and its vertex falls at the origin O (Figure 95). Choose P as any point on the terminal side of angle $(A + B)$, which is in standard position. From P drop a perpendicular to the initial side of B at Q. Draw PM and QS perpendicular to the x-axis and draw QR perpendicular to PM. Then angle RPQ equals angle A because they are acute angles with their sides respectively perpendicular. Using the definition of the sine of an angle, we have

$$\sin (A + B) = \frac{MP}{OP} = \frac{MR + RP}{OP} = \frac{SQ + RP}{OP}$$

$$= \frac{SQ}{OP} + \frac{RP}{OP}$$

$$= \frac{SQ}{*} \cdot \frac{*}{OP} + \frac{RP}{\dagger} \cdot \frac{\dagger}{OP}$$

$$= \frac{SQ}{OQ} \cdot \frac{OQ}{OP} + \frac{RP}{QP} \cdot \frac{QP}{OP}$$

$$= \sin A \cos B + \cos A \sin B$$

Using the definition of the cosine of an angle, we have

---

* Since SQ and OP lie in different triangles SOQ and POQ, we multiply top and bottom of the fraction by OQ, the common side of the two triangles.

† The common side of triangles RQP and OQP is QP.

$$\cos{(A + B)} = \frac{OM}{OP} = \frac{OS - MS}{OP} = \frac{OS - RQ}{OP}$$

$$= \frac{OS}{OP} - \frac{RQ}{OP}$$

$$= \frac{OS}{OQ} \cdot \frac{OQ}{OP} - \frac{RQ}{QP} \cdot \frac{QP}{OP}$$

$$= \cos A \cos B - \sin A \sin B$$

These proofs are not general because we considered only the case in which A and B are positive acute angles with a sum of less than 90°. A general proof is given in Section 39.

94 SOLVING OBLIQUE TRIANGLES: SAS AND SSS

94   SOLVING OBLIQUE TRIANGLES: SAS AND SSS   If we want a high degree of accuracy in the computed parts of an oblique triangle, it is usually desirable to avoid using the law of cosines in solving the *SAS* and *SSS* cases. The following outline should assist the student in choosing a suitable plan of attack.

	Without logs	With logs
SAA	Law of sines	Law of sines
SSA	Law of sines	Law of sines
SAS	Law of cosines	Law of cosines (or Law of tangents)
SSS	Law of cosines	Law of cosines (or Half-angle formulas)

95 THE LAW OF TANGENTS

In any triangle, the difference of two sides is to their sum as the tangent of half the difference of the opposite angles is to the tangent of half their sum;

$$\frac{a - b}{a + b} = \frac{\tan \frac{1}{2}(A - B)}{\tan \frac{1}{2}(A + B)} \qquad \frac{b - a}{b + a} = \frac{\tan \frac{1}{2}(B - A)}{\tan \frac{1}{2}(B + A)}$$

$$\frac{b - c}{b + c} = \frac{\tan \frac{1}{2}(B - C)}{\tan \frac{1}{2}(B + C)} \qquad \frac{c - b}{c + b} = \frac{\tan \frac{1}{2}(C - B)}{\tan \frac{1}{2}(C + B)}$$

$$\frac{c - a}{c + a} = \frac{\tan \frac{1}{2}(C - A)}{\tan \frac{1}{2}(C + A)} \qquad \frac{a - c}{a + c} = \frac{\tan \frac{1}{2}(A - C)}{\tan \frac{1}{2}(A + C)}$$

Proof   Let the two given sides be a and b, with $a > b$. By the law of sines,

$$\frac{a}{b} = \frac{\sin A}{\sin B}$$

Subtract 1 from each side; add 1 to each side:

$$\frac{a}{b} - 1 = \frac{\sin A}{\sin B} - 1 \qquad \frac{a}{b} + 1 = \frac{\sin A}{\sin B} + 1$$

Hence

$$\frac{a - b}{b} = \frac{\sin A - \sin B}{\sin B} \qquad \frac{a + b}{b} = \frac{\sin A + \sin B}{\sin B}$$

Divide the first equation by the second:

$$\frac{a - b}{a + b} = \frac{\sin A - \sin B}{\sin A + \sin B}$$

Apply formulas (16) and (15) of Section 45:

$$\frac{a - b}{a + b} = \frac{2 \cos \frac{1}{2}(A + B) \sin \frac{1}{2}(A - B)}{2 \sin \frac{1}{2}(A + B) \cos \frac{1}{2}(A - B)}$$

$$= \cot \frac{1}{2}(A + B) \tan \frac{1}{2}(A - B) = \frac{\tan \frac{1}{2}(A - B)}{\tan \frac{1}{2}(A + B)}$$

This proves the first of the six formulas.

If a and b are interchanged, then their opposite angles, A and B, must be interchanged and we get the second formula:

$$\frac{b - a}{b + a} = \frac{\tan \frac{1}{2}(B - A)}{\tan \frac{1}{2}(B + A)}$$

The remaining formulas may be obtained by cyclic permutation (Section 76).

96   APPLICATIONS OF THE LAW OF TANGENTS: SAS   Suppose that the given parts are the sides a and b and the included angle C. If $a > b$, we use the following form of the law of tangents:

$$\tan \frac{1}{2} (A - B) = \frac{a - b}{a + b} \tan \frac{1}{2} (A + B)$$

Knowing a and b, we can find $(a - b)$ and $(a + b)$. Since C is given, we can find $\frac{1}{2}(A + B)$ by halving the relation $A + B = 180° - C$. Thus the three quantities on the right side of the equation are easily obtained from the given data. By applying the law of tangents, we find the value of $\frac{1}{2}(A - B)$. *Knowing $\frac{1}{2}(A + B)$ and $\frac{1}{2}(A - B)$, we add to get A, and subtract to get B.* The sixth part, c, can be found with the law of sines. The problem can be checked by finding c again with another form of the law of sines.

EXAMPLE   Solve the triangle ABC, given $a = 6810$, $b = 4828$, $C = 55° 8'$.

Solution   Since $a > b$, use the first form of the law of tangents:

$$\tan \frac{1}{2} (A - B) = \frac{a - b}{a + b} \tan \frac{1}{2} (A + B)$$

$$a = \quad 6810$$
$$b = \quad 4828$$

$$\overline{}$$

$$a - b = \quad 1982$$
$$a + b = 11638$$
$$A + B = 180° - 55° 8'$$
$$\qquad\quad = 124° 52'$$
$$\tfrac{1}{2}(A + B) = 62° 26' \longrightarrow \tan \tfrac{1}{2}(A - B) = \frac{1982}{11638} \tan 62° 26'$$

$$\log 1982 = \quad 3.29710$$
$$\log \tan 62° 26' = 10.28229 - 10$$
$$\overline{}\text{A}$$
$$\log \text{num.} = 13.57939 - 10$$
$$\log 11638 = \quad 4.06588$$
$$\overline{}\text{S}$$
$$\log \tan \tfrac{1}{2}(A - B) = \quad 9.51351 - 10$$
$$\tfrac{1}{2}(A - B) = 18° 4' \longleftarrow \quad \tfrac{1}{2}(A - B) = 18° 4'$$

$$\overline{\phantom{\tfrac{1}{2}(A-B)=18}}$$

$$A = 80° 30'$$
$$B = 44° 22'$$

To find c,

$$c = \frac{a \sin C}{\sin A} = \frac{6810 \sin 55° 8'}{\sin 80° 30'}$$

$$\log 6810 = 3.83315$$
$$\log \sin 55° \ 8' = 9.91407 - 10$$
$$\overline{\qquad\qquad}\text{A}$$
$$\log \text{num.} = 13.74722 - 10$$
$$\log \sin 80° \ 30' = 9.99400 - 10$$
$$\overline{\qquad\qquad}\text{S}$$
$$\log c = 3.75322$$
$$c = 5665$$

To check,

$$c = \frac{b \sin C}{\sin B} = \frac{4828 \sin 55° \ 8'}{\sin 44° \ 22'}$$

$$\log 4828 = 3.68377$$
$$\log \sin 55° \ 8' = 9.91407 - 10$$
$$\overline{\qquad\qquad}\text{A}$$
$$\log \text{num.} = 13.59784 - 10$$
$$\log \sin 44° \ 22' = 9.84463 - 10$$
$$\overline{\qquad\qquad}\text{S}$$
$$\log c = 3.75321$$
$$c = 5665$$

EXERCISE 50

Make a complete outline of the logarithmic solution and check of the oblique triangle in which the following parts are known.

1 a, c, B, with $c > a$ 2 b, c, A, with $c > b$
3 a, b, C, with $b > a$ 4 a, c, B, with $a > c$

Solve the following triangles. Check as directed by the instructor.

5 $a = 7982, b = 1475, C = 100° \ 0'$
6 $a = 33.22, b = 17.28, C = 40° \ 20'$
7 $a = 406.5, b = 301.5, C = 81° \ 18'$
8 $a = 4567, b = 3333, C = 72° \ 0'$
9 $a = 9.125, c = 7.625, B = 37° \ 18'$
10 $a = 1950, c = 1942, B = 60° \ 18'$
11 $b = 56{,}789, c = 34{,}567, A = 123° \ 45.6'$
12 $b = 82.606, c = 87.654, A = 59° \ 29.8'$
13 $a = 2.7183, b = 3.1416, C = 57° \ 17.8'$
14 $a = 36{,}475, b = 48{,}596, C = 80° \ 0.0'$

15 The difference between the two shortest sides of a triangle is 123.4 ft. Two angles of the triangle are 80° 0′ and 60° 0′. Find the two shortest sides of the triangle.

16 In order to find the width *RS* of a swamp, a point *T* is chosen on dry land. By measurement it is found that *TR* = 1513 ft., *TS* = 1925 ft., and ∠*STR* = 43° 56′. Find *RS*.

17 The diagonals of a parallelogram are 49.12 ft. and 75.68 ft. They intersect at an angle of 55° 56′. Find the smaller side of the parallelogram.

18 San Francisco, Calif., is 340 miles N 41° W from Los Angeles, Calif. Reno, Nev., is 390 miles N 12° W from Los Angeles. Find the bearing and the distance of Reno from San Francisco. *

97 THE HALF-ANGLE FORMULAS

In any triangle *ABC*,

$$\tan \frac{A}{2} = \frac{r}{s-a} \qquad \tan \frac{B}{2} = \frac{r}{s-b} \qquad \tan \frac{C}{2} = \frac{r}{s-c}$$

where

$$r = \sqrt{\frac{(s-a)(s-b)(s-c)}{s}} \quad \text{and} \quad s = \frac{1}{2}(a+b+c)$$

Proof In Section 79 we proved that the area of a triangle with sides *a*, *b*, *c* is $K = \sqrt{s(s-a)(s-b)(s-c)}$, where $s = \frac{1}{2}(a+b+c)$. In triangle *ABC*, let *AO*, *BO*, and *CO* be the bisectors of angles *A*, *B*, and *C*, respectively (Figure 96). Then, by geometry, *O* is the center of the inscribed circle. *Let r be the radius of the inscribed circle.* It is obvious that

$$\text{area } \triangle ABC = \text{area } \triangle AOB + \text{area } \triangle BOC$$
$$+ \text{ area } \triangle COA$$
$$= \tfrac{1}{2}cr + \tfrac{1}{2}ar + \tfrac{1}{2}br$$
$$= r \cdot \tfrac{1}{2}(c + a + b) = rs$$

Equate the two values of the area:

$$rs = \sqrt{s(s-a)(s-b)(s-c)}$$
$$r = \sqrt{\frac{(s-a)(s-b)(s-c)}{s}}$$

---

* Ignore the curvature of the earth and assume only two-place accuracy. Get angles to the nearest degree. Get distances to the nearest multiple of 10 miles.

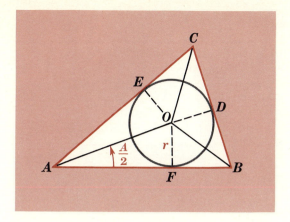

FIGURE 96

In $\triangle ABC$, by geometry, $AF = AE$, $BF = BD$, and $CD = CE$. Why? Since the sum of these six segments equals the perimeter, $2s$,

$$2AF + 2BD + 2CD = 2s$$
$$AF = s - (BD + CD) = s - a$$

Similarly

$$BD = s - b \qquad \text{and} \qquad CE = s - c$$

Since AO bisects angle A and $OF \perp AF$,

$$\tan \frac{A}{2} = \frac{FO}{AF} = \frac{r}{s - a}$$

Similarly

$$\tan \frac{B}{2} = \frac{r}{s - b} \qquad \text{and} \qquad \tan \frac{C}{2} = \frac{r}{s - c}$$

98   APPLICATIONS OF THE HALF-ANGLE FORMULAS: SSS   When the three sides are given, we first compute s, then $(s - a)$, $(s - b)$, $(s - c)$. Knowing these four quantities, we find r by use of logarithms. Then the half-angle formulas give us $A/2$, $B/2$, $C/2$. When these values are doubled, we get the three angles of the triangle. The half-angle formulas may be used to check other solutions.

EXAMPLE   Solve the triangle ABC, given $a = 76.54$, $b = 60.06$, $c = 54.32$.

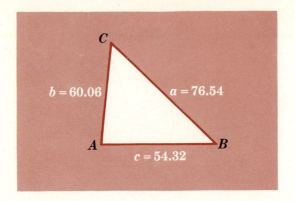

FIGURE 97

Solution   a =   76.54
              b =   60.06
              c =   54.32
         ————————— A
    2s = 190.92
     s =   95.46

         s − a = 18.92   log (s − a) = 1.27692
         s − b = 35.40   log (s − b) = 1.54900
         s − c = 41.14   log (s − c) = 1.61426
    ———————————————— A   ———————————— A
    Check*: s = 95.46   log num. = 4.44018
                                    log s = 1.97982
                                    ——————————— S
    $$r = \sqrt{\frac{(s-a)(s-b)(s-c)}{s}}$$ log radicand = 2.46036
                                              2 ——————————
                                              log r = 1.23018

    $$\tan \frac{A}{2} = \frac{r}{s-a}$$ $$\tan \frac{B}{2} = \frac{r}{s-b}$$

        log r = 1.23018   log r = 1.23018
    log (s − a) = 1.27692   log (s − b) = 1.54900
    ———————————————— S   ———————————————— S

    $$\log \tan \frac{A}{2} = 9.95326 - 10$$ $$\log \tan \frac{B}{2} = 9.68118 - 10$$

    $$\frac{A}{2} = 41° \, 55.4'$$ $$\frac{B}{2} = 25° \, 38.3'$$

    $$A = 83° \, 50.8'$$ $$B = 51° \, 16.6'$$

* (s − a) + (s − b) + (s − c) = 3s − (a + b + c) = s

$$\tan \frac{C}{2} = \frac{r}{s-c}$$

$$\log r = 1.23018$$
$$\log (s - c) = 1.61426$$
$$\overline{}\text{S}$$
$$\log \tan \frac{C}{2} = 9.61592 - 10$$

$$\frac{C}{2} = 22° \ 26.4'$$

$$C = 44° \ 52.8'$$

Rounding off these results to the nearest minute for four-figure accuracy, we get

$A = 83° \ 51'$
$B = 51° \ 17'$
$C = 44° \ 53'$

Check:
$A + B + C = 180° \ 1'$
The check is satisfactory.

EXERCISE 51

1 Make a complete outline of the logarithmic solution of an oblique triangle in which the three sides are known.

Use logarithms to solve the following triangles.

2 $a = 5.03, \ b = 4.02, \ c = 3.01$
3 $a = 16.20, \ b = 19.20, \ c = 19.50$
4 $a = 5050, \ b = 5220, \ c = 9944$
5 $a = 3113, \ b = 4114, \ c = 5115$
6 $a = 1954, \ b = 1492, \ c = 1776$
7 $a = 7002, \ b = 2996, \ c = 5002$
8 $a = 38.066, \ b = 38.044, \ c = 38.022$
9 $a = 501.05, \ b = 660.33, \ c = 412.34$
10 $a = 6.5432, \ b = 3.1416, \ c = 8.3838$
11 $a = 46,138, \ b = 77,769, \ c = 98,073$

12 Make an attempt to solve the following triangle: $a = 6785, \ b = 3798,$ $c = 2986.$ Explain.
13 Lubbock, Tex., is 320 miles from Stillwater, Okla. Dallas, Tex., is 300 miles S 78° E from Lubbock. Stillwater is 230 miles from Dallas. What is the bearing of Stillwater from Lubbock? From Dallas? *

* Ignore the curvature of the earth and assume only two-place accuracy. Get angles to the nearest degree. Get distances to the nearest multiple of 10 miles.

14   In a triangle of area 3030 sq. ft., a circle of radius 20.2 ft. is inscribed. One angle of the triangle is 44° 0'. Find the side opposite it.

15   Prove that $AF = AE$ in Figure 96.

99   THE MIL AS A UNIT OF ANGULAR MEASURE   Certain branches of the United States Army employ the mil system for measuring angles. The unit in this system is the **mil,** which is defined as $\frac{1}{1600}$ of a right angle.

The mil is a convenient unit for certain kinds of rapid calculation. The reason lies in the fact that one mil subtends an arc equal to *approximately* $\frac{1}{1000}$ of the radius. If the angle is small, the arc and chord are almost equal and we have the approximation

$$c = \frac{r\,\theta}{1000} \qquad (\theta \; in \; \text{mils}) \qquad\qquad (1)$$

where the chord c and the radius r are measured in the same units.

The following relationships exist among the three units of angular measure.

$$90° = 1600 \text{ mils}$$

1° = 17.778 mils	1 mil = 0.056250°
1' = 0.29630 mil	1 mil = 3.3750'
1 radian = 1018.6 mils	1 mil = 0.00098175 radian

EXAMPLE 1   An automobile of standard width (5.5 ft.) is facing an observer. How far away is the auto if it subtends a horizontal angle of 5 mils?

Solution   Using $c = \dfrac{r\theta}{1000}$, we have $5.5 = \dfrac{r \cdot 5}{1000}$. Hence $r = 1100$ ft.

EXAMPLE 2   A circular pill box 300 yd. from an observer subtends a horizontal angle of 16 mils. Find the diameter of the pill box.

Solution   Using $c = \dfrac{r\theta}{1000}$, we get $c = \dfrac{300 \cdot 16}{1000} = 4.8$ yd.

EXERCISE 52

Express in degrees and minutes. (Round off to the nearest minute.)

1   80 mils   2   11 mils
3   30.5 mils   4   44 mils

Express in mils. (Round off to the nearest mil.)

5   3°   6   4° 20′
7   7°   8   1° 40′

Express in radians. (Round off to three-figure accuracy.)

9   120 mils   10   300 mils
11   250 mils   12   440 mils

Express in mils. (Round off to the nearest mil.)

13   0.2 radian   14   0.066 radian
15   0.051 radian   16   0.3 radian

17   Find the height of the Space Needle in Seattle, Wash., if it subtends an angle of 50 mils at a distance of 12,100 ft.

18   A billboard 16 ft. high subtends an angle of 40 mils at the eye of an observer. How far is the billboard from the observer?

19   Find in mils the angle subtended by a 12-ft. pole at a distance of 200 ft.

20   At a distance of 5000 ft., the Pacific Trade Center in Honolulu subtends an angle of 72 mils. How high is the building?

21   A car 16 ft. long subtends a horizontal angle of 30 mils at the eye of an observer. How far away is the car if it is perpendicular to his line of vision?

22   The width of a boy's thumb is $\frac{3}{4}$ in. Find in mils the angle subtended by the thumb's width at a distance of 21 in.

23   The sun at a distance of 92,900,000 miles subtends an angle of 9.3 mils. What is the diameter of the sun? (Write result with three-figure accuracy.)

24   A 50-ft. water tower subtends an angle of 80 mils at the eye of an observer on level ground. How far is the observer from the tower?

25   Prove Equation (1) in Section 99.

ANSWERS[*]

13 202°   **14** 129°   **15** 345°

17 6600°

EXERCISE 3, Pages 14–15

1 0.94, −0.34, −2.7   **2** −0.77, 0.64, −1.2

3 0.98, 0.17, 5.6   **5** −0.64, −0.77, 0.84

6 0.34, −0.94, −0.36   **7** −0.17, 0.98, −0.18

9 0.17, 0.98, 0.18   **10** −0.98, −0.17, 5.6

11 0.64, −0.77, −0.84

13 0, 0.17, 0.34, 0.50, 0.64, 0.77, 0.87, 0.94, 0.98, 1

14 1, 0.98, 0.94, 0.87, 0.77, 0.64, 0.50, 0.34, 0.17, 0

15 0, 1, 0, *does not exist*, 1, *does not exist*

17 −1, 0, *does not exist*, 0, *does not exist*, −1

18 0, −1, 0, *does not exist*, −1, *does not exist*

19 $-\frac{8}{17}, -\frac{15}{17}, \frac{8}{15}, \frac{15}{8}, -\frac{17}{15}, -\frac{17}{8}$

21 $-\frac{12}{13}, \frac{5}{13}, -\frac{12}{5}, -\frac{5}{12}, \frac{13}{5}, -\frac{13}{12}$

22 $\frac{24}{25}, \frac{7}{25}, \frac{24}{7}, \frac{7}{24}, \frac{25}{7}, \frac{25}{24}$

23 $\dfrac{2\sqrt{5}}{5}, -\dfrac{\sqrt{5}}{5}, -2, -\dfrac{1}{2}, -\sqrt{5}, \dfrac{\sqrt{5}}{2}$

25 $\dfrac{3\sqrt{58}}{58}, \dfrac{7\sqrt{58}}{58}, \dfrac{3}{7}, \dfrac{7}{3}, \dfrac{\sqrt{58}}{7}, \dfrac{\sqrt{58}}{3}$

26 $-\dfrac{4\sqrt{97}}{97}, \dfrac{9\sqrt{97}}{97}, -\dfrac{4}{9}, -\dfrac{9}{4}, \dfrac{\sqrt{97}}{9}, -\dfrac{\sqrt{97}}{4}$

27 $-\dfrac{\sqrt{11}}{6}, \dfrac{5}{6}, -\dfrac{\sqrt{11}}{5}, -\dfrac{5\sqrt{11}}{11}, \dfrac{6}{5}, -\dfrac{6\sqrt{11}}{11}$

29 $\dfrac{\sqrt{7}}{3}, -\dfrac{\sqrt{2}}{3}, -\dfrac{\sqrt{14}}{2}, -\dfrac{\sqrt{14}}{7}, -\dfrac{3\sqrt{2}}{2}, \dfrac{3\sqrt{7}}{7}$

30 $-\dfrac{\sqrt{10}}{4}, -\dfrac{\sqrt{6}}{4}, \dfrac{\sqrt{15}}{3}, \dfrac{\sqrt{15}}{5}, -\dfrac{2\sqrt{6}}{3}, -\dfrac{2\sqrt{10}}{5}$

31 III   **33** III   **34** II

35 II   **37** IV   **38** IV

39 Impossible   **41** Possible   **42** Possible

43 Possible   **45** Impossible   **46** Impossible

47 Close to 0°   **49** Close to 0°   **50** Close to 90°

51 Close to 90°   **53** IV   **54** I

55 II

EXERCISE 4, Page 17

1 $\cos\theta = \dfrac{\sqrt{7}}{4}, \tan\theta = \dfrac{3\sqrt{7}}{7}, \cot\theta = \dfrac{\sqrt{7}}{3}, \sec\theta = \dfrac{4\sqrt{7}}{7},$

$$\csc \theta = \frac{4}{3}$$

2 $\sin \theta = \dfrac{\sqrt{15}}{4}$, $\tan \theta = -\sqrt{15}$, $\cot \theta = -\dfrac{\sqrt{15}}{15}$,

 $\sec \theta = -4$, $\csc \theta = \dfrac{4\sqrt{15}}{15}$

3 $\sin \theta = -\dfrac{3\sqrt{13}}{13}$, $\cos \theta = \dfrac{2\sqrt{13}}{13}$, $\cot \theta = -\dfrac{2}{3}$,

 $\sec \theta = \dfrac{\sqrt{13}}{2}$, $\csc \theta = -\dfrac{\sqrt{13}}{3}$

5 $\sin \theta = -\frac{7}{25}$, $\tan \theta = -\frac{7}{24}$, $\cot \theta = -\frac{24}{7}$,
   $\sec \theta = \frac{25}{24}$, $\csc \theta = -\frac{25}{7}$

6 $\sin \theta = \frac{15}{17}$, $\cos \theta = \frac{8}{17}$, $\cot \theta = \frac{8}{15}$, $\sec \theta = \frac{17}{8}$, $\csc \theta = \frac{17}{15}$

7 $\cos \theta = -\frac{3}{5}$, $\tan \theta = -\frac{4}{3}$, $\cot \theta = -\frac{3}{4}$, $\sec \theta = -\frac{5}{3}$, $\csc \theta = \frac{5}{4}$

9 $\sin \theta = -\dfrac{2}{3}$, $\cos \theta = -\dfrac{\sqrt{5}}{3}$, $\cot \theta = \dfrac{\sqrt{5}}{2}$, $\sec \theta = -\dfrac{3\sqrt{5}}{5}$,

 $\csc \theta = -\dfrac{3}{2}$

10 $\cos \theta = \dfrac{9\sqrt{85}}{85}$, $\tan \theta = -\dfrac{2}{9}$, $\cot \theta = -\dfrac{9}{2}$, $\sec \theta = \dfrac{\sqrt{85}}{9}$,

 $\csc \theta = -\dfrac{\sqrt{85}}{2}$

11 $\sin \theta = -\dfrac{7}{8}$, $\tan \theta = \dfrac{7\sqrt{15}}{15}$, $\cot \theta = \dfrac{\sqrt{15}}{7}$, $\sec \theta = -\dfrac{8\sqrt{15}}{15}$,

 $\csc \theta = -\dfrac{8}{7}$

13 $\sin \theta = \dfrac{2\sqrt{2}}{3}$, $\cos \theta = -\dfrac{1}{3}$, $\tan \theta = -2\sqrt{2}$, $\cot \theta = -\dfrac{\sqrt{2}}{4}$,

 $\csc \theta = \dfrac{3\sqrt{2}}{4}$

14 $\sin \theta = -\dfrac{\sqrt{26}}{26}$, $\cos \theta = -\dfrac{5\sqrt{26}}{26}$, $\tan \theta = \dfrac{1}{5}$,

 $\sec \theta = -\dfrac{\sqrt{26}}{5}$, $\csc \theta = -\sqrt{26}$

15 $\sin \theta = \dfrac{1}{a}$, $\cos \theta = \dfrac{\sqrt{a^2 - 1}}{a}$, $\tan \theta = \dfrac{\sqrt{a^2 - 1}}{a^2 - 1}$,

 $\cot \theta = \sqrt{a^2 - 1}$, $\sec \theta = \dfrac{a\sqrt{a^2 - 1}}{a^2 - 1}$

EXERCISE 6, Page 23

1	3	**2**	$\frac{15}{16}$	**3**	$5\sqrt{3}$	**5**	$\frac{3}{4}$
6	$\frac{1}{8}$	**7**	$\frac{7}{8}$	**9**	True	**10**	False
11	True	**13**	False	**14**	True	**15**	True

EXERCISE 7, Page 25

1	0.4488	**2**	0.9787	**3**	0.3378	**5**	0.4566
6	0.8290	**7**	0.1703	**9**	6.561	**10**	0.8847
11	0.8208	**13**	57° 40′	**14**	79° 10′	**15**	26° 0′
17	9° 30′	**18**	84° 50′	**19**	40° 40′	**21**	66° 0′
22	43° 20′	**23**	87° 30′				

EXERCISE 8, Pages 27–28

1	0.8482	**2**	0.7896	**3**	2.271	**5**	0.3759
6	0.8973	**7**	0.2636	**9**	3.322	**10**	0.8674
11	0.9850	**13**	0.1051	**14**	0.5460	**15**	0.7879
17	12° 22′	**18**	63° 26′	**19**	21° 3′	**21**	14° 17′
22	86° 27′	**23**	6° 34′	**25**	72° 12′	**26**	38° 42′
27	65° 8′	**29**	65° 56′	**30**	36° 52′	**31**	75° 41′

EXERCISE 9, Page 30

1 (a) 37.42, (b) 37.4, (c) 37
2 (a) 0.02806, (b) 0.0281, (c) 0.028
3 (a) 0.5372, (b) 0.537, (c) 0.54
5 (a) 72° 48′, (b) 72° 50′, (c) 73°
6 (a) 16° 27′, (b) 16° 30′, (c) 16°
7 (a) 85° 18′, (b) 85° 20′, (c) 85°
9 61.25 to 61.35, exclusive
10 8.395 to 8.405, inclusive
11 925.5 to 926.5, inclusive
13 0.48275 to 0.48285, inclusive
14 3690.5 to 3691.5, exclusive
15 73.145 to 73.155, exclusive

EXERCISE 10, Pages 33–34

1 $B = 58° 40′$, $b = 342$, $c = 400$
2 $A = 65° 10′$, $a = 8.17$, $c = 9.00$

3   $A = 20°$, $b = 7.1$, $c = 7.6$
5   $A = 44° 9'$, $a = 5578$, $b = 5746$
6   $B = 18°$, $a = 94$, $b = 31$
7   $A = 35° 30'$, $B = 54° 30'$, $c = 0.7370$
9   $A = 66° 30'$, $B = 23° 30'$, $a = 7.06$
10 $A = 23° 14'$, $B = 66° 46'$, $a = 4599$ or 4600, depending upon the method used
11 $A = 6° 20'$, $B = 83° 40'$, $b = 298$
13 $A = 59°$, $B = 31°$, $c = 93$
14 $A = 38° 10'$, $B = 51° 50'$, $c = 0.636$
15 $B = 60° 10'$, $a = 256$, $b = 447$
17 $B = 22° 20'$, $a = 1.85$, $c = 2.00$
18 $A = 79°$, $B = 11°$, $a = 98$
19 $A = 24° 30'$, $B = 65° 30'$, $a = 1244$

EXERCISE 11, Pages 36–39

1   640 ft.
2   1454 ft.
3   57°
5   1053 ft.
6   Only 70 ft., but to him it may seem like worlds
7   1575 ft.
9   N 66° E, S 66° W
10 N 68° E, S 68° W
11 310 miles, 160 miles
13 127 ft.
14 4.7 miles
15 288 ft.
17 N 6° 20′ E
18 12:54 P.M.
19 9:38 A.M.
21 30.2 in.
22 68° 0′, 68° 0′, 44° 0′
23 6.276 ft., 6.283 ft.
25 751 ft.
27 $a(1 + \cot \phi \tan \theta)$
29 938 ft.
30 2750 ft.
31 41 miles, 48 miles

EXERCISE 12, Pages 44–45

1   $\tan 8A$
2   $\cot 320°$
3   $\cos 4A$
5   $\pm \cot A$
6   $\cos 140°$
7   $\sin 100°$
9   $\sec 260°$
10 $\sin 48°$
11 True
13 False
14 True
15 False
17 False
18 True
19 True
21 True
22 True
23 False
25 True
26 False
27 False
29 True
30 False
31 True

EXERCISE 13, Pages 47–48

1   $1 + \sin \theta \cos \theta$
2   1
3   $6 \sin^2 \theta$
5   1
6   $\sin \theta$
7   0

9 $\cot 147° = -\sqrt{\csc^2 147° - 1}$ **10** $\csc 246° = \dfrac{1}{\sin 246°}$

11 $\tan 278° = -\sqrt{\sec^2 278° - 1}$ **13** $\sec 3\theta = \dfrac{1}{\cos 3\theta}$

14 $\cos 5A = \pm\sqrt{1 - \sin^2 5A}$

15 $\sin \theta = \pm\dfrac{\tan \theta}{\sqrt{1 + \tan^2 \theta}}$, $\cos \theta = \pm\dfrac{1}{\sqrt{1 + \tan^2 \theta}}$, $\cot \theta = \dfrac{1}{\tan \theta}$,

$\sec \theta = \pm\sqrt{1 + \tan^2 \theta}$, $\csc \theta = \pm\dfrac{\sqrt{1 + \tan^2 \theta}}{\tan \theta}$

EXERCISE 14, Pages 53–57

33 An identity

34 Not an identity. Left side $= \cot \theta/(1 + \tan \theta) \neq 1/(1 + \tan \theta)$. True only if $\cot \theta = 1$, that is, $\theta = 45, 225°$ and angles coterminal with them.

35 An identity

37 Not an identity. Left side $= \sec \theta \neq \csc \theta$. True only if $\sec \theta = \csc \theta$, that is, if $\theta = 45°, 225°$ and angles coterminal with them.

38 An identity

39 Not an identity. Left side $= \sin \theta/(1 - \cos \theta) \neq 1/(1 - \cos \theta)$. True only if $\sin \theta = 1$, that is, $\theta = 90°$ and angles coterminal with it.

45 $90°, 270°$

46 $0°, 90°, 180°, 270°$

47 $90°, 180°, 270°$

63 $a = 18, b = 20, c = 7$

EXERCISE 15, Page 59

1	$10°$	**2**	$70°$	**3**	$32°$	**5**	$82° 15'$
6	$27° 25'$	**7**	$71°$	**9**	$65°$	**10**	$42°$
11	$8°$						

EXERCISE 16, Pages 61–62

1 $\sin 300° = -\dfrac{\sqrt{3}}{2}$, $\cos 300° = \dfrac{1}{2}$

2 $\sin 315° = -\dfrac{\sqrt{2}}{2}$, $\cos 315° = \dfrac{\sqrt{2}}{2}$

3 $\sin 225° = -\dfrac{\sqrt{2}}{2}$, $\cos 225° = -\dfrac{\sqrt{2}}{2}$

5 $\sin 135° = \dfrac{\sqrt{2}}{2}$, $\cos 135° = -\dfrac{\sqrt{2}}{2}$

6   $\sin 210° = -\dfrac{1}{2}, \cos 210° = -\dfrac{\sqrt{3}}{2}$

7   $\sin 150° = \dfrac{1}{2}, \cos 150° = -\dfrac{\sqrt{3}}{2}$

9   $\sin 570° = -\dfrac{1}{2}, \cos 570° = -\dfrac{\sqrt{3}}{2}$

10 $\sin 840° = \dfrac{\sqrt{3}}{2}, \cos 840° = -\dfrac{1}{2}$

11 $\sin 660° = -\dfrac{\sqrt{3}}{2}, \cos 660° = \dfrac{1}{2}$

13 0.7431	14 0.7002	15 −0.9397
17 −0.3987	18 −0.5022	19 −0.9787
21 −20.46	22 −0.1181	23 3.001
25 True	26 True	27 False
29 False	30 False	31 True
33 125°, 235°, 305°	34 17°, 197°, 343°	35 42°, 138°, 318°

EXERCISE 17, Page 64

1   $\sin(-30°) = -\dfrac{1}{2}, \cos(-30°) = \dfrac{\sqrt{3}}{2}, \tan(-30°) = -\dfrac{\sqrt{3}}{3}$

2   $\sin(-45°) = -\dfrac{\sqrt{2}}{2}, \cos(-45°) = \dfrac{\sqrt{2}}{2}, \tan(-45°) = -1$

3   $\sin(-60°) = -\dfrac{\sqrt{3}}{2}, \cos(-60°) = \dfrac{1}{2}, \tan(-60°) = -\sqrt{3}$

5 True	6 False	7 True	9 False
10 True	11 False	13 False	14 True
15 False			

18 (a), (d), (f) even; (b), (e), (g) odd; (c), (h) neither

EXERCISE 18, Pages 68–69

1 140°	2 108°	3 162°
5 330°	6 405°	7 300°
9 −270°	10 −70°	11 1800°

13 $\dfrac{666°}{\pi} \rightarrow 212° \, 0'$ 14 $\dfrac{720°}{\pi} \rightarrow 229° \, 11'$ 15 $\dfrac{108°}{\pi} \rightarrow 34° \, 23'$

17 $\dfrac{\pi}{4}$ 18 $\dfrac{5\pi}{6}$ 19 $\dfrac{3\pi}{4}$

21 $\dfrac{2\pi}{3}$ 22 $\dfrac{3\pi}{2}$ 23 $\dfrac{7\pi}{6}$

25 $\dfrac{10\pi}{3}$ 26 $\dfrac{\pi}{18}$ 27 $-\dfrac{\pi}{8}$

29 $\dfrac{16\pi}{9}$ 30 $\dfrac{35\pi}{9}$ 31 $\dfrac{40\pi}{9}$

33 0.3872   34 1.4265   35 1.0001

37 $\dfrac{\sqrt{2}}{2}$ 38 $\dfrac{\sqrt{3}}{2}$ 39 $-\dfrac{1}{2}$

41 $-\dfrac{\sqrt{3}}{2}$ 42 $-\dfrac{\sqrt{3}}{2}$ 43 $-\dfrac{\sqrt{2}}{2}$

45 $\dfrac{2\sqrt{3}}{3}$ 46 $-\sqrt{2}$ 47 $\sqrt{3}$

49 *Does not exist*   50 0   51 -1

53 0.1132   54 0.2896   55 0.4913

57 $\dfrac{7\pi}{6}, 20\pi$ 58 $\dfrac{5\pi}{12}, \dfrac{2\pi}{3}$ 59 $\dfrac{7\pi}{10}$

EXERCISE 19, Pages 71–74

1 (a) 108 ft., (b) 36.1 ft.   2 (a) 37.5 in., (b) 22.7 in.
3 655 ft.   5 275° 1′   6 16.8 ft.
7 2500 miles, 3800 miles   9 5700 miles
10 39° N   11 38° N   13 290 ft.
14 2200 miles   15 28 ft.   17 4.92
18 46° 40′, 66° 40′, 66° 40′   19 10.5 in.

EXERCISE 20, Page 75

1 $\dfrac{2400}{\pi}$ rpm → 764 rpm   2 $\dfrac{3600}{\pi}$ rpm → 1150 rpm

3 $\dfrac{1320}{\pi}$ rpm → 420 rpm   5 $\dfrac{11}{\pi}$ ft. → 3.50 ft.

6 330π mph → 1040 mph   7 839 mph

9 $\dfrac{25\pi}{6}$ mph → 13.1 mph   10 $\dfrac{45}{2\pi}$ in. → 7.16 in.

EXERCISE 21, Pages 84–86

9 180°   11 30°, 330°   13 0°, 180°
14 135°, 225°   15 225°, 315°   17 120°, 240°

18 $60°, 120°$ **19** $30°, 210°$ **21** $45°, 135°$

22 $135°, 315°$ **23** $\dfrac{3\pi}{2}$ **25** $\dfrac{5\pi}{6}, \dfrac{7\pi}{6}$

26 $\dfrac{\pi}{6}, \dfrac{7\pi}{6}$ **27** $\dfrac{\pi}{3}, \dfrac{5\pi}{3}$ **29** $\dfrac{3\pi}{4}, \dfrac{7\pi}{4}$

30 No solution   **31** $22°, 202°$ **33** $196°, 344°$

34 $55°, 305°$ **35** $109°, 251°$ **37** $34°, 214°$

38 $91°, 271°$ **39** **(a)** $4, 2$ **(b)** $7, 3$ **(c)** $8, 5$

EXERCISE 22, Pages 92–94

1 $\sin 105° = \dfrac{\sqrt{6} + \sqrt{2}}{4} \doteq 0.966$

2 $\cos 285° = \dfrac{\sqrt{6} - \sqrt{2}}{4} \doteq 0.259$

3 $\cos 165° = -\dfrac{\sqrt{6} + \sqrt{2}}{4} \doteq -0.966$

6 -1

7 $\sin \theta$

9 1

17 $\cos 55° = \cos 22° \cos 33° - \sin 22° \sin 33°$

18 $\sin 74° = \sin 70° \cos 4° + \cos 70° \sin 4°$

21 $\dfrac{7\sqrt{91} + 3\sqrt{51}}{100}$

22 $\frac{171}{221}$

23 **(a)** $\frac{156}{205}$, **(b)** $\frac{133}{205}$, **(c)** Q I

EXERCISE 23, Pages 96–98

1 $\tan 345° = \dfrac{1 - \sqrt{3}}{1 + \sqrt{3}} = \sqrt{3} - 2 \doteq -0.268$

2 $\tan 75° = \dfrac{\sqrt{3} + 1}{\sqrt{3} - 1} = 2 + \sqrt{3} \doteq 3.732$

3 $\cos 195° = -\dfrac{\sqrt{6} + \sqrt{2}}{4} \doteq -0.966$

5 -1 **6** $\frac{1}{2}$ **7** $\tan 3\theta$

13 **(a)** $\frac{84}{85}$, **(b)** $\frac{13}{85}$, **(c)** $\frac{36}{77}$, **(d)** $\frac{84}{13}$

14 **(a)** $-\frac{204}{325}$, **(b)** $\frac{253}{325}$, **(c)** $\frac{36}{323}$, **(d)** $-\frac{204}{253}$

15 $\dfrac{5\sqrt{35} - \sqrt{11}}{36}$

17 False   **18** True   **19** True

21 $\sin 11° = \sin 88° \cos 77° - \cos 88° \sin 77°$

22 $\cos 20° = \cos 70° \cos 50° + \sin 70° \sin 50°$

23 $\tan 36° = \dfrac{\tan 76° - \tan 40°}{1 + \tan 76° \tan 40°}$

37 $17 \sin (\theta + 61° 56')$

38 $\sqrt{2} \sin (\theta + 135°)$

39 $5 \sin (\theta + 306° 52')$

41 $5\sqrt{3} \sin \theta - 5 \cos \theta$

42 $-\sqrt{2} \sin \theta - \sqrt{2} \cos \theta$

EXERCISE 24, Pages 101–104

3 $\sin 22\frac{1}{2}° = \frac{1}{2}\sqrt{2 - \sqrt{2}}, \ \cos 22\frac{1}{2}° = \frac{1}{2}\sqrt{2 + \sqrt{2}}$

5 $\cos 16\theta$ **6** $9 \sin 20°$ **7** $\frac{1}{2} \sin 24°$

9 $\cos 140°$ **10** $\pm \sin 7\theta$ **11** 1

13 $\sin \dfrac{A}{2} = \dfrac{4}{5}, \ \sin 2A = -\dfrac{336}{625}$

14 $\cos \dfrac{A}{2} = -\dfrac{3}{4}, \ \cos 2A = -\dfrac{31}{32}$

15 $\cos 3B = -\dfrac{\sqrt{6}}{6}, \ \cos 12B = -\dfrac{1}{9}$

17 $\sin 155° = \sqrt{\dfrac{1 - \cos 310°}{2}}$

18 $\cos 175° = -\sqrt{\dfrac{1 + \cos 350°}{2}}$

19 $\sin 280° = -2 \sin 140° \sqrt{1 - \sin^2 140°}$

21 $\cos 4B = 1 - 2 \sin^2 2B$

22 $\sin C = 2 \sin \dfrac{C}{2} \cos \dfrac{C}{2}$

23 $\sin 3D = \pm \sqrt{\dfrac{1 - \cos 6D}{2}}$

25 True   **26** False   **27** True

29 False   **30** True   **31** True

33 True   **34** True   **35** False

67 (c) 90° is a nonpermissible value of A because $\tan 90°$ does not exist. The identity does *not* hold true for $A = 90°$.

EXERCISE 25, Pages 106–107

1 $\frac{1}{2} \sin 275° + \frac{1}{2} \sin 141°$ **2** $5 \cos 67° + 5 \cos 25°$

3   $-5 - 10 \cos 226°$ 5 $\cos 11\theta + \cos 5\theta$
6   $\frac{1}{2} \cos \theta - \frac{1}{2} \cos 11\theta$ 7 $2 \sin 13\theta \cos 2\theta$
9   $-\cos 70°$ 10 $\sqrt{3} \sin (A + 60°)$

EXERCISE 26, Pages 112–114

1   $155°, 245°$ 2 $20°, 80°$
3   $30°, 150°, 210°, 330°$ 5 $60°, 120°, 240°, 300°$
6   $45°, 135°, 225°, 315°$ 7 $120°, 240°, 300°$
9   $60°, 120°, 150°, 210°$ 10 $30°, 225°, 315°, 330°$
11   $60°, 90°, 120°, 240°, 270°, 300°$
13   $60°, 180°, 300°$ 14 $60°, 270°, 300°$
15   $90°, 135°, 270°, 315°$ 17 $120°, 300°$
18   $120°, 240°$ 19   No solution
21   No solution   22 $0°, 135°, 180°, 315°$
23   $0°, 135°, 180°, 225°$ 25 $90°, 210°, 330°$
26   $45°, 135°, 225°, 315°$ 27 $0°, 120°, 240°$
29   $30°, 90°, 150°, 270°$ 30 $0°, 180°, 210°, 330°$
31   $20°, 80°, 140°, 200°, 260°, 320°$
33   The given equation is an identity. It holds true for all permissible values of θ, that is, all θ except $0°, 90°, 180°, 270°$, and angles coterminal with them.
34   $25°, 115°, 205°, 295°$ 35 $196° 36', 270°, 343° 24'$
37   $70° 32', 120°, 240°, 289° 28'$
38   No solution   39 $78° 28', 281° 32'$
41   $33° 41', 213° 41'$
42   $14° 29', 165° 31', 210°, 330°$
43   $180°, 300°$ 45 $330°$
46   $90°, 180°$ 47 $60°, 300°$
49   $119° 33', 346° 43'$ 50 $88° 4', 328° 4'$
51   $82° 37', 322° 37'$
53   $10°, 70°, 100°, 190°, 250°, 280°$
54   $10°, 70°, 130°, 150°, 190°, 250°, 310°, 330°$
55   $15°, 75°, 90°, 195°, 255°, 270°$
57   $0°, 180°$ 58 $135°, 225°$
59   $120°, 240°$ 61 $120°, 240°$
62   $45°, 116° 34', 225°, 296° 34'$

EXERCISE 27, Pages 122–124

1   $\dfrac{\pi}{3}, 4$ 2 $\dfrac{2\pi}{3}, 5$ 3 $8, 1$

5 $\dfrac{8\pi}{3}$, ∞ 6 7π, 1   7 $\dfrac{\pi}{4}$, $\dfrac{1}{2}$

9 $\dfrac{2\pi}{3}$, 1, 2 units to right   10 8, 5, 4 units to left

11 $\dfrac{\pi}{2}$, 7, $\dfrac{\pi}{12}$ units to left

EXERCISE 28, Pages 127–128

1	2	2	-1	3	1
5	0	6	$\frac{1}{2}$	7	-1
9	$\frac{1}{3}$	10	0	11	$\frac{1}{6}$
13	$-\frac{3}{2}$	14	$-\frac{2}{3}$	15	-3
17	3	18	2	19	$\frac{1}{16}$
21	365	22	$\frac{1}{36}$	23	2
25	10	26	$\sqrt[3]{6}$	27	5
29	$\log_{625} 125 = \frac{3}{4}$	30	$\log_p r = q$	31	$\log_7 1 = 0$
33	$(\frac{1}{9})^{-1/2} = 3$	34	$243^{4/5} = 81$	35	$216^{2/3} = 36$
37	True	38	True	39	True

EXERCISE 29, Pages 129–130

1	-0.18	2	0.78	3	2.33
5	1.80	6	0.15	7	0.08
9	0.35	10	-3.70	11	4.62
13	0.13	14	1.08	15	-1.55

17 $\log (a^2 - 1)$ 18 $\log \dfrac{a - b}{b}$ 19 $\log \pi \sqrt{\dfrac{l}{g}}$

21 $\log \dfrac{x^5}{z^2 \sqrt[3]{y}}$ 22 $\log \sqrt{\dfrac{(s - a)(s - b)(s - c)}{s}}$

23 $\log a^4 \sqrt[6]{\dfrac{c^4}{b^5}}$ 25 True   26 False

27	False	29	False	30	False
31	True	33	True	34	True
35	False	37	False	38	True
39	True	41	True		

EXERCISE 30, Page 134

1	4.53857	2	$9.53857 - 10$	3	$8.88666 - 10$
5	$7.88666 - 10$	6	3.88666	7	6.53857

9	1.53857	10	5.53857 — 10	11	9.88666 — 10
13	0.01234	14	1.234	15	66,070,000
17	660.7	18	0.006607	19	0.000 1234
21	0.000 000 000 01234	22	123,400	23	66.07
25	0.6607	26	0.000 6607	27	0.001234

EXERCISE 31, Page 136

1	0.77240	2	1.88773	3	5.85794 — 10
5	5.64992	6	8.96028 — 10	7	4.57031
9	6.95022 — 10	10	6.00303	11	0.53301
13	1098	14	0.000 005432	15	246,800
17	0.000 06363	18	27,310	19	0.001414
21	6,200,000	22	0.000 8915	23	337.1

EXERCISE 32, Pages 138–139

1	9.72440 — 10	2	2.57523	3	1.06648
5	6.65009	6	7.28072 — 10	7	6.92000 — 10
9	8.44820 — 10	10	4.90256	11	5.99454
13	3.79527	14	0.63565	15	8.33118 — 10
17	16.836	18	0.33428	19	0.000 024957
21	8.9146	22	4788.7	23	77,664
25	37,032	26	0.027008	27	0.12008
29	0.000 60007	30	10.173	31	1397.9

EXERCISE 33, Pages 141–143

1	60,340	2	6.820	3	0.08851
5	1.210	6	27.18	7	4.974
9	3,487,000,000	10	0.000 3614	11	6,000,000
13	2923	14	0.8566	15	0.02498
17	0.05470	18	0.5570	19	0.4488
21	−0.000 01569	22	−0.000 8331	23	−5.480
25	1.632	26	−9.828	27	0.2006
29	0.000 9507	30	0.006337	31	138.6
33	0.000 088726	34	1.6220	35	0.014465
37	0.42711	38	0.32955	39	0.044302
41	0.0090950	42	872,740	43	82.308
45	125,000	46	0.01818	47	0.000 3342
49	7.208	50	0.000 702	51	0.005033
53	1.1087	54	40,400	55	104
57	18,255	58	16.99		

EXERCISE 34, Pages 147–148

1	$1, -4$	**2**	-1	**3**	-2
5	4	**6**	17	**7**	$x = 3 + \sqrt{9 + b^y}$

9 $\dfrac{7 \log a + 9 \log b}{6 \log a - 8 \log b}$ **10** $\dfrac{\log b + \log c}{\log a - \log c}$ **11** -4.004

13	2.526	**14**	0.07248	**15**	0.6014
17	1.017	**18**	0.8193	**19**	8.915
21	-0.5460	**22**	-3.187	**23**	6.113

EXERCISE 35, Pages 150–151

1	$9.99920 - 10$	**2**	$9.95937 - 10$	**3**	$9.97497 - 10$
5	0.09483	**6**	0.19283	**7**	$9.17290 - 10$
9	$9.48066 - 10$	**10**	$9.93158 - 10$	**11**	$9.98434 - 10$
13	$9.50265 - 10$	**14**	$9.28775 - 10$	**15**	0.51128
17	$24°\ 4.0'$	**18**	$71°\ 20.0'$	**19**	$63°\ 18.0'$
21	$64°\ 26.6'$	**22**	$3°\ 20.1'$	**23**	$25°\ 32.2'$
25	$81°\ 40.0'$	**26**	$9°\ 14.4'$	**27**	$24°\ 38.7'$
29	$18°\ 24.3'$	**30**	$84°\ 1.0'$	**31**	$45°\ 56.2'$
33	$72°\ 9.6'$	**34**	Impossible	**35**	$32°\ 24.6'$

EXERCISE 36, Page 154

5 $B = 57°\ 19',\ a = 2671,\ b = 4164$

6 $B = 28°\ 43',\ b = 186.5,\ c = 388.3$

7 $A = 63°\ 11',\ b = 34.08,\ c = 75.54$

9 $A = 19°\ 10',\ B = 70°\ 50',\ c = 0.2056$

10 $A = 37°\ 3',\ B = 52°\ 57',\ a = 3.155$

11 $A = 17°\ 33',\ B = 72°\ 27',\ b = 812.6$

13 $A = 20°\ 55.2',\ a = 27{,}865,\ c = 78{,}038$

14 $B = 67°\ 25.2',\ a = 22{,}346,\ b = 53{,}734$

15 $A = 7°\ 36.0',\ a = 1.1381,\ b = 8.5292$

17 $A = 40°\ 8.0',\ B = 49°\ 52.0',\ a = 5.9211$

18 $A = 41°\ 41.3',\ B = 48°\ 18.7',\ c = 85.874$ or 85.876, depending upon the method used

19 $A = 65°\ 45.0',\ B = 24°\ 15.0',\ c = 82{,}980$

EXERCISE 37, Pages 157–159

1	2650 ft.	**2**	1350 ft.
3	555 ft.	**5**	**(a)** 378.0 lb., **(b)** 2971 lb.
6	9032 lb.	**7**	**(a)** $16°\ 9.7'$, **(b)** 1691.0 lb.
9	N $15°\ 52'$ E, 4115 ft. per min.		

10 N 77° 40′ E, 103 mph   **11** N 6° 40′ E, 151 mph
13 32.1 mph   **14** 2.00 min.
15 18.0 mph   **17** N 2° 20′ E
18 481 ft. 4.0 in., 755 ft. 8.8 in.
19 61.14 lb. due west, 85.10 lb. due south
22 $C = 73° 12′, a = 62.37, c = 81.42$
23 $B = 80° 30′, C = 44° 22′, a = 5665$
25 $C = 40° 20′, b = 17.28, c = 22.96$
26 $B = 14° 41′, C = 25° 37′, a = 3690$

EXERCISE 38, Pages 164–165

1 $A = 77° 10′, a = 770, c = 779$
2 $C = 43°, a = 29, b = 18$
7 $B = 28° 54′, a = 3986, c = 3759$
9 $A = 40° 20′, b = 33.21, c = 17.28$
10 $B = 72° 34′, a = 8.472, c = 5.298$
11 $A = 35° 30′, a = 254, b = 325$
13 $C = 58° 58.3′, a = 9242.2, c = 8404.2$
14 $B = 85° 35.0′, a = 17,149, b = 52,554$
15 $A = 59° 46.5′, b = 0.090544, c = 0.090340$
17 3605 ft.
18 140 miles, 130 miles
19 17 mph

EXERCISE 39, Pages 169–170

1 $B = 78° 57′, C = 46° 23′, c = 14.38;$
　 $B′ = 101° 3′, C′ = 24° 17′, c′ = 8.167$
2 $B = 47° 10′, C = 59° 38′, c = 0.7338$
3 No triangle
5 $B = 90° 0′, C = 78° 59′, c = 6857$
6 No triangle
7 $B = 69° 50′, C = 68° 47′, c = 7.006;$
　 $B′ = 110° 10′, C′ = 28° 27′, c′ = 3.580$
9 No triangle
10 $B = 62° 13.4′, C = 78° 55.0′, b = 200.85;$
　 $B′ = 40° 3.4′, C′ = 101° 5.0′, b′ = 146.09$
11 $A = 20° 35.3′, C = 44° 57.7′, a = 15,131$
13 There are two Lexingtons that satisfy the conditions of the problem. Lexington, Ky., is 170 miles from Evansville; Lexington, Va., is 440 miles from Evansville.
14 N 36° 30′ W, 1.41 min.
15 2:13 P.M. and 3:19 P.M.

EXERCISE 40, Pages 173–174

1 102°	2 149°	3 59°	5 26
6 5.34	7 29.39	9 6.5	10 99

11 $A = 22°, B = 38°, C = 120°$

13 51° and 18°   14 16° and 164°

15 N 24° E, 270 miles   17 77° 20′

18 42° 20′   19 2.51

21 $b = 35.5, A = 43° 20′, C = 117° 10′$

22 $c = 38.82, A = 36° 46′, B = 79° 34′$

EXERCISE 41, Page 177

1 3.1	2 501	3 45	5 252
6 $12\sqrt{17} \doteq 49$	7 185.7	9 897.3	10 3.432
11 6.014		14 3.37 acres	

15 12° 2′ or 167° 58′   17 $8 + \frac{1}{5}\sqrt{85}, 8 - \frac{1}{5}\sqrt{85}$

EXERCISE 42, Pages 178–181

1 87°	2 3.5
3 649.1	5 51° 53′ or 128° 7′
6 4552	7 31
9 57 yd. or 71 yd.	10 98°, 22°, 60°
11 N 29° W	13 260 mph, 270 mph
14 5°	

15 N 32° E, N 85° W or S 32° E, S 85° W

17 87°, 93°   18 N 4° E, 126 mph

19 45.86 in., 52.21 in.   21 405,000,000 miles

22 $y = \dfrac{a \sin (\beta - \alpha)}{\cos \beta}$ 23 24 miles

25 *Hint:* Use the law of sines.

EXERCISE 43, Pages 185–186

1 $-\dfrac{\pi}{6}$	2 $\dfrac{\pi}{3}$	3 $\dfrac{\pi}{3}$	5 $\dfrac{\pi}{6}$
6 $\dfrac{2\pi}{3}$	7 0	9 $\dfrac{\pi}{4}$	10 $\dfrac{\pi}{4}$
11 π	13 $\dfrac{\pi}{2}$	14 0	15 $-\dfrac{\pi}{4}$
17 $\dfrac{\pi}{4}$	18 π	19 $\dfrac{\pi}{6}$	21 $-\dfrac{\pi}{3}$

22 0.8494   23 −0.5876   25 3.0107   26 −0.4451
27 1.3526

EXERCISE 44, Pages 188–190

1 $\dfrac{\sqrt{1 - u^2}}{u}$

2 $\dfrac{u}{\sqrt{u^2 + 1}}$

3 $\dfrac{1}{u}$

5 $\frac{5}{13}$

6 $-2\sqrt{2}$

7 $\dfrac{1}{\sqrt{5}}$

9 $uv - \sqrt{1 - u^2}\sqrt{1 - v^2}$

10 $u\sqrt{1 - v^2} - v\sqrt{1 - u^2}$

11 $\sqrt{1 - u^2}\sqrt{1 - v^2} + uv$

13 $-\dfrac{\sqrt{33}}{7}$

14 $-\frac{2}{3}$

15 $\frac{1}{2}(\sqrt{1 - u^2} + u\sqrt{3})$

17 $2u^2 - 1$

18 $2u\sqrt{1 - u^2}$

19 $16u\sqrt{1 - 64u^2}$

21 $2u$

22 $6u$

23 $\dfrac{3\pi}{8}$

25 $-\dfrac{\pi}{5}$

26 $\dfrac{2\pi}{9}$

27 $-\dfrac{\pi}{2}$

29 $\dfrac{2\pi}{3}$

30 $-\dfrac{\pi}{4}$

31 $\dfrac{1}{\sqrt{u^2 + 1}}, u$

33 $\dfrac{\sqrt{4 - u^2}}{2}, \dfrac{u}{2}$

34 $\dfrac{\sqrt{u^2 + 4u + 13}}{3}, \dfrac{u + 2}{3}$

35 $\dfrac{1 - 2u^2}{2u\sqrt{1 - u^2}}$

46 True   47 True   49 True   50 True
51 True   53 False if $u < 0$ 54 False
55 False; right side should be $u/\sqrt{1 - u^2}$
57 False   59 $1, \frac{1}{2}$ 61 $-1 < x < 1$
62 0, 1

EXERCISE 46, Page 195

1 $12 - 3i$ 2 $5 - 13i$ 3 $-6 + 3i$
5 $46 - 3i$ 6 $4\sqrt{7} + 19i$ 7 $14 - 8i\sqrt{5}$

9   $-33 + 56i$ 10 $21 - 20i$ 11 i

13   $-i$ 14   1   15 $\frac{56}{37} + \frac{3}{37}i$

17   $\dfrac{2}{7} + \dfrac{9\sqrt{6}}{7}i$ 18 $-\frac{17}{25} - \frac{44}{25}i$ 19 $5 - 7i$

21   $x = 7, y = -8$ 22 $x = -2, y = 32$

23   $x = 9, y = 5; x = -5, y = -9$ 25   0

EXERCISE 47, Pages 199–200

1   $6 - 2i$ 2 $5 + 3i$ 3 $2 + 2i$

5   $-3 + 4i$ 6 $-7 - 6i$ 7 $-1 + 6i$

9   $6\sqrt{2}(\cos 45° + i\sin 45°)$ 10 $2(\cos 315° + i\sin 315°)$

11   $5\sqrt{2}(\cos 135° + i\sin 135°)$ 13 $5(\cos 270° + i\sin 270°)$

14   $3(\cos 180° + i\sin 180°)$ 15 $7(\cos 0° + i\sin 0°)$

17   $2(\cos 120° + i\sin 120°)$ 18 $17(\cos 61° 56' + i\sin 61° 56')$

19   $5(\cos 216° 52' + i\sin 216° 52')$

21   $-3.420 - 9.397i$ 22 $-3 + 3i\sqrt{3}$ 23 $\sqrt{3} - i$

EXERCISE 48, Pages 203–205

1   $7 - 7i\sqrt{3}$ 2 $-3 - 3i\sqrt{3}$ 3 $-10\sqrt{2} + 10i\sqrt{2}$

5   6   6 $12\sqrt{3} + 12i$ 7 $8\sqrt{3} + 24i$

9   $15\sqrt{3} + 15i$ 10 $-14 - 14i\sqrt{3}$ 11 i

13   $-16\sqrt{3} + 16i$ 14 $-243i$ 15 $-32 + 32i$

17   $64\sqrt{2} + 64i\sqrt{2}$ 18 $32i$ 19 $-128 + 128i\sqrt{3}$

21   $-8431 - 5378i$, using four-place tables. The correct
value is $-8432 - 5376i$.   22 $-648 + 648i\sqrt{3}$

23   $2.9631 + 0.4692i, -1.8879 + 2.3313i, -1.0752 - 2.8008i$

25   $3\sqrt{2} + 3i\sqrt{2}, -3\sqrt{2} - 3i\sqrt{2}$ 26 $-1 + i, 1 - i$

27   $-1 + i\sqrt{3}, 1 - i\sqrt{3}$ 29 $2, -1 + i\sqrt{3}, -1 - i\sqrt{3}$

30   $\sqrt{2} + i\sqrt{2}, -1.9318 + 0.5176i, 0.5176 - 1.9318i$

31   $1.8478 + 0.7654i, -0.7654 + 1.8478i,$
     $-1.8478 - 0.7654i, 0.7654 - 1.8478i$

33   $1 + i\sqrt{3}, -\sqrt{3} + i, -1 - i\sqrt{3}, \sqrt{3} - i$

34   $2\sqrt{2} + 2i\sqrt{2}, -2\sqrt{2} + 2i\sqrt{2}, -2\sqrt{2} - 2i\sqrt{2}, 2\sqrt{2} - 2i\sqrt{2}$

35   $2, \sqrt{2} + i\sqrt{2}, 2i, -\sqrt{2} + i\sqrt{2}, -2, -\sqrt{2} - i\sqrt{2}, -2i, \sqrt{2} - i\sqrt{2}$

37   $1.6180 + 1.1756i, -0.6180 + 1.9022i, -2,$
     $-0.6180 - 1.9022i, 1.6180 - 1.1756i$

38   $4, 2 + 2i\sqrt{3}, -2 + 2i\sqrt{3}, -4, -2 - 2i\sqrt{3}, 2 - 2i\sqrt{3}$

EXERCISE 49, Page 210

3   0.09983342

EXERCISE 50, Pages 215–216

5   $A = 70° 0'$, $B = 10° 0'$, $c = 8365$
6   $A = 110° 31'$, $B = 29° 9'$, $c = 22.96$
7   $A = 59° 9'$, $B = 39° 33'$, $c = 468.0$
9   $A = 86° 13'$, $C = 56° 29'$, $b = 5.542$
10   $A = 60° 3'$, $C = 59° 39'$, $b = 1955$
11   $B = 35° 31.6'$, $C = 20° 42.8'$, $a = 81,253$ or $81,250$,
     depending upon the method used
13   $A = 53° 49.2'$, $B = 68° 53.0'$, $c = 2.8339$
14   $A = 40° 21.8'$, $B = 59° 38.2'$, $c = 55,465$
15   478.7 ft., 355.3 ft.
17   31.53 ft.
18   N 49° E, 190 miles

EXERCISE 51, Pages 219–220

2   $A = 90° 10'$, $B = 53° 0'$, $C = 36° 50'$
3   $A = 49° 29'$, $B = 64° 18'$, $C = 66° 13'$
5   $A = 37° 29'$, $B = 53° 32'$, $C = 88° 59'$
6   $A = 72° 52'$, $B = 46° 51'$, $C = 60° 17'$
7   $A = 120° 6'$, $B = 21° 44'$, $C = 38° 10'$
9   $A = 49° 18.8'$, $B = 92° 4.4'$, $C = 38° 36.8'$
10   $A = 44° 51.2'$, $B = 19° 47.6'$, $C = 115° 21.2'$
11   $A = 27° 26.6'$, $B = 50° 58.2'$, $C = 101° 35.2'$
13   N 59° E, N 5° W
14   100 ft.

EXERCISE 52, Page 221

1	4° 30'	2	0° 37'	3	1° 43'
5	53 mils	6	77 mils	7	124 mils
9	0.118 radian	10	0.295 radian	11	0.245 radian
13	204 mils	14	67 mils	15	52 mils
17	605 ft.	18	400 ft.	19	60 mils
21	533 ft.	22	36 mils	23	864,000 miles

LOGARITHMIC AND TRIGONOMETRIC TABLES

VALUES OF THE TRIGONOMETRIC FUNCTIONS
TO FOUR PLACES

TABLE 1

Radians	Degrees	Sin	Tan	Cot	Cos		
.0000	0° 00′	.0000	.0000	——	1.0000	90° 00′	1.5708
029	10	029	029	343.8	000	89° 50′	679
058	20	058	058	171.9	000	40	650
.0087	30	.0087	.0087	114.6	1.0000	30	1.5621
116	40	116	116	85.94	.9999	20	592
145	50	145	145	68.75	999	10	563
.0175	1° 00′	.0175	.0175	57.29	.9998	89° 00′	1.5533
204	10	204	204	49.10	998	88° 50′	504
233	20	233	233	42.96	997	40	475
.0262	30	.0262	.0262	38.19	.9997	30	1.5446
291	40	291	291	34.37	996	20	417
320	50	320	320	31.24	995	10	388
.0349	2° 00′	.0349	.0349	28.64	.9994	88° 00′	1.5359
378	10	378	378	26.43	993	87° 50′	330
407	20	407	407	24.54	992	40	301
.0436	30	.0436	.0437	22.90	.9990	30	1.5272
465	40	465	466	21.47	989	20	243
495	50	494	495	20.21	988	10	213
.0524	3° 00′	.0523	.0524	19.08	.9986	87° 00′	1.5184
553	10	552	553	18.07	985	86° 50′	155
582	20	581	582	17.17	983	40	126
.0611	30	.0610	.0612	16.35	.9981	30	1.5097
640	40	640	641	15.60	980	20	068
669	50	669	670	14.92	978	10	039
.0698	4° 00′	.0698	.0699	14.30	.9976	86° 00′	1.5010
727	10	727	729	13.73	974	85° 50′	981
756	20	756	758	13.20	971	40	952
.0785	30	.0785	.0787	12.71	.9969	30	1.4923
814	40	814	816	12.25	967	20	893
844	50	843	846	11.83	964	10	864
.0873	5° 00′	.0872	.0875	11.43	.9962	85° 00′	1.4835
902	10	901	904	11.06	959	84° 50′	806
931	20	929	934	10.71	957	40	777
.0960	30	.0958	.0963	10.39	.9954	30	1.4748
989	40	987	992	10.08	951	20	719
.1018	50	.1016	.1022	9.788	948	10	690
.1047	6° 00′	.1045	.1051	9.514	.9945	84° 00′	1.4661
076	10	074	080	9.255	942	83° 50′	632
105	20	103	110	9.010	939	40	603
.1134	30	.1132	.1139	8.777	.9936	30	1.4573
164	40	161	169	8.556	932	20	544
193	50	190	198	8.345	929	10	515
.1222	7° 00′	.1219	.1228	8.144	.9925	83° 00′	1.4486
251	10	248	257	7.953	922	82° 50′	457
280	20	276	287	7.770	918	40	428
.1309	30	.1305	.1317	7.596	.9914	30	1.4399
338	40	334	346	7.429	911	20	370
367	50	363	376	7.269	907	10	341
.1396	8° 00′	.1392	.1405	7.115	.9903	82° 00′	1.4312
425	10	421	435	6.968	899	81° 50′	283
454	20	449	465	6.827	894	40	254
.1484	30	.1478	.1495	6.691	.9890	30	1.4224
513	40	507	524	6.561	886	20	195
542	50	536	554	6.435	881	10	166
.1571	9° 00′	.1564	.1584	6.314	.9877	81° 00′	1.4137
		Cos	Cot	Tan	Sin	Degrees	Radians

TABLE 1

Radians	Degrees	Sin	Tan	Cot	Cos		
.1571	9° 00'	.1564	.1584	6.314	.9877	81° 00'	1.4137
600	10	593	614	197	872	80° 50	108
629	20	622	644	084	868	40	079
.1658	30	.1650	.1673	5.976	.9863	30	1.4050
687	40	679	703	871	858	20	1.4021
716	50	708	733	769	853	10	992
.1745	10° 00'	.1736	.1763	5.671	.9848	80° 00'	1.3963
774	10	765	793	576	843	79° 50'	934
804	20	794	823	485	838	40	904
.1833	30	.1822	.1853	5.396	.9833	30	1.3875
862	40	851	883	309	827	20	846
891	50	880	914	226	822	10	817
.1920	11° 00'	.1908	.1944	5.145	.9816	79° 00'	1.3788
949	10	937	974	066	811	78° 50'	759
978	20	965	.2004	4.989	805	40	730
.2007	30	.1994	.2035	4.915	.9799	30	1.3701
036	40	.2022	065	843	793	20	672
065	50	051	095	773	787	10	643
.2094	12° 00'	.2079	.2126	4.705	.9781	78° 00'	1.3614
123	10	108	156	638	775	77° 50'	584
153	20	136	186	574	769	40	555
.2182	30	.2164	.2217	4.511	.9763	30	1.3526
211	40	193	247	449	757	20	497
240	50	221	278	390	750	10	468
.2269	13° 00'	.2250	.2309	4.331	.9744	77° 00'	1.3439
298	10	278	339	275	737	76° 50'	410
327	20	306	370	219	730	40	381
.2356	30	.2334	.2401	4.165	.9724	30	1.3352
385	40	363	432	113	717	20	323
414	50	391	462	061	710	10	294
.2443	14° 00'	.2419	.2493	4.011	.9703	76° 00'	1.3265
473	10	447	524	3.962	696	75° 50'	235
502	20	476	555	914	689	40	206
.2531	30	.2504	.2586	3.867	.9681	30	1.3177
560	40	532	617	821	674	20	148
589	50	560	648	776	667	10	119
.2618	15° 00'	.2588	.2679	3.732	.9659	75° 00'	1.3090
647	10	616	711	689	652	74° 50'	061
676	20	644	742	647	644	40	032
.2705	30	.2672	.2773	3.606	.9636	30	1.3003
734	40	700	805	566	628	20	974
763	50	728	836	526	621	10	945
.2793	16° 00'	.2756	.2867	3.487	.9613	74° 00'	1.2915
822	10	784	899	450	605	73° 50'	886
851	20	812	931	412	596	40	857
.2880	30	.2840	.2962	3.376	.9588	30	1.2828
909	40	868	994	340	580	20	799
938	50	896	.3026	305	572	10	770
.2967	17° 00'	.2924	.3057	3.271	.9563	73° 00'	1.2741
996	10	952	089	237	555	72° 50'	712
.3025	20	979	121	204	546	40	683
.3054	30	.3007	.3153	3.172	.9537	30	1.2654
083	40	035	185	140	528	20	625
113	50	062	217	108	520	10	595
.3142	18° 00'	.3090	.3249	3.078	.9511	72° 00'	1.2566
		Cos	Cot	Tan	Sin	Degrees	Radians

TABLE 1

Radians	Degrees	Sin	Tan	Cot	Cos		
.3142	18° 00′	.3090	.3249	3.078	.9511	72° 00′	1.2566
171	10	118	281	047	502	71° 50′	537
200	20	145	314	018	492	40	508
.3229	30	.3173	.3346	2.989	.9483	30	1.2479
258	40	201	378	960	474	20	450
287	50	228	411	932	465	10	421
.3316	19° 00′	.3256	.3443	2.904	.9455	71° 00′	1.2392
345	10	283	476	877	446	70° 50′	363
374	20	311	508	850	436	40	334
.3403	30	.3338	.3541	2.824	.9426	30	1.2305
432	40	365	574	798	417	20	275
462	50	393	607	773	407	10	246
.3491	20° 00′	.3420	.3640	2.747	.9397	70° 00′	1.2217
520	10	448	673	723	387	69° 50′	188
549	20	475	706	699	377	40	159
.3578	30	.3502	.3739	2.675	.9367	30	1.2130
607	40	529	772	651	356	20	101
636	50	557	805	628	346	10	072
.3665	21° 00′	.3584	.3839	2.605	.9336	69° 00′	1.2043
694	10	611	872	583	325	68° 50′	1.2014
723	20	638	906	560	315	40	985
.3752	30	.3665	.3939	2.539	.9304	30	1.1956
782	40	692	973	517	293	20	926
811	50	719	.4006	496	283	10	897
.3840	22° 00′	.3746	.4040	2.475	.9272	68° 00′	1.1868
869	10	773	074	455	261	67° 50′	839
898	20	800	108	434	250	40	810
.3927	30	.3827	.4142	2.414	.9239	30	1.1781
956	40	854	176	394	228	20	752
985	50	881	210	375	216	10	723
.4014	23° 00′	.3907	.4245	2.356	.9205	67° 00′	1.1694
043	10	934	279	337	194	66° 50′	665
072	20	961	314	318	182	40	636
.4102	30	.3987	.4348	2.300	.9171	30	1.1606
131	40	.4014	383	282	159	20	577
160	50	041	417	264	147	10	548
.4189	24° 00′	.4067	.4452	2.246	.9135	66° 00′	1.1519
218	10	094	487	229	124	65° 50′	490
247	20	120	522	211	112	40	461
.4276	30	.4147	.4557	2.194	.9100	30	1.1432
305	40	173	592	177	088	20	403
334	50	200	628	161	075	10	374
.4363	25° 00′	.4226	.4663	2.145	.9063	65° 00′	1.1345
392	10	253	699	128	051	64° 50′	316
422	20	279	734	112	038	40	286
.4451	30	.4305	.4770	2.097	.9026	30	1.1257
480	40	331	806	081	013	20	228
509	50	358	841	066	001	10	199
.4538	26° 00′	.4384	.4877	2.050	.8988	64° 00′	1.1170
567	10	410	913	035	975	63° 50′	141
596	20	436	950	020	962	40	112
.4625	30	.4462	.4986	2.006	.8949	30	1.1083
654	40	488	.5022	1.991	936	20	054
683	50	514	059	977	923	10	1.1025
.4712	27° 00′	.4540	.5095	1.963	.8910	63° 00′	1.0996
		Cos	Cot	Tan	Sin	Degrees	Radians

TABLE 1

Radians	Degrees	Sin	Tan	Cot	Cos		
.4712	27° 00′	.4540	.5095	1.963	.8910	63° 00′	1.0996
741	10	566	132	949	897	62° 50′	966
771	20	592	169	935	884	40	937
.4800	30	.4617	.5206	1.921	.8870	30	1.0908
829	40	643	243	907	857	20	879
858	50	669	280	894	843	10	850
.4887	28° 00′	.4695	.5317	1.881	.8829	62° 00′	1.0821
916	10	720	354	868	816	61° 50′	792
945	20	746	392	855	802	40	763
.4974	30	.4772	.5430	1.842	.8788	30	1.0734
.5003	40	797	467	829	774	20	705
032	50	823	505	816	760	10	676
.5061	29° 00′	.4848	.5543	1.804	.8746	61° 00′	1.0647
091	10	874	581	792	732	60° 50′	617
120	20	899	619	780	718	40	588
.5149	30	.4924	.5658	1.767	.8704	30	1.0559
178	40	950	696	756	689	20	530
207	50	975	735	744	675	10	501
.5236	30° 00′	.5000	.5774	1.732	.8660	60° 00′	1.0472
265	10	025	812	720	646	59° 50′	443
294	20	050	851	709	631	40	414
.5323	30	.5075	.5890	1.698	.8616	30	1.0385
352	40	100	930	686	601	20	356
381	50	125	969	675	587	10	327
.5411	31° 00′	.5150	.6009	1.664	.8572	59° 00′	1.0297
440	10	175	048	653	557	58° 50′	268
469	20	200	088	643	542	40	239
.5498	30	.5225	.6128	1.632	.8526	30	1.0210
527	40	250	168	621	511	20	181
556	50	275	208	611	496	10	152
.5585	32° 00′	.5299	.6249	1.600	.8480	58° 00′	1.0123
614	10	324	289	590	465	57° 50′	094
643	20	348	330	580	450	40	065
.5672	30	.5373	.6371	1.570	.8434	30	1.0036
701	40	398	412	560	418	20	1.0007
730	50	422	453	550	403	10	977
.5760	33° 00′	.5446	.6494	1.540	.8387	57° 00′	.9948
789	10	471	536	530	371	56° 50′	919
818	20	495	577	520	355	40	890
.5847	30	.5519	.6619	1.511	.8339	30	.9861
876	40	544	661	501	323	20	832
905	50	568	703	1.492	307	10	803
.5934	34° 00′	.5592	.6745	1.483	.8290	56° 00′	.9774
963	10	616	787	473	274	55° 50′	745
992	20	640	830	464	258	40	716
.6021	30	.5664	.6873	1.455	.8241	30	.9687
050	40	688	916	446	225	20	657
080	50	712	959	437	208	10	628
.6109	35° 00′	.5736	.7002	1.428	.8192	55° 00′	.9599
138	10	760	046	419	175	54° 50′	570
167	20	783	089	411	158	40	541
.6196	30	.5807	.7133	1.402	.8141	30	.9512
225	40	831	177	393	124	20	483
254	50	854	221	385	107	10	454
.6283	36° 00′	.5878	.7265	1.376	.8090	54° 00′	.9425
		Cos	Cot	Tan	Sin	Degrees	Radians

TABLE 1

Radians	Degrees	Sin	Tan	Cot	Cos		
.6283	36° 00′	.5878	.7265	1.376	.8090	54° 00′	.9425
312	10	901	310	368	073	53° 50′	396
341	20	925	355	360	056	40	367
.6370	30	.5948	.7400	1.351	.8039	30	.9338
400	40	972	445	343	021	20	308
429	50	995	490	335	004	10	279
.6458	37° 00′	.6018	.7536	1.327	.7986	53° 00′	.9250
487	10	041	581	319	969	52° 50′	221
516	20	065	627	311	951	40	192
.6545	30	.6088	.7673	1.303	.7934	30	.9163
574	40	111	720	295	916	20	134
603	50	134	766	288	898	10	105
.6632	38° 00′	.6157	.7813	1.280	.7880	52° 00′	.9076
661	10	180	860	272	862	51° 50′	047
690	20	202	907	265	844	40	.9018
.6720	30	.6225	.7954	1.257	.7826	30	.8988
749	40	248	.8002	250	808	20	959
778	50	271	050	242	790	10	930
.6807	39° 00′	.6293	.8098	1.235	.7771	51° 00′	.8901
836	10	316	146	228	753	50° 50′	872
865	20	338	195	220	735	40	843
.6894	30	.6361	.8243	1.213	.7716	30	.8814
923	40	383	292	206	698	20	785
952	50	406	342	199	679	10	756
.6981	40° 00′	.6428	.8391	1.192	.7660	50° 00′	.8727
.7010	10	450	441	185	642	49° 50′	698
039	20	472	491	178	623	40	668
.7069	30	.6494	.8541	1.171	.7604	30	.8639
098	40	517	591	164	585	20	610
127	50	539	642	157	566	10	581
.7156	41° 00′	.6561	.8693	1.150	.7547	49° 00′	.8552
185	10	583	744	144	528	48° 50′	523
214	20	604	796	137	509	40	494
.7243	30	.6626	.8847	1.130	.7490	30	.8465
272	40	648	899	124	470	20	436
301	50	670	952	117	451	10	407
.7330	42° 00′	.6691	.9004	1.111	.7431	48° 00′	.8378
359	10	713	057	104	412	47° 50′	348
389	20	734	110	098	392	40	319
.7418	30	.6756	.9163	1.091	.7373	30	.8290
447	40	777	217	085	353	20	261
476	50	799	271	079	333	10	232
.7505	43° 00′	.6820	.9325	1.072	.7314	47° 00′	.8203
534	10	841	380	066	294	46° 50′	174
563	20	862	435	060	274	40	145
.7592	30	.6884	.9490	1.054	.7254	30	.8116
621	40	905	545	048	234	20	087
650	50	926	601	042	214	10	058
.7679	44° 00′	.6947	.9657	1.036	.7193	46° 00′	.8029
709	10	967	713	030	173	45° 50′	999
738	20	988	770	024	153	40	970
.7767	30	.7009	.9827	1.018	.7133	30	.7941
796	40	030	884	012	112	20	912
825	50	050	942	006	092	10	883
.7854	45° 00′	.7071	1.000	1.000	.7071	45° 00′	.7854
		Cos	Cot	Tan	Sin	Degrees	Radians

TABLE 2

MANTISSAS OF COMMON LOGARITHMS OF NUMBERS
TO FIVE DECIMAL PLACES

100-150

N.	0	1	2	3	4	5	6	7	8	9
100	00 000	043	087	130	173	217	260	303	346	389
101	432	475	518	561	604	647	689	732	775	817
102	860	903	945	988	*030	*072	*115	*157	*199	*242
103	01 284	326	368	410	452	494	536	578	620	662
104	703	745	787	828	870	912	953	995	*036	*078
105	02 119	160	202	243	284	325	366	407	449	490
106	531	572	612	653	694	735	776	816	857	898
107	938	979	*019	*060	*100	*141	*181	*222	*262	*302
108	03 342	383	423	463	503	543	583	623	663	703
109	743	782	822	862	902	941	981	*021	*060	*100
110	04 139	179	218	258	297	336	376	415	454	493
111	532	571	610	650	689	727	766	805	844	883
112	922	961	999	*038	*077	*115	*154	*192	*231	*269
113	05 308	346	385	423	461	500	538	576	614	652
114	690	729	767	805	843	881	918	956	994	*032
115	06 070	108	145	183	221	258	296	333	371	408
116	446	483	521	558	595	633	670	707	744	781
117	819	856	893	930	967	*004	*041	*078	*115	*151
118	07 188	225	262	298	335	372	408	445	482	518
119	555	591	628	664	700	737	773	809	846	882
120	918	954	990	*027	*063	*099	*135	*171	*207	*243
121	08 279	314	350	386	422	458	493	529	565	600
122	636	672	707	743	778	814	849	884	920	955
123	991	*026	*061	*096	*132	*167	*202	*237	*272	*307
124	09 342	377	412	447	482	517	552	587	621	656
125	691	726	760	795	830	864	899	934	968	*003
126	10 037	072	106	140	175	209	243	278	312	346
127	380	415	449	483	517	551	585	619	653	687
128	721	755	789	823	857	890	924	958	992	*025
129	11 059	093	126	160	193	227	261	294	327	361
130	394	428	461	494	528	561	594	628	661	694
131	727	760	793	826	860	893	926	959	992	*024
132	12 057	090	123	156	189	222	254	287	320	352
133	385	418	450	483	516	548	581	613	646	678
134	710	743	775	808	840	872	905	937	969	*001
135	13 033	066	098	130	162	194	226	258	290	322
136	354	386	418	450	481	513	545	577	609	640
137	672	704	735	767	799	830	862	893	925	956
138	988	*019	*051	*082	*114	*145	*176	*208	*239	*270
139	14 301	333	364	395	426	457	489	520	551	582
140	613	644	675	706	737	768	799	829	860	891
141	922	953	983	*014	*045	*076	*106	*137	*168	*198
142	15 229	259	290	320	351	381	412	442	473	503
143	534	564	594	625	655	685	715	746	776	806
144	836	866	897	927	957	987	*017	*047	*077	*107
145	16 137	167	197	227	256	286	316	346	376	406
146	435	465	495	524	554	584	613	643	673	702
147	732	761	791	820	850	879	909	938	967	997
148	17 026	056	085	114	143	173	202	231	260	289
149	319	348	377	406	435	464	493	522	551	580
150	609	638	667	696	725	754	782	811	840	869
N.	0	1	2	3	4	5	6	7	8	9

Prop. Pts.

	44	43	42
1	4.4	4.3	4.2
2	8.8	8.6	8.4
3	13.2	12.9	12.6
4	17.6	17.2	16.8
5	22.0	21.5	21.0
6	26.4	25.8	25.2
7	30.8	30.1	29.4
8	35.2	34.4	33.6
9	39.6	38.7	37.8

	41	40	39
1	4.1	4.0	3.9
2	8.2	8.0	7.8
3	12.3	12.0	11.7
4	16.4	16.0	15.6
5	20.5	20.0	19.5
6	24.6	24.0	23.4
7	28.7	28.0	27.3
8	32.8	32.0	31.2
9	36.9	36.0	35.1

	38	37	36
1	3.8	3.7	3.6
2	7.6	7.4	7.2
3	11.4	11.1	10.8
4	15.2	14.8	14.4
5	19.0	18.5	18.0
6	22.8	22.2	21.6
7	26.6	25.9	25.2
8	30.4	29.6	28.8
9	34.2	33.3	32.4

	35	34	33
1	3.5	3.4	3.3
2	7.0	6.8	6.6
3	10.5	10.2	9.9
4	14.0	13.6	13.2
5	17.5	17.0	16.5
6	21.0	20.4	19.8
7	24.5	23.8	23.1
8	28.0	27.2	26.4
9	31.5	30.6	29.7

	32	31	30
1	3.2	3.1	3.0
2	6.4	6.2	6.0
3	9.6	9.3	9.0
4	12.8	12.4	12.0
5	16.0	15.5	15.0
6	19.2	18.6	18.0
7	22.4	21.7	21.0
8	25.6	24.8	24.0
9	28.8	27.9	27.0

150-200

N.	0	1	2	3	4	5	6	7	8	9
150	17 609	638	667	696	725	754	782	811	840	869
151	898	926	955	984	*013	*041	*070	*099	*127	*156
152	18 184	213	241	270	298	327	355	384	412	441
153	469	498	526	554	583	611	639	667	696	724
154	752	780	808	837	865	893	921	949	977	*005
155	19 033	061	089	117	145	173	201	229	257	285
156	312	340	368	396	424	451	479	507	535	562
157	590	618	645	673	700	728	756	783	811	838
158	866	893	921	948	976	*003	*030	*058	*085	*112
159	20 140	167	194	222	249	276	303	330	358	385
160	412	439	466	493	520	548	575	602	629	656
161	683	710	737	763	790	817	844	871	898	925
162	952	978	*005	*032	*059	*085	*112	*139	*165	*192
163	21 219	245	272	299	325	352	378	405	431	458
164	484	511	537	564	590	617	643	669	696	722
165	748	775	801	827	854	880	906	932	958	985
166	22 011	037	063	089	115	141	167	194	220	246
167	272	298	324	350	376	401	427	453	479	505
168	531	557	583	608	634	660	686	712	737	763
169	789	814	840	866	891	917	943	968	994	*019
170	23 045	070	096	121	147	172	198	223	249	274
171	300	325	350	376	401	426	452	477	502	528
172	553	578	603	629	654	679	704	729	754	779
173	805	830	855	880	905	930	955	980	*005	*030
174	24 055	080	105	130	155	180	204	229	254	279
175	304	329	353	378	403	428	452	477	502	527
176	551	576	601	625	650	674	699	724	748	773
177	797	822	846	871	895	920	944	969	993	*018
178	25 042	066	091	115	139	164	188	212	237	261
179	285	310	334	358	382	406	431	455	479	503
180	527	551	575	600	624	648	672	696	720	744
181	768	792	816	840	864	888	912	935	959	983
182	26 007	031	055	079	102	126	150	174	198	221
183	245	269	293	316	340	364	387	411	435	458
184	482	505	529	553	576	600	623	647	670	694
185	717	741	764	788	811	834	858	881	905	928
186	951	975	998	*021	*045	*068	*091	*114	*138	*161
187	27 184	207	231	254	277	300	323	346	370	393
188	416	439	462	485	508	531	554	577	600	623
189	646	669	692	715	738	761	784	807	830	852
190	875	898	921	944	967	989	*012	*035	*058	*081
191	28 103	126	149	171	194	217	240	262	285	307
192	330	353	375	398	421	443	466	488	511	533
193	556	578	601	623	646	668	691	713	735	758
194	780	803	825	847	870	892	914	937	959	981
195	29 003	026	048	070	092	115	137	159	181	203
196	226	248	270	292	314	336	358	380	403	425
197	447	469	491	513	535	557	579	601	623	645
198	667	688	710	732	754	776	798	820	842	863
199	885	907	929	951	973	994	*016	*038	*060	*081
200	30 103	125	146	168	190	211	233	255	276	298

Prop. Pts.

	29	28
1	2.9	2.8
2	5.8	5.6
3	8.7	8.4
4	11.6	11.2
5	14.5	14.0
6	17.4	16.8
7	20.3	19.6
8	23.2	22.4
9	26.1	25.2

	27	26
1	2.7	2.6
2	5.4	5.2
3	8.1	7.8
4	10.8	10.4
5	13.5	13.0
6	16.2	15.6
7	18.9	18.2
8	21.6	20.8
9	24.3	23.4

	25
1	2.5
2	5.0
3	7.5
4	10.0
5	12.5
6	15.0
7	17.5
8	20.0
9	22.5

	24	23
1	2.4	2.3
2	4.8	4.6
3	7.2	6.9
4	9.6	9.2
5	12.0	11.5
6	14.4	13.8
7	16.8	16.1
8	19.2	18.4
9	21.6	20.7

	22	21
1	2.2	2.1
2	4.4	4.2
3	6.6	6.3
4	8.8	8.4
5	11.0	10.5
6	13.2	12.6
7	15.4	14.7
8	17.6	16.8
9	19.8	18.9

TABLE

2

200-250

N.	0	1	2	3	4	5	6	7	8	9
200	30 103	125	146	168	190	211	233	255	276	298
201	320	341	363	384	406	428	449	471	492	514
202	535	557	578	600	621	643	664	685	707	728
203	750	771	792	814	835	856	878	899	920	942
204	963	984	*006	*027	*048	*069	*091	*112	*133	*154
205	31 175	197	218	239	260	281	302	323	345	366
206	387	408	429	450	471	492	513	534	555	576
207	597	618	639	660	681	702	723	744	765	785
208	806	827	848	869	890	911	931	952	973	994
209	32 015	035	056	077	098	118	139	160	181	201
210	222	243	263	284	305	325	346	366	387	408
211	428	449	469	490	510	531	552	572	593	613
212	634	654	675	695	715	736	756	777	797	818
213	838	858	879	899	919	940	960	980	*001	*021
214	33 041	062	082	102	122	143	163	183	203	224
215	244	264	284	304	325	345	365	385	405	425
216	445	465	486	506	526	546	566	586	606	626
217	646	666	686	706	726	746	766	786	806	826
218	846	866	885	905	925	945	965	985	*005	*025
219	34 044	064	084	104	124	143	163	183	203	223
220	242	262	282	301	321	341	361	380	400	420
221	439	459	479	498	518	537	557	577	596	616
222	635	655	674	694	713	733	753	772	792	811
223	830	850	869	889	908	928	947	967	986	*005
224	35 025	044	064	083	102	122	141	160	180	199
225	218	238	257	276	295	315	334	353	372	392
226	411	430	449	468	488	507	526	545	564	583
227	603	622	641	660	679	698	717	736	755	774
228	793	813	832	851	870	889	908	927	946	965
229	984	*003	*021	*040	*059	*078	*097	*116	*135	*154
230	36 173	192	211	229	248	267	286	305	324	342
231	361	380	399	418	436	455	474	493	511	530
232	549	568	586	605	624	642	661	680	698	717
233	736	754	773	791	810	829	847	866	884	903
234	922	940	959	977	996	*014	*033	*051	*070	*088
235	37 107	125	144	162	181	199	218	236	254	273
236	291	310	328	346	365	383	401	420	438	457
237	475	493	511	530	548	566	585	603	621	639
238	658	676	694	712	731	749	767	785	803	822
239	840	858	876	894	912	931	949	967	985	*003
240	38 021	039	057	075	093	112	130	148	166	184
241	202	220	238	256	274	292	310	328	346	364
242	382	399	417	435	453	471	489	507	525	543
243	561	578	596	614	632	650	668	686	703	721
244	739	757	775	792	810	828	846	863	881	899
245	917	934	952	970	987	*005	*023	*041	*058	*076
246	39 094	111	129	146	164	182	199	217	235	252
247	270	287	305	322	340	358	375	393	410	428
248	445	463	480	498	515	533	550	568	585	602
249	620	637	655	672	690	707	724	742	759	777
250	794	811	829	846	863	881	898	915	933	950
N.	0	1	2	3	4	5	6	7	8	9

Prop. Pts.

	22	21
1	2.2	2.1
2	4.4	4.2
3	6.6	6.3
4	8.8	8.4
5	11.0	10.5
6	13.2	12.6
7	15.4	14.7
8	17.6	16.8
9	19.8	18.9

	20
1	2.0
2	4.0
3	6.0
4	8.0
5	10.0
6	12.0
7	14.0
8	16.0
9	18.0

	19
1	1.9
2	3.8
3	5.7
4	7.6
5	9.5
6	11.4
7	13.3
8	15.2
9	17.1

	18
1	1.8
2	3.6
3	5.4
4	7.2
5	9.0
6	10.8
7	12.6
8	14.4
9	16.2

	17
1	1.7
2	3.4
3	5.1
4	6.8
5	8.5
6	10.2
7	11.9
8	13.6
9	15.3

TABLE 2

250-300

N.	0	1	2	3	4	5	6	7	8	9
250	39 794	811	829	846	863	881	898	915	933	950
251	967	985	*002	*019	*037	*054	*071	*088	*106	*123
252	40 140	157	175	192	209	226	243	261	278	295
253	312	329	346	364	381	398	415	432	449	466
254	483	500	518	535	552	569	586	603	620	637
255	654	671	688	705	722	739	756	773	790	807
256	824	841	858	875	892	909	926	943	960	976
257	993	*010	*027	*044	*061	*078	*095	*111	*128	*145
258	41 162	179	196	212	229	246	263	280	296	313
259	330	347	363	380	397	414	430	447	464	481
260	497	514	531	547	564	581	597	614	631	647
261	664	681	697	714	731	747	764	780	797	814
262	830	847	863	880	896	913	929	946	963	979
263	996	*012	*029	*045	*062	*078	*095	*111	*127	*144
264	42 160	177	193	210	226	243	259	275	292	308
265	325	341	357	374	390	406	423	439	455	472
266	488	504	521	537	553	570	586	602	619	635
267	651	667	684	700	716	732	749	765	781	797
268	813	830	846	862	878	894	911	927	943	959
269	975	991	*008	*024	*040	*056	*072	*088	*104	*120
270	43 136	152	169	185	201	217	233	249	265	281
271	297	313	329	345	361	377	393	409	425	441
272	457	473	489	505	521	537	553	569	584	600
273	616	632	648	664	680	696	712	727	743	759
274	775	791	807	823	838	854	870	886	902	917
275	933	949	965	981	996	*012	*028	*044	*059	*075
276	44 091	107	122	138	154	170	185	201	217	232
277	248	264	279	295	311	326	342	358	373	389
278	404	420	436	451	467	483	498	514	529	545
279	560	576	592	607	623	638	654	669	685	700
280	716	731	747	762	778	793	809	824	840	855
281	871	886	902	917	932	948	963	979	994	*010
282	45 025	040	056	071	086	102	117	133	148	163
283	179	194	209	225	240	255	271	286	301	317
284	332	347	362	378	393	408	423	439	454	469
285	484	500	515	530	545	561	576	591	606	621
286	637	652	667	682	697	712	728	743	758	773
287	788	803	818	834	849	864	879	894	909	924
288	939	954	969	984	*000	*015	*030	*045	*060	*075
289	46 090	105	120	135	150	165	180	195	210	225
290	240	255	270	285	300	315	330	345	359	374
291	389	404	419	434	449	464	479	494	509	523
292	538	553	568	583	598	613	627	642	657	672
293	687	702	716	731	746	761	776	790	805	820
294	835	850	864	879	894	909	923	938	953	967
295	982	997	*012	*026	*041	*056	*070	*085	*100	*114
296	47 129	144	159	173	188	202	217	232	246	261
297	276	290	305	319	334	349	363	378	392	407
298	422	436	451	465	480	494	509	524	538	553
299	567	582	596	611	625	640	654	669	683	698
300	712	727	741	756	770	784	799	813	828	842
N.	0	1	2	3	4	5	6	7	8	9

Prop. Pts.

18
1 1.8
2 3.6
3 5.4
4 7.2
5 9.0
6 10.8
7 12.6
8 14.4
9 16.2

17
1 1.7
2 3.4
3 5.1
4 6.8
5 8.5
6 10.2
7 11.9
8 13.6
9 15.3

$\log e = 0.43429$

16
1 1.6
2 3.2
3 4.8
4 6.4
5 8.0
6 9.6
7 11.2
8 12.8
9 14.4

15
1 1.5
2 3.0
3 4.5
4 6.0
5 7.5
6 9.0
7 10.5
8 12.0
9 13.5

14
1 1.4
2 2.8
3 4.2
4 5.6
5 7.0
6 8.4
7 9.8
8 11.2
9 12.6

TABLE
2

300-350

TABLE **2**

N.	0	1	2	3	4	5	6	7	8	9
300	47 712	727	741	756	770	784	799	813	828	842
301	857	871	885	900	914	929	943	958	972	986
302	48 001	015	029	044	058	073	087	101	116	130
303	144	159	173	187	202	216	230	244	259	273
304	287	302	316	330	344	359	373	387	401	416
305	430	444	458	473	487	501	515	530	544	558
306	572	586	601	615	629	643	657	671	686	700
307	714	728	742	756	770	785	799	813	827	841
308	855	869	883	897	911	926	940	954	968	982
309	996	*010	*024	*038	*052	*066	*080	*094	*108	*122
310	49 136	150	164	178	192	206	220	234	248	262
311	276	290	304	318	332	346	360	374	388	402
312	415	429	443	457	471	485	499	513	527	541
313	554	568	582	596	610	624	638	651	665	679
314	693	707	721	734	748	762	776	790	803	817
315	831	845	859	872	886	900	914	927	941	955
316	969	982	996	*010	*024	*037	*051	*065	*079	*092
317	50 106	120	133	147	161	174	188	202	215	229
318	243	256	270	284	297	311	325	338	352	365
319	379	393	406	420	433	447	461	474	488	501
320	515	529	542	556	569	583	596	610	623	637
321	651	664	678	691	705	718	732	745	759	772
322	786	799	813	826	840	853	866	880	893	907
323	920	934	947	961	974	987	*001	*014	*028	*041
324	51 055	068	081	095	108	121	135	148	162	175
325	188	202	215	228	242	255	268	282	295	308
326	322	335	348	362	375	388	402	415	428	441
327	455	468	481	495	508	521	534	548	561	574
328	587	601	614	627	640	654	667	680	693	706
329	720	733	746	759	772	786	799	812	825	838
330	851	865	878	891	904	917	930	943	957	970
331	983	996	*009	*022	*035	*048	*061	*075	*088	*101
332	52 114	127	140	153	166	179	192	205	218	231
333	244	257	270	284	297	310	323	336	349	362
334	375	388	401	414	427	440	453	466	479	492
335	504	517	530	543	556	569	582	595	608	621
336	634	647	660	673	686	699	711	724	737	750
337	763	776	789	802	815	827	840	853	866	879
338	892	905	917	930	943	956	969	982	994	*007
339	53 020	033	046	058	071	084	097	110	122	135
340	148	161	173	186	199	212	224	237	250	263
341	275	288	301	314	326	339	352	364	377	390
342	403	415	428	441	453	466	479	491	504	517
343	529	542	555	567	580	593	605	618	631	643
344	656	668	681	694	706	719	732	744	757	769
345	782	794	807	820	832	845	857	870	882	895
346	908	920	933	945	958	970	983	995	*008	*020
347	54 033	045	058	070	083	095	108	120	133	145
348	158	170	183	195	208	220	233	245	258	270
349	283	295	307	320	332	345	357	370	382	394
350	407	419	432	444	456	469	481	494	506	518
N.	**0**	**1**	**2**	**3**	**4**	**5**	**6**	**7**	**8**	**9**

Prop. Pts.

15	
1	1.5
2	3.0
3	4.5
4	6.0
5	7.5
6	9.0
7	10.5
8	12.0
9	13.5

$\log \pi = 0.49715$

14	
1	1.4
2	2.8
3	4.2
4	5.6
5	7.0
6	8.4
7	9.8
8	11.2
9	12.6

13	
1	1.3
2	2.6
3	3.9
4	5.2
5	6.5
6	7.8
7	9.1
8	10.4
9	11.7

12	
1	1.2
2	2.4
3	3.6
4	4.8
5	6.0
6	7.2
7	8.4
8	9.6
9	10.8

350–400

N.	0	1	2	3	4	5	6	7	8	9	Prop. Pts.
350	54 407	419	432	444	456	469	481	494	506	518	
351	531	543	555	568	580	593	605	617	630	642	
352	654	667	679	691	704	716	728	741	753	765	
353	777	790	802	814	827	839	851	864	876	888	
354	900	913	925	937	949	962	974	986	998	*011	
355	55 023	035	047	060	072	084	096	108	121	133	
356	145	157	169	182	194	206	218	230	242	255	
357	267	279	291	303	315	328	340	352	364	376	
358	388	400	413	425	437	449	461	473	485	497	
359	509	522	534	546	558	570	582	594	606	618	
360	630	642	654	666	678	691	703	715	727	739	
361	751	763	775	787	799	811	823	835	847	859	
362	871	883	895	907	919	931	943	955	967	979	
363	991	*003	*015	*027	*038	*050	*062	*074	*086	*098	
364	56 110	122	134	146	158	170	182	194	205	217	
365	229	241	253	265	277	289	301	312	324	336	
366	348	360	372	384	396	407	419	431	443	455	
367	467	478	490	502	514	526	538	549	561	573	
368	585	597	608	620	632	644	656	667	679	691	
369	703	714	726	738	750	761	773	785	797	808	
370	820	832	844	855	867	879	891	902	914	926	
371	937	949	961	972	984	996	*008	*019	*031	*043	
372	57 054	066	078	089	101	113	124	136	148	159	
373	171	183	194	206	217	229	241	252	264	276	
374	287	299	310	322	334	345	357	368	380	392	
375	403	415	426	438	449	461	473	484	496	507	
376	519	530	542	553	565	576	588	600	611	623	
377	634	646	657	669	680	692	703	715	726	738	
378	749	761	772	784	795	807	818	830	841	852	
379	864	875	887	898	910	921	933	944	955	967	
380	978	990	*001	*013	*024	*035	*047	*058	*070	*081	
381	58 092	104	115	127	138	149	161	172	184	195	
382	206	218	229	240	252	263	274	286	297	309	
383	320	331	343	354	365	377	388	399	410	422	
384	433	444	456	467	478	490	501	512	524	535	
385	546	557	569	580	591	602	614	625	636	647	
386	659	670	681	692	704	715	726	737	749	760	
387	771	782	794	805	816	827	838	850	861	872	
388	883	894	906	917	928	939	950	961	973	984	
389	995	*006	*017	*028	*040	*051	*062	*073	*084	*095	
390	59 106	118	129	140	151	162	173	184	195	207	
391	218	229	240	251	262	273	284	295	306	318	
392	329	340	351	362	373	384	395	406	417	428	
393	439	450	461	472	483	494	506	517	528	539	
394	550	561	572	583	594	605	616	627	638	649	
395	660	671	682	693	704	715	726	737	748	759	
396	770	780	791	802	813	824	835	846	857	868	
397	879	890	901	912	923	934	945	956	966	977	
398	988	999	*010	*021	*032	*043	*054	*065	*076	*086	
399	60 097	108	119	130	141	152	163	173	184	195	
400	206	217	228	239	249	260	271	282	293	304	
N.	0	1	2	3	4	5	6	7	8	9	Prop. Pts.

Prop. Pts.

13
1	1.3
2	2.6
3	3.9
4	5.2
5	6.5
6	7.8
7	9.1
8	10.4
9	11.7

12
1	1.2
2	2.4
3	3.6
4	4.8
5	6.0
6	7.2
7	8.4
8	9.6
9	10.8

11
1	1.1
2	2.2
3	3.3
4	4.4
5	5.5
6	6.6
7	7.7
8	8.8
9	9.9

10
1	1.0
2	2.0
3	3.0
4	4.0
5	5.0
6	6.0
7	7.0
8	8.0
9	9.0

TABLE

2

400–450

N.	0	1	2	3	4	5	6	7	8	9	Prop. Pts.
400	60 206	217	228	239	249	260	271	282	293	304	
401	314	325	336	347	358	369	379	390	401	412	
402	423	433	444	455	466	477	487	498	509	520	
403	531	541	552	563	574	584	595	606	617	627	
404	638	649	660	670	681	692	703	713	724	735	
405	746	756	767	778	788	799	810	821	831	842	
406	853	863	874	885	895	906	917	927	938	949	
407	959	970	981	991	*002	*013	*023	*034	*045	*055	
408	61 066	077	087	098	109	119	130	140	151	162	
409	172	183	194	204	215	225	236	247	257	268	
410	278	289	300	310	321	331	342	352	363	374	
411	384	395	405	416	426	437	448	458	469	479	
412	490	500	511	521	532	542	553	563	574	584	
413	595	606	616	627	637	648	658	669	679	690	
414	700	711	721	731	742	752	763	773	784	794	
415	805	815	826	836	847	857	868	878	888	899	
416	909	920	930	941	951	962	972	982	993	*003	
417	62 014	024	034	045	055	066	076	086	097	107	
418	118	128	138	149	159	170	180	190	201	211	
419	221	232	242	252	263	273	284	294	304	315	
420	325	335	346	356	366	377	387	397	408	418	
421	428	439	449	459	469	480	490	500	511	521	
422	531	542	552	562	572	583	593	603	613	624	
423	634	644	655	665	675	685	696	706	716	726	
424	737	747	757	767	778	788	798	808	818	829	
425	839	849	859	870	880	890	900	910	921	931	
426	941	951	961	972	982	992	*002	*012	*022	*033	
427	63 043	053	063	073	083	094	104	114	124	134	
428	144	155	165	175	185	195	205	215	225	236	
429	246	256	266	276	286	296	306	317	327	337	
430	347	357	367	377	387	397	407	417	428	438	
431	448	458	468	478	488	498	508	518	528	538	
432	548	558	568	579	589	599	609	619	629	639	
433	649	659	669	679	689	699	709	719	729	739	
434	749	759	769	779	789	799	809	819	829	839	
435	849	859	869	879	889	899	909	919	929	939	
436	949	959	969	979	988	998	*008	*018	*028	*038	
437	64 048	058	068	078	088	098	108	118	128	137	
438	147	157	167	177	187	197	207	217	227	237	
439	246	256	266	276	286	296	306	316	326	335	
440	345	355	365	375	385	395	404	414	424	434	
441	444	454	464	473	483	493	503	513	523	532	
442	542	552	562	572	582	591	601	611	621	631	
443	640	650	660	670	680	689	699	709	719	729	
444	738	748	758	768	777	787	797	807	816	826	
445	836	846	856	865	875	885	895	904	914	924	
446	933	943	953	963	972	982	992	*002	*011	*021	
447	65 031	040	050	060	070	079	089	099	108	118	
448	128	137	147	157	167	176	186	196	205	215	
449	225	234	244	254	263	273	283	292	302	312	
450	321	331	341	350	360	369	379	389	398	408	
N.	0	1	2	3	4	5	6	7	8	9	Prop. Pts.

Prop. Pts.

11
1 1.1
2 2.2
3 3.3
4 4.4
5 5.5
6 6.6
7 7.7
8 8.8
9 9.9

10
1 1.0
2 2.0
3 3.0
4 4.0
5 5.0
6 6.0
7 7.0
8 8.0
9 9.0

9
1 0.9
2 1.8
3 2.7
4 3.6
5 4.5
6 5.4
7 6.3
8 7.2
9 8.1

TABLE

2

450-500

N.	0	1	2	3	4	5	6	7	8	9
450	65 321	331	341	350	360	369	379	389	398	408
451	418	427	437	447	456	466	475	485	495	504
452	514	523	533	543	552	562	571	581	591	600
453	610	619	629	639	648	658	667	677	686	696
454	706	715	725	734	744	753	763	772	782	792
455	801	811	820	830	839	849	858	868	877	887
456	896	906	916	925	935	944	954	963	973	982
457	992	*001	*011	*020	*030	*039	*049	*058	*068	*077
458	66 087	096	106	115	124	134	143	153	162	172
459	181	191	200	210	219	229	238	247	257	266
460	276	285	295	304	314	323	332	342	351	361
461	370	380	389	398	408	417	427	436	445	455
462	464	474	483	492	502	511	521	530	539	549
463	558	567	577	586	596	605	614	624	633	642
464	652	661	671	680	689	699	708	717	727	736
465	745	755	764	773	783	792	801	811	820	829
466	839	848	857	867	876	885	894	904	913	922
467	932	941	950	960	969	978	987	997	*006	*015
468	67 025	034	043	052	062	071	080	089	099	108
469	117	127	136	145	154	164	173	182	191	201
470	210	219	228	237	247	256	265	274	284	293
471	302	311	321	330	339	348	357	367	376	385
472	394	403	413	422	431	440	449	459	468	477
473	486	495	504	514	523	532	541	550	560	569
474	578	587	596	605	614	624	633	642	651	660
475	669	679	688	697	706	715	724	733	742	752
476	761	770	779	788	797	806	815	825	834	843
477	852	861	870	879	888	897	906	916	925	934
478	943	952	961	970	979	988	997	*006	*015	*024
479	68 034	043	052	061	070	079	088	097	106	115
480	124	133	142	151	160	169	178	187	196	205
481	215	224	233	242	251	260	269	278	287	296
482	305	314	323	332	341	350	359	368	377	386
483	395	404	413	422	431	440	449	458	467	476
484	485	494	502	511	520	529	538	547	556	565
485	574	583	592	601	610	619	628	637	646	655
486	664	673	681	690	699	708	717	726	735	744
487	753	762	771	780	789	797	806	815	824	833
488	842	851	860	869	878	886	895	904	913	922
489	931	940	949	958	966	975	984	993	*002	*011
490	69 020	028	037	046	055	064	073	082	090	099
491	108	117	126	135	144	152	161	170	179	188
492	197	205	214	223	232	241	249	258	267	276
493	285	294	302	311	320	329	338	346	355	364
494	373	381	390	399	408	417	425	434	443	452
495	461	469	478	487	496	504	513	522	531	539
496	548	557	566	574	583	592	601	609	618	627
497	636	644	653	662	671	679	688	697	705	714
498	723	732	740	749	758	767	775	784	793	801
499	810	819	827	836	845	854	862	871	880	888
500	897	906	914	923	932	940	949	958	966	975

Prop. Pts.

	10
1	1.0
2	2.0
3	3.0
4	4.0
5	5.0
6	6.0
7	7.0
8	8.0
9	9.0

	9
1	0.9
2	1.8
3	2.7
4	3.6
5	4.5
6	5.4
7	6.3
8	7.2
9	8.1

	8
1	0.8
2	1.6
3	2.4
4	3.2
5	4.0
6	4.8
7	5.6
8	6.4
9	7.2

TABLE 2

500-550

TABLE 2

N.	0	1	2	3	4	5	6	7	8	9
500	69 897	906	914	923	932	940	949	958	966	975
501	984	992	*001	*010	*018	*027	*036	*044	*053	*062
502	70 070	079	088	096	105	114	122	131	140	148
503	157	165	174	183	191	200	209	217	226	234
504	243	252	260	269	278	286	295	303	312	321
505	329	338	346	355	364	372	381	389	398	406
506	415	424	432	441	449	458	467	475	484	492
507	501	509	518	526	535	544	552	561	569	578
508	586	595	603	612	621	629	638	646	655	663
509	672	680	689	697	706	714	723	731	740	749
510	757	766	774	783	791	800	808	817	825	834
511	842	851	859	868	876	885	893	902	910	919
512	927	935	944	952	961	969	978	986	995	*003
513	71 012	020	029	037	046	054	063	071	079	088
514	096	105	113	122	130	139	147	155	164	172
515	181	189	198	206	214	223	231	240	248	257
516	265	273	282	290	299	307	315	324	332	341
517	349	357	366	374	383	391	399	408	416	425
518	433	441	450	458	466	475	483	492	500	508
519	517	525	533	542	550	559	567	575	584	592
520	600	609	617	625	634	642	650	659	667	675
521	684	692	700	709	717	725	734	742	750	759
522	767	775	784	792	800	809	817	825	834	842
523	850	858	867	875	883	892	900	908	917	925
524	933	941	950	958	966	975	983	991	999	*008
525	72 016	024	032	041	049	057	066	074	082	090
526	099	107	115	123	132	140	148	156	165	173
527	181	189	198	206	214	222	230	239	247	255
528	263	272	280	288	296	304	313	321	329	337
529	346	354	362	370	378	387	395	403	411	419
530	428	436	444	452	460	469	477	485	493	501
531	509	518	526	534	542	550	558	567	575	583
532	591	599	607	616	624	632	640	648	656	665
533	673	681	689	697	705	713	722	730	738	746
534	754	762	770	779	787	795	803	811	819	827
535	835	843	852	860	868	876	884	892	900	908
536	916	925	933	941	949	957	965	973	981	989
537	997	*006	*014	*022	*030	*038	*046	*054	*062	*070
538	73 078	086	094	102	111	119	127	135	143	151
539	159	167	175	183	191	199	207	215	223	231
540	239	247	255	263	272	280	288	296	304	312
541	320	328	336	344	352	360	368	376	384	392
542	400	408	416	424	432	440	448	456	464	472
543	480	488	496	504	512	520	528	536	544	552
544	560	568	576	584	592	600	608	616	624	632
545	640	648	656	664	672	679	687	695	703	711
546	719	727	735	743	751	759	767	775	783	791
547	799	807	815	823	830	838	846	854	862	870
548	878	886	894	902	910	918	926	933	941	949
549	957	965	973	981	989	997	*005	*013	*020	*028
550	74 036	044	052	060	068	076	084	092	099	107
N.	0	1	2	3	4	5	6	7	8	9

Prop. Pts.

9
1	0.9
2	1.8
3	2.7
4	3.6
5	4.5
6	5.4
7	6.3
8	7.2
9	8.1

8
1	0.8
2	1.6
3	2.4
4	3.2
5	4.0
6	4.8
7	5.6
8	6.4
9	7.2

7
1	0.7
2	1.4
3	2.1
4	2.8
5	3.5
6	4.2
7	4.9
8	5.6
9	6.3

550-600

N.	0	1	2	3	4	5	6	7	8	9	Prop. Pts.
550	74 036	044	052	060	068	076	084	092	099	107	
551	115	123	131	139	147	155	162	170	178	186	
552	194	202	210	218	225	233	241	249	257	265	
553	273	280	288	296	304	312	320	327	335	343	
554	351	359	367	374	382	390	398	406	414	421	
555	429	437	445	453	461	468	476	484	492	500	
556	507	515	523	531	539	547	554	562	570	578	
557	586	593	601	609	617	624	632	640	648	656	
558	663	671	679	687	695	702	710	718	726	733	
559	741	749	757	764	772	780	788	796	803	811	
560	819	827	834	842	850	858	865	873	881	889	
561	896	904	912	920	927	935	943	950	958	966	
562	974	981	989	997	*005	*012	*020	*028	*035	*043	
563	75 051	059	066	074	082	089	097	105	113	120	
564	128	136	143	151	159	166	174	182	189	197	
565	205	213	220	228	236	243	251	259	266	274	
566	282	289	297	305	312	320	328	335	343	351	
567	358	366	374	381	389	397	404	412	420	427	
568	435	442	450	458	465	473	481	488	496	504	
569	511	519	526	534	542	549	557	565	572	580	
570	587	595	603	610	618	626	633	641	648	656	
571	664	671	679	686	694	702	709	717	724	732	
572	740	747	755	762	770	778	785	793	800	808	
573	815	823	831	838	846	853	861	868	876	884	
574	891	899	906	914	921	929	937	944	952	959	
575	967	974	982	989	997	*005	*012	*020	*027	*035	
576	76 042	050	057	065	072	080	087	095	103	110	
577	118	125	133	140	148	155	163	170	178	185	
578	193	200	208	215	223	230	238	245	253	260	
579	268	275	283	290	298	305	313	320	328	335	
580	343	350	358	365	373	380	388	395	403	410	
581	418	425	433	440	448	455	462	470	477	485	
582	492	500	507	515	522	530	537	545	552	559	
583	567	574	582	589	597	604	612	619	626	634	
584	641	649	656	664	671	678	686	693	701	708	
585	716	723	730	738	745	753	760	768	775	782	
586	790	797	805	812	819	827	834	842	849	856	
587	864	871	879	886	893	901	908	916	923	930	
588	938	945	953	960	967	975	982	989	997	*004	
589	77 012	019	026	034	041	048	056	063	070	078	
590	085	093	100	107	115	122	129	137	144	151	
591	159	166	173	181	188	195	203	210	217	225	
592	232	240	247	254	262	269	276	283	291	298	
593	305	313	320	327	335	342	349	357	364	371	
594	379	386	393	401	408	415	422	430	437	444	
595	452	459	466	474	481	488	495	503	510	517	
596	525	532	539	546	554	561	568	576	583	590	
597	597	605	612	619	627	634	641	648	656	663	
598	670	677	685	692	699	706	714	721	728	735	
599	743	750	757	764	772	779	786	793	801	808	
600	815	822	830	837	844	851	859	866	873	880	
N.	0	1	2	3	4	5	6	7	8	9	Prop. Pts.

Prop. Pts.

8
1 | 0.8
2 | 1.6
3 | 2.4
4 | 3.2
5 | 4.0
6 | 4.8
7 | 5.6
8 | 6.4
9 | 7.2

7
1 | 0.7
2 | 1.4
3 | 2.1
4 | 2.8
5 | 3.5
6 | 4.2
7 | 4.9
8 | 5.6
9 | 6.3

TABLE
2

600-650

N.	0	1	2	3	4	5	6	7	8	9	Prop. Pts.
600	77 815	822	830	837	844	851	859	866	873	880	
601	887	895	902	909	916	924	931	938	945	952	
602	960	967	974	981	988	996	*003	*010	*017	*025	
603	78 032	039	046	053	061	068	075	082	089	097	
604	104	111	118	125	132	140	147	154	161	168	
605	176	183	190	197	204	211	219	226	233	240	
606	247	254	262	269	276	283	290	297	305	312	
607	319	326	333	340	347	355	362	369	376	383	
608	390	398	405	412	419	426	433	440	447	455	
609	462	469	476	483	490	497	504	512	519	526	
610	533	540	547	554	561	569	576	583	590	597	
611	604	611	618	625	633	640	647	654	661	668	
612	675	682	689	696	704	711	718	725	732	739	
613	746	753	760	767	774	781	789	796	803	810	
614	817	824	831	838	845	852	859	866	873	880	
615	888	895	902	909	916	923	930	937	944	951	
616	958	965	972	979	986	993	*000	*007	*014	*021	
617	79 029	036	043	050	057	064	071	078	085	092	
618	099	106	113	120	127	134	141	148	155	162	
619	169	176	183	190	197	204	211	218	225	232	
620	239	246	253	260	267	274	281	288	295	302	
621	309	316	323	330	337	344	351	358	365	372	
622	379	386	393	400	407	414	421	428	435	442	
623	449	456	463	470	477	484	491	498	505	511	
624	518	525	532	539	546	553	560	567	574	581	
625	588	595	602	609	616	623	630	637	644	650	
626	657	664	671	678	685	692	699	706	713	720	
627	727	734	741	748	754	761	768	775	782	789	
628	796	803	810	817	824	831	837	844	851	858	
629	865	872	879	886	893	900	906	913	920	927	
630	934	941	948	955	962	969	975	982	989	996	
631	80 003	010	017	024	030	037	044	051	058	065	
632	072	079	085	092	099	106	113	120	127	134	
633	140	147	154	161	168	175	182	188	195	202	
634	209	216	223	229	236	243	250	257	264	271	
635	277	284	291	298	305	312	318	325	332	339	
636	346	353	359	366	373	380	387	393	400	407	
637	414	421	428	434	441	448	455	462	468	475	
638	482	489	496	502	509	516	523	530	536	543	
639	550	557	564	570	577	584	591	598	604	611	
640	618	625	632	638	645	652	659	665	672	679	
641	686	693	699	706	713	720	726	733	740	747	
642	754	760	767	774	781	787	794	801	808	814	
643	821	828	835	841	848	855	862	868	875	882	
644	889	895	902	909	916	922	929	936	943	949	
645	956	963	969	976	983	990	996	*003	*010	*017	
646	81 023	030	037	043	050	057	064	070	077	084	
647	090	097	104	111	117	124	131	137	144	151	
648	158	164	171	178	184	191	198	204	211	218	
649	224	231	238	245	251	258	265	271	278	285	
650	291	298	305	311	318	325	331	338	345	351	
N.	0	1	2	3	4	5	6	7	8	9	Prop. Pts.

TABLE 2

Prop. Pts.

8
1 | 0.8
2 | 1.6
3 | 2.4
4 | 3.2
5 | 4.0
6 | 4.8
7 | 5.6
8 | 6.4
9 | 7.2

7
1 | 0.7
2 | 1.4
3 | 2.1
4 | 2.8
5 | 3.5
6 | 4.2
7 | 4.9
8 | 5.6
9 | 6.3

6
1 | 0.6
2 | 1.2
3 | 1.8
4 | 2.4
5 | 3.0
6 | 3.6
7 | 4.2
8 | 4.8
9 | 5.4

650-700

N.	0	1	2	3	4	5	6	7	8	9
650	81 291	298	305	311	318	325	331	338	345	351
651	358	365	371	378	385	391	398	405	411	418
652	425	431	438	445	451	458	465	471	478	485
653	491	498	505	511	518	525	531	538	544	551
654	558	564	571	578	584	591	598	604	611	617
655	624	631	637	644	651	657	664	671	677	684
656	690	697	704	710	717	723	730	737	743	750
657	757	763	770	776	783	790	796	803	809	816
658	823	829	836	842	849	856	862	869	875	882
659	889	895	902	908	915	921	928	935	941	948
660	954	961	968	974	981	987	994	*000	*007	*014
661	82 020	027	033	040	046	053	060	066	073	079
662	086	092	099	105	112	119	125	132	138	145
663	151	158	164	171	178	184	191	197	204	210
664	217	223	230	236	243	249	256	263	269	276
665	282	289	295	302	308	315	321	328	334	341
666	347	354	360	367	373	380	387	393	400	406
667	413	419	426	432	439	445	452	458	465	471
668	478	484	491	497	504	510	517	523	530	536
669	543	549	556	562	569	575	582	588	595	601
670	607	614	620	627	633	640	646	653	659	666
671	672	679	685	692	698	705	711	718	724	730
672	737	743	750	756	763	769	776	782	789	795
673	802	808	814	821	827	834	840	847	853	860
674	866	872	879	885	892	898	905	911	918	924
675	930	937	943	950	956	963	969	975	982	988
676	995	*001	*008	*014	*020	*027	*033	*040	*046	*052
677	83 059	065	072	078	085	091	097	104	110	117
678	123	129	136	142	149	155	161	168	174	181
679	187	193	200	206	213	219	225	232	238	245
680	251	257	264	270	276	283	289	296	302	308
681	315	321	327	334	340	347	353	359	366	372
682	378	385	391	398	404	410	417	423	429	436
683	442	448	455	461	467	474	480	487	493	499
684	506	512	518	525	531	537	544	550	556	563
685	569	575	582	588	594	601	607	613	620	626
686	632	639	645	651	658	664	670	677	683	689
687	696	702	708	715	721	727	734	740	746	753
688	759	765	771	778	784	790	797	803	809	816
689	822	828	835	841	847	853	860	866	872	879
690	885	891	897	904	910	916	923	929	935	942
691	948	954	960	967	973	979	985	992	998	*004
692	84 011	017	023	029	036	042	048	055	061	067
693	073	080	086	092	098	105	111	117	123	130
694	136	142	148	155	161	167	173	180	186	192
695	198	205	211	217	223	230	236	242	248	255
696	261	267	273	280	286	292	298	305	311	317
697	323	330	336	342	348	354	361	367	373	379
698	386	392	398	404	410	417	423	429	435	442
699	448	454	460	466	473	479	485	491	497	504
700	510	516	522	528	535	541	547	553	559	566
N.	0	1	2	3	4	5	6	7	8	9

Prop. Pts.

7	
1	0.7
2	1.4
3	2.1
4	2.8
5	3.5
6	4.2
7	4.9
8	5.6
9	6.3

6	
1	0.6
2	1.2
3	1.8
4	2.4
5	3.0
6	3.6
7	4.2
8	4.8
9	5.4

TABLE

2

700-750

N.	0	1	2	3	4	5	6	7	8	9
700	84 510	516	522	528	535	541	547	553	559	566
701	572	578	584	590	597	603	609	615	621	628
702	634	640	646	652	658	665	671	677	683	689
703	696	702	708	714	720	726	733	739	745	751
704	757	763	770	776	782	788	794	800	807	813
705	819	825	831	837	844	850	856	862	868	874
706	880	887	893	899	905	911	917	924	930	936
707	942	948	954	960	967	973	979	985	991	997
708	85 003	009	016	022	028	034	040	046	052	058
709	065	071	077	083	089	095	101	107	114	120
710	126	132	138	144	150	156	163	169	175	181
711	187	193	199	205	211	217	224	230	236	242
712	248	254	260	266	272	278	285	291	297	303
713	309	315	321	327	333	339	345	352	358	364
714	370	376	382	388	394	400	406	412	418	425
715	431	437	443	449	455	461	467	473	479	485
716	491	497	503	509	516	522	528	534	540	546
717	552	558	564	570	576	582	588	594	600	606
718	612	618	625	631	637	643	649	655	661	667
719	673	679	685	691	697	703	709	715	721	727
720	733	739	745	751	757	763	769	775	781	788
721	794	800	806	812	818	824	830	836	842	848
722	854	860	866	872	878	884	890	896	902	908
723	914	920	926	932	938	944	950	956	962	968
724	974	980	986	992	998	*004	*010	*016	*022	*028
725	86 034	040	046	052	058	064	070	076	082	088
726	094	100	106	112	118	124	130	136	141	147
727	153	159	165	171	177	183	189	195	201	207
728	213	219	225	231	237	243	249	255	261	267
729	273	279	285	291	297	303	308	314	320	326
730	332	338	344	350	356	362	368	374	380	386
731	392	398	404	410	415	421	427	433	439	445
732	451	457	463	469	475	481	487	493	499	504
733	510	516	522	528	534	540	546	552	558	564
734	570	576	581	587	593	599	605	611	617	623
735	629	635	641	646	652	658	664	670	676	682
736	688	694	700	705	711	717	723	729	735	741
737	747	753	759	764	770	776	782	788	794	800
738	806	812	817	823	829	835	841	847	853	859
739	864	870	876	882	888	894	900	906	911	917
740	923	929	935	941	947	953	958	964	970	976
741	982	988	994	999	*005	*011	*017	*023	*029	*035
742	87 040	046	052	058	064	070	075	081	087	093
743	099	105	111	116	122	128	134	140	146	151
744	157	163	169	175	181	186	192	198	204	210
745	216	221	227	233	239	245	251	256	262	268
746	274	280	286	291	297	303	309	315	320	326
747	332	338	344	349	355	361	367	373	379	384
748	390	396	402	408	413	419	425	431	437	442
749	448	454	460	466	471	477	483	489	495	500
750	506	512	518	523	529	535	541	547	552	558
N.	0	1	2	3	4	5	6	7	8	9

TABLE 2

Prop. Pts.

7
1 0.7
2 1.4
3 2.1
4 2.8
5 3.5
6 4.2
7 4.9
8 5.6
9 6.3

6
1 0.6
2 1.2
3 1.8
4 2.4
5 3.0
6 3.6
7 4.2
8 4.8
9 5.4

5
1 0.5
2 1.0
3 1.5
4 2.0
5 2.5
6 3.0
7 3.5
8 4.0
9 4.5

750-800

N.	0	1	2	3	4	5	6	7	8	9
750	87 506	512	518	523	529	535	541	547	552	558
751	564	570	576	581	587	593	599	604	610	616
752	622	628	633	639	645	651	656	662	668	674
753	679	685	691	697	703	708	714	720	726	731
754	737	743	749	754	760	766	772	777	783	789
755	795	800	806	812	818	823	829	835	841	846
756	852	858	864	869	875	881	887	892	898	904
757	910	915	921	927	933	938	944	950	955	961
758	967	973	978	984	990	996	*001	*007	*013	*018
759	88 024	030	036	041	047	053	058	064	070	076
760	081	087	093	098	104	110	116	121	127	133
761	138	144	150	156	161	167	173	178	184	190
762	195	201	207	213	218	224	230	235	241	247
763	252	258	264	270	275	281	287	292	298	304
764	309	315	321	326	332	338	343	349	355	360
765	366	372	377	383	389	395	400	406	412	417
766	423	429	434	440	446	451	457	463	468	474
767	480	485	491	497	502	508	513	519	525	530
768	536	542	547	553	559	564	570	576	581	587
769	593	598	604	610	615	621	627	632	638	643
770	649	655	660	666	672	677	683	689	694	700
771	705	711	717	722	728	734	739	745	750	756
772	762	767	773	779	784	790	795	801	807	812
773	818	824	829	835	840	846	852	857	863	868
774	874	880	885	891	897	902	908	913	919	925
775	930	936	941	947	953	958	964	969	975	981
776	986	992	997	*003	*009	*014	*020	*025	*031	*037
777	89 042	048	053	059	064	070	076	081	087	092
778	098	104	109	115	120	126	131	137	143	148
779	154	159	165	170	176	182	187	193	198	204
780	209	215	221	226	232	237	243	248	254	260
781	265	271	276	282	287	293	298	304	310	315
782	321	326	332	337	343	348	354	360	365	371
783	376	382	387	393	398	404	409	415	421	426
784	432	437	443	448	454	459	465	470	476	481
785	487	492	498	504	509	515	520	526	531	537
786	542	548	553	559	564	570	575	581	586	592
787	597	603	609	614	620	625	631	636	642	647
788	653	658	664	669	675	680	686	691	697	702
789	708	713	719	724	730	735	741	746	752	757
790	763	768	774	779	785	790	796	801	807	812
791	818	823	829	834	840	845	851	856	862	867
792	873	878	883	889	894	900	905	911	916	922
793	927	933	938	944	949	955	960	966	971	977
794	982	988	993	998	*004	*009	*015	*020	*026	*031
795	90 037	042	048	053	059	064	069	075	080	086
796	091	097	102	108	113	119	124	129	135	140
797	146	151	157	162	168	173	179	184	189	195
798	200	206	211	217	222	227	233	238	244	249
799	255	260	266	271	276	282	287	293	298	304
800	309	314	320	325	331	336	342	347	352	358
N.	0	1	2	3	4	5	6	7	8	9

Prop. Pts.

6
1 | 0.6
2 | 1.2
3 | 1.8
4 | 2.4
5 | 3.0
6 | 3.6
7 | 4.2
8 | 4.8
9 | 5.4

5
1 | 0.5
2 | 1.0
3 | 1.5
4 | 2.0
5 | 2.5
6 | 3.0
7 | 3.5
8 | 4.0
9 | 4.5

TABLE

2

800-850

N.	0	1	2	3	4	5	6	7	8	9	Prop. Pts.
800	90 309	314	320	325	331	336	342	347	352	358	
801	363	369	374	380	385	390	396	401	407	412	
802	417	423	428	434	439	445	450	455	461	466	
803	472	477	482	488	493	499	504	509	515	520	
804	526	531	536	542	547	553	558	563	569	574	
805	580	585	590	596	601	607	612	617	623	628	
806	634	639	644	650	655	660	666	671	677	682	
807	687	693	698	703	709	714	720	725	730	736	
808	741	747	752	757	763	768	773	779	784	789	
809	795	800	806	811	816	822	827	832	838	843	
810	849	854	859	865	870	875	881	886	891	897	
811	902	907	913	918	924	929	934	940	945	950	**6**
812	956	961	966	972	977	982	988	993	998	*004	1 0.6
813	91 009	014	020	025	030	036	041	046	052	057	2 1.2
814	062	068	073	078	084	089	094	100	105	110	3 1.8 4 2.4
815	116	121	126	132	137	142	148	153	158	164	5 3.0
816	169	174	180	185	190	196	201	206	212	217	6 3.6
817	222	228	233	238	243	249	254	259	265	270	7 4.2
818	275	281	286	291	297	302	307	312	318	323	8 4.8
819	328	334	339	344	350	355	360	365	371	376	9 5.4
820	381	387	392	397	403	408	413	418	424	429	
821	434	440	445	450	455	461	466	471	477	482	
822	487	492	498	503	508	514	519	524	529	535	
823	540	545	551	556	561	566	572	577	582	587	
824	593	598	603	609	614	619	624	630	635	640	
825	645	651	656	661	666	672	677	682	687	693	
826	698	703	709	714	719	724	730	735	740	745	
827	751	756	761	766	772	777	782	787	793	798	
828	803	808	814	819	824	829	834	840	845	850	
829	855	861	866	871	876	882	887	892	897	903	
830	908	913	918	924	929	934	939	944	950	955	
831	960	965	971	976	981	986	991	997	*002	*007	**5**
832	92 012	018	023	028	033	038	044	049	054	059	1 0.5
833	065	070	075	080	085	091	096	101	106	111	2 1.0 3 1.5
834	117	122	127	132	137	143	148	153	158	163	4 2.0
835	169	174	179	184	189	195	200	205	210	215	5 2.5
836	221	226	231	236	241	247	252	257	262	267	6 3.0
837	273	278	283	288	293	298	304	309	314	319	7 3.5
838	324	330	335	340	345	350	355	361	366	371	8 4.0
839	376	381	387	392	397	402	407	412	418	423	9 4.5
840	428	433	438	443	449	454	459	464	469	474	
841	480	485	490	495	500	505	511	516	521	526	
842	531	536	542	547	552	557	562	567	572	578	
843	583	588	593	598	603	609	614	619	624	629	
844	634	639	645	650	655	660	665	670	675	681	
845	686	691	696	701	706	711	716	722	727	732	
846	737	742	747	752	758	763	768	773	778	783	
847	788	793	799	804	809	814	819	824	829	834	
848	840	845	850	855	860	865	870	875	881	886	
849	891	896	901	906	911	916	921	927	932	937	
850	942	947	952	957	962	967	973	978	983	988	
N.	0	1	2	3	4	5	6	7	8	9	Prop. Pts.

TABLE 2

850-900

N.	0	1	2	3	4	5	6	7	8	9	Prop. Pts.
850	92 942	947	952	957	962	967	973	978	983	988	
851	993	998	*003	*008	*013	*018	*024	*029	*034	*039	
852	£3 044	049	054	059	064	069	075	080	085	090	
853	095	100	105	110	115	120	125	131	136	141	
854	146	151	156	161	166	171	176	181	186	192	
855	197	202	207	212	217	222	227	232	237	242	
856	247	252	258	263	268	273	278	283	288	293	
857	298	303	308	313	318	323	328	334	339	344	
858	349	354	359	364	369	374	379	384	389	394	
859	399	404	409	414	420	425	430	435	440	445	
860	450	455	460	465	470	475	480	485	490	495	
861	500	505	510	515	520	526	531	536	541	546	
862	551	556	561	566	571	576	581	586	591	596	
863	601	606	611	616	621	626	631	636	641	646	
864	651	656	661	666	671	676	682	687	692	697	
865	702	707	712	717	722	727	732	737	742	747	
866	752	757	762	767	772	777	782	787	792	797	
867	802	807	812	817	822	827	832	837	842	847	
868	852	857	862	867	872	877	882	887	892	897	
869	902	907	912	917	922	927	932	937	942	947	
870	952	957	962	967	972	977	982	987	992	997	
871	94 002	007	012	017	022	027	032	037	042	047	
872	052	057	062	067	072	077	082	086	091	096	
873	101	106	111	116	121	126	131	136	141	146	
874	151	156	161	166	171	176	181	186	191	196	
875	201	206	211	216	221	226	231	236	240	245	
876	250	255	260	265	270	275	280	285	290	295	
877	300	305	310	315	320	325	330	335	340	345	
878	349	354	359	364	369	374	379	384	389	394	
879	399	404	409	414	419	424	429	433	438	443	
880	448	453	458	463	468	473	478	483	488	493	
881	498	503	507	512	517	522	527	532	537	542	
882	547	552	557	562	567	571	576	581	586	591	
883	596	601	606	611	616	621	626	630	635	640	
884	645	650	655	660	665	670	675	680	685	689	
885	694	699	704	709	714	719	724	729	734	738	
886	743	748	753	758	763	768	773	778	783	787	
887	792	797	802	807	812	817	822	827	832	836	
888	841	846	851	856	861	866	871	876	880	885	
889	890	895	900	905	910	915	919	924	929	934	
890	939	944	949	954	959	963	968	973	978	983	
891	988	993	998	*002	*007	*012	*017	*022	*027	*032	
892	95 036	041	046	051	056	061	066	071	075	080	
893	085	090	095	100	105	109	114	119	124	129	
894	134	139	143	148	153	158	163	168	173	177	
895	182	187	192	197	202	207	211	216	221	226	
896	231	236	240	245	250	255	260	265	270	274	
897	279	284	289	294	299	303	308	313	318	323	
898	328	332	337	342	347	352	357	361	366	371	
899	376	381	386	390	395	400	405	410	415	419	
900	424	429	434	439	444	448	453	458	463	468	
N.	0	1	2	3	4	5	6	7	8	9	Prop. Pts.

Prop. Pts.

6	
1	0.6
2	1.2
3	1.8
4	2.4
5	3.0
6	3.6
7	4.2
8	4.8
9	5.4

5	
1	0.5
2	1.0
3	1.5
4	2.0
5	2.5
6	3.0
7	3.5
8	4.0
9	4.5

4	
1	0.4
2	0.8
3	1.2
4	1.6
5	2.0
6	2.4
7	2.8
8	3.2
9	3.6

TABLE

2

900-950

N.	0	1	2	3	4	5	6	7	8	9	Prop. Pts.
900	95 424	429	434	439	444	448	453	458	463	468	
901	472	477	482	487	492	497	501	506	511	516	
902	521	525	530	535	540	545	550	554	559	564	
903	569	574	578	583	588	593	598	602	607	612	
904	617	622	626	631	636	641	646	650	655	660	
905	665	670	674	679	684	689	694	698	703	708	
906	713	718	722	727	732	737	742	746	751	756	
907	761	766	770	775	780	785	789	794	799	804	
908	809	813	818	823	828	832	837	842	847	852	
909	856	861	866	871	875	880	885	890	895	899	
910	904	909	914	918	923	928	933	938	942	947	
911	952	957	961	966	971	976	980	985	990	995	**5**
912	999	*004	*009	*014	*019	*023	*028	*033	*038	*042	1 0.5
913	96 047	052	057	061	066	071	076	080	085	090	2 1.0
											3 1.5
914	095	099	104	109	114	118	123	128	133	137	4 2.0
915	142	147	152	156	161	166	171	175	180	185	5 2.5
916	190	194	199	204	209	213	218	223	227	232	6 3.0
											7 3.5
917	237	242	246	251	256	261	265	270	275	280	8 4.0
918	284	289	294	298	303	308	313	317	322	327	9 4.5
919	332	336	341	346	350	355	360	365	369	374	
920	379	384	388	393	398	402	407	412	417	421	
921	426	431	435	440	445	450	454	459	464	468	
922	473	478	483	487	492	497	501	506	511	515	
923	520	525	530	534	539	544	548	553	558	562	
924	567	572	577	581	586	591	595	600	605	609	
925	614	619	624	628	633	638	642	647	652	656	
926	661	666	670	675	680	685	689	694	699	703	
927	708	713	717	722	727	731	736	741	745	750	
928	755	759	764	769	774	778	783	788	792	797	
929	802	806	811	816	820	825	830	834	839	844	
930	848	853	858	862	867	872	876	881	886	890	
931	895	900	904	909	914	918	923	928	932	937	**4**
932	942	946	951	956	960	965	970	974	979	984	1 0.4
933	988	993	997	*002	*007	*011	*016	*021	*025	*030	2 0.8
											3 1.2
934	97 035	039	044	049	053	058	063	067	072	077	4 1.6
935	081	086	090	095	100	104	109	114	118	123	5 2.0
936	128	132	137	142	146	151	155	160	165	169	6 2.4
											7 2.8
937	174	179	183	188	192	197	202	206	211	216	8 3.2
938	220	225	230	234	239	243	248	253	257	262	9 3.6
939	267	271	276	280	285	290	294	299	304	308	
940	313	317	322	327	331	336	340	345	350	354	
941	359	364	368	373	377	382	387	391	396	400	
942	405	410	414	419	424	428	433	437	442	447	
943	451	456	460	465	470	474	479	483	488	493	
944	497	502	506	511	516	520	525	529	534	539	
945	543	548	552	557	562	566	571	575	580	585	
946	589	594	598	603	607	612	617	621	626	630	
947	635	640	644	649	653	658	663	667	672	676	
948	681	685	690	695	699	704	708	713	717	722	
949	727	731	736	740	745	749	754	759	763	768	
950	772	777	782	786	791	795	800	804	809	813	
N.	0	1	2	3	4	5	6	7	8	9	Prop. Pts.

TABLE

2

950-1000

N.	0	1	2	3	4	5	6	7	8	9	Prop. Pts.
950	97 772	777	782	786	791	795	800	804	809	813	
951	818	823	827	832	836	841	845	850	855	859	
952	864	868	873	877	882	886	891	896	900	905	
953	909	914	918	923	928	932	937	941	946	950	
954	955	959	964	968	973	978	982	987	991	996	
955	98 000	005	009	014	019	023	028	032	037	041	
956	046	050	055	059	064	068	073	078	082	087	
957	091	096	100	105	109	114	118	123	127	132	
958	137	141	146	150	155	159	164	168	173	177	
959	182	186	191	195	200	204	209	214	218	223	
960	227	232	236	241	245	250	254	259	263	268	
961	272	277	281	286	290	295	299	304	308	313	
962	318	322	327	331	336	340	345	349	354	358	**5**
963	363	367	372	376	381	385	390	394	399	403	1 0.5
964	408	412	417	421	426	430	435	439	444	448	2 1.0 3 1.5
965	453	457	462	466	471	475	480	484	489	493	4 2.0
966	498	502	507	511	516	520	525	529	534	538	5 2.5
967	543	547	552	556	561	565	570	574	579	583	6 3.0 7 3.5
968	588	592	597	601	605	610	614	619	623	628	8 4.0
969	632	637	641	646	650	655	659	664	668	673	9 4.5
970	677	682	686	691	695	700	704	709	713	717	
971	722	726	731	735	740	744	749	753	758	762	
972	767	771	776	780	784	789	793	798	802	807	
973	811	816	820	825	829	834	838	843	847	851	
974	856	860	865	869	874	878	883	887	892	896	
975	900	905	909	914	918	923	927	932	936	941	
976	945	949	954	958	963	967	972	976	981	985	
977	989	994	998	*003	*007	*012	*016	*021	*025	*029	
978	99 034	038	043	047	052	056	061	065	069	074	
979	078	083	087	092	096	100	105	109	114	118	
980	123	127	131	136	140	145	149	154	158	162	
981	167	171	176	180	185	189	193	198	202	207	
982	211	216	220	224	229	233	238	242	247	251	**4**
983	255	260	264	269	273	277	282	286	291	295	1 0.4
984	300	304	308	313	317	322	326	330	335	339	2 0.8 3 1.2
985	344	348	352	357	361	366	370	374	379	383	4 1.6
986	388	392	396	401	405	410	414	419	423	427	5 2.0
987	432	436	441	445	449	454	458	463	467	471	6 2.4 7 2.8
988	476	480	484	489	493	498	502	506	511	515	8 3.2
989	520	524	528	533	537	542	546	550	555	559	9 3.6
990	564	568	572	577	581	585	590	594	599	603	
991	607	612	616	621	625	629	634	638	642	647	
992	651	656	660	664	669	673	677	682	686	691	
993	695	699	704	708	712	717	721	726	730	734	
994	739	743	747	752	756	760	765	769	774	778	
995	782	787	791	795	800	804	808	813	817	822	
996	826	830	835	839	843	848	852	856	861	865	
997	870	874	878	883	887	891	896	900	904	909	
998	913	917	922	926	930	935	939	944	948	952	
999	957	961	965	970	974	978	983	987	991	996	
1000	00 000	004	009	013	017	022	026	030	035	039	
N.	0	1	2	3	4	5	6	7	8	9	Prop. Pts.

TABLE

2

TABLE **3**

LOGARITHMS OF TRIGONOMETRIC FUNCTIONS
TO FIVE DECIMAL PLACES
(Subtract 10 from Each Entry)

Special interpolation in Table 3   In the first three columns on pages 271, 272, and 273, the tabular differences are so large that linear interpolation gives inaccurate results.   Since every entry for the logarithm of the cosine of an angle near 90° can be thought of as the logarithm of the sine of an angle near 0°, it is sufficient if we consider only the sines of small angles.   In like manner, since the tangent and cotangent are both reciprocals and cofunctions, it is sufficient to consider interpolation for tangents of very small angles, for we have

$$\log \tan A = -\log \cot A = -\log \tan (90° - A)$$
$$= \log \cot (90° - A)$$

If A and B are sufficiently small angles, both expressed in minutes, it can be shown that the following relations are accurate to five places of decimals.

(a) *log sin A = log sin B + log A − log B*
(b) *log tan A = log tan B + log A − log B*

In particular, these relations will be reliable to five places of decimals if both A and B are less than 3° and $A - B$ is less than 1'.   Since these conditions are true for interpolation problems for the sine and tangent in the first columns on pages 271, 272, and 273, these formulas may be used.

EXAMPLE 1   Find $\log \cos 89° \; 27.5'$. This is the same as finding $\log \sin 0° \; 32.5'$.

Let $A = 32.5'$, and let $B = 32'$.   Then

$$\log \sin 32.5' = \log \sin 32' + \log 32.5 - \log 32$$

whence $\log \cos 89° \; 27.5' = 7.97560 - 10$

EXAMPLE 2   Find $\log \tan 1° \; 15.2'$.   Let $A = 1° \; 15.2' = 75.2'$, and let $B = 1° \; 15' = 75'$.   Then

$$\log \tan 1° \; 15.2' = \log \tan 1° \; 15' + \log 75.2 - \log 75$$
$$= 8.34002 - 10$$

NOTE.   Ordinary linear interpolation in the first example would have given the result $7.97555 - 10$, with an error of 5 in the last place.   Linear interpolation in the last example would have given the result $8.34001 - 10$, with an error of 1 in the last place.

0°

′	L Sin	d	L Tan	c d	L Cot	L Cos	
0	————		————			0.00 000	60
1	6.46 373		6.46 373		13.53 627	0.00 000	59
2	6.76 476	30103	6.76 476	30103	13.23 524	0.00 000	58
3	6.94 085	17609	6.94 085	17609	13.05 915	0.00 000	57
4	7.06 579	12494	7.06 579	12494	12.93 421	0.00 000	56
5	7.16 270	9691	7.16 270	9691	12.83 730	0.00 000	55
6	7.24 188	7918	7.24 188	7918	12.75 812	0.00 000	54
7	7.30 882	6694	7.30 882	6694	12.69 118	0.00 000	53
8	7.36 682	5800	7 36 682	5800	12.63 318	0.00 000	52
9	7.41 797	5115	7.41 797	5115	12.58 203	0.00 000	51
10	7.46 373	4576	7.46 373	4576	12.53 627	0.00 000	50
11	7.50 512	4139	7.50 512	4139	12.49 488	0.00 000	49
12	7.54 291	3779	7.54 291	3779	12.45 709	0.00 000	48
13	7.57 767	3476	7.57 767	3476	12.42 233	0.00 000	47
14	7.60 985	3218	7.60 986	3219	12.39 014	0.00 000	46
15	7.63 982	2997	7.63 982	2996	12.36 018	0.00 000	45
16	7.66 784	2802	7.66 785	2803	12.33 215	0.00 000	44
17	7.69 417	2633	7.69 418	2633	12.30 582	9.99 999	43
18	7.71 900	2483	7.71 900	2482	12.28 100	9.99 999	42
19	7.74 248	2348	7.74 248	2348	12.25 752	9.99 999	41
20	7.76 475	2227	7.76 476	2228	12.23 524	9.99 999	40
21	7.78 594	2119	7.78 595	2119	12.21 405	9.99 999	39
22	7.80 615	2021	7.80 615	2020	12.19 385	9.99 999	38
23	7.82 545	1930	7.82 546	1931	12.17 454	9.99 999	37
24	7.84 393	1848	7.84 394	1848	12.15 606	9.99 999	36
25	7.86 166	1773	7.86 167	1773	12.13 833	9.99 999	35
26	7.87 870	1704	7.87 871	1704	12.12 129	9.99 999	34
27	7.89 509	1639	7.89 510	1639	12.10 490	9.99 999	33
28	7.91 088	1579	7.91 089	1579	12.08 911	9.99 999	32
29	7.92 612	1524	7.92 613	1524	12.07 387	9.99 998	31
30	7.94 084	1472	7.94 086	1473	12.05 914	9.99 998	30
31	7.95 508	1424	7.95 510	1424	12.04 490	9.99 998	29
32	7.96 887	1379	7.96 889	1379	12.03 111	9.99 998	28
33	7.98 223	1336	7.98 225	1336	12.01 775	9.99 998	27
34	7.99 520	1297	7.99 522	1297	12.00 478	9.99 998	26
35	8.00 779	1259	8.00 781	1259	11.99 219	9.99 998	25
36	8.02 002	1223	8.02 004	1223	11.97 996	9.99 998	24
37	8.03 192	1190	8.03 194	1190	11.96 806	9.99 997	23
38	8.04 350	1158	8.04 353	1159	11.95 647	9.99 997	22
39	8.05 478	1128	8.05 481	1128	11.94 519	9.99 997	21
40	8.06 578	1100	8.06 581	1100	11.93 419	9.99 997	20
41	8.07 650	1072	8.07 653	1072	11.92 347	9.99 997	19
42	8.08 696	1046	8.08 700	1047	11.91 300	9.99 997	18
43	8.09 718	1022	8.09 722	1022	11.90 278	9.99 997	17
44	8.10 717	999	8.10 720	998	11.89 280	9.99 996	16
45	8.11 693	976	8.11 696	976	11.88 304	9.99 996	15
46	8.12 647	954	8.12 651	955	11.87 349	9.99 996	14
47	8.13 581	934	8.13 585	934	11.86 415	9.99 996	13
48	8.14 495	914	8.14 500	915	11.85 500	9.99 996	12
49	8.15 391	896	8.15 395	895	11.84 605	9.99 996	11
50	8.16 268	877	8.16 273	878	11.83 727	9.99 995	10
51	8.17 128	860	8.17 133	860	11.82 867	9.99 995	9
52	8.17 971	843	8.17 976	843	11.82 024	9.99 995	8
53	8.18 798	827	8.18 804	828	11.81 196	9.99 995	7
54	8.19 610	812	8.19 616	812	11.80 384	9.99 995	6
55	8.20 407	797	8.20 413	797	11.79 587	9.99 994	5
56	8.21 189	782	8.21 195	782	11.78 805	9.99 994	4
57	8.21 958	769	8.21 964	769	11.78 036	9.99 994	3
58	8.22 713	755	8.22 720	756	11.77 280	9.99 994	2
59	8.23 456	743	8.23 462	742	11.76 538	9.99 994	1
60	8.24 186	730	8.24 192	730	11.75 808	9.99 993	0
	L Cos	d	L Cot	c d	L Tan	L Sin	′

89°

Since the tabular differences in the first three columns of this page, and on each of the two pages following, are so large and change so rapidly in value that ordinary linear interpolation does not give results accurate to five places of decimals, special methods of interpolation are necessary. A brief account of these special methods is given on page 270 of the tables.

TABLE

3

1°

'	L Sin	d	L Tan	c d	L Cot	L Cos	'
0	8.24 186	717	8.24 192	718	11.75 808	9.99 993	60
1	8.24 903	706	8.24 910	706	11.75 090	9.99 993	59
2	8.25 609	695	8.25 616	696	11.74 384	9.99 993	58
3	8.26 304	684	8.26 312	684	11.73 688	9.99 993	57
4	8.26 988	673	8.26 996	673	11.73 004	9.99 992	56
5	8.27 661	663	8.27 669	663	11.72 331	9.99 992	55
6	8.28 324	653	8.28 332	654	11.71 668	9.99 992	54
7	8.28 977	644	8.28 986	643	11.71 014	9.99 992	53
8	8.29 621	634	8.29 629	634	11.70 371	9.99 992	52
9	8.30 255	624	8.30 263	625	11.69 737	9.99 991	51
10	8.30 879	616	8.30 888	617	11.69 112	9.99 991	50
11	8.31 495	608	8.31 505	607	11.68 495	9.99 991	49
12	8.32 103	599	8.32 112	599	11.67 888	9.99 990	48
13	8.32 702	590	8.32 711	591	11.67 289	9.99 990	47
14	8.33 292	583	8.33 302	584	11.66 698	9.99 990	46
15	8.33 875	575	8.33 886	575	11.66 114	9.99 990	45
16	8.34 450	568	8.34 461	568	11.65 539	9.99 989	44
17	8.35 018	560	8.35 029	561	11.64 971	9.99 989	43
18	8.35 578	553	8.35 590	553	11.64 410	9.99 989	42
19	8.36 131	547	8.36 143	546	11.63 857	9.99 989	41
20	8.36 678	539	8.36 689	540	11.63 311	9.99 988	40
21	8.37 217	533	8.37 229	533	11.62 771	9.99 988	39
22	8.37 750	526	8.37 762	527	11.62 238	9.99 988	38
23	8.83 276	520	8.38 289	520	11.61 711	9.99 987	37
24	8.38 796	514	8.38 809	514	11.61 191	9.99 987	36
25	8.39 310	508	8.39 323	509	11.60 677	9.99 987	35
26	8.39 818	502	8.39 832	502	11.60 168	9.99 986	34
27	8.40 320	496	8.40 334	496	11.59 666	9.99 986	33
28	8.40 816	491	8.40 830	491	11.59 170	9.99 986	32
29	8.41 307	485	8.41 321	486	11.58 679	9.99 985	31
30	8.41 792	480	8.41 807	480	11.58 193	9.99 985	30
31	8.42 272	474	8.42 287	475	11.57 713	9.99 985	29
32	8.42 746	470	8.42 762	470	11.57 238	9.99 984	28
33	8.43 216	464	8.43 232	464	11.56 768	9.99 984	27
34	8.43 680	459	8.43 696	460	11.56 304	9.99 984	26
35	8.44 139	455	8.44 156	455	11.55 844	9.99 983	25
36	8.44 594	450	8.44 611	450	11.55 389	9.99 983	24
37	8.45 044	445	8.45 061	446	11.54 939	9.99 983	23
38	8.45 489	441	8.45 507	441	11.54 493	9.99 982	22
39	8.45 930	436	8.45 948	437	11.54 052	9.99 982	21
40	8.46 366	433	8.46 385	432	11.53 615	9.99 982	20
41	8.46 799	427	8.46 817	428	11.53 183	9.99 981	19
42	8.47 226	424	8.47 245	424	11.52 755	9.99 981	18
43	8.47 650	419	8.47 669	420	11.52 331	9.99 981	17
44	8.48 069	416	8.48 089	416	11.51 911	9.99 980	16
45	8.48 485	411	8.48 505	412	11.51 495	9.99 980	15
46	8.48 896	408	8.48 917	408	11.51 083	9.99 979	14
47	8.49 304	404	8.49 325	404	11.50 675	9.99 979	13
48	8.49 708	400	8.49 729	401	11.50 271	9.99 979	12
49	8.50 108	396	8.50 130	397	11.49 870	9.99 978	11
50	8.50 504	393	8.50 527	393	11.49 473	9.99 978	10
51	8.50 897	390	8.50 920	390	11.49 080	9.99 977	9
52	8.51 287	386	8.51 310	386	11.48 690	9.99 977	8
53	8.51 673	382	8.51 696	383	11.48 304	9.99 977	7
54	8.52 055	379	8.52 079	380	11.47 921	9.99 976	6
55	8.52 434	376	8.52 459	376	11.47 541	9.99 976	5
56	8.52 810	373	8.52 835	373	11.47 165	9.99 975	4
57	8.53 183	369	8.53 208	370	11.46 792	9.99 975	3
58	8.53 552	367	8.53 578	367	11.46 422	9.99 974	2
59	8.53 919	363	8.53 945	363	11.46 055	9.99 974	1
60	8.54 282		8.54 308		11.45 692	9.99 974	0
	L Cos	d	L Cot	c d	L Tan	L Sin	'

88°

If ordinary linear interpolation is not sufficiently accurate, use the special methods described on page 270 of the tables.

TABLE
3

2°

′	L Sin	d	L Tan	c d	L Cot	L Cos	
0	8.54 282	360	8.54 308	361	11.45 692	9.99 974	60
1	8.54 642	357	8.54 669	358	11.45 331	9.99 973	59
2	8.54 999	355	8.55 027	355	11.44 973	9.99 973	58
3	8.55 354	351	8.55 382	352	11.44 618	9.99 972	57
4	8.55 705	349	8.55 734	349	11.44 266	9.99 972	56
5	8.56 054	346	8.56 083	346	11.43 917	9.99 971	55
6	8.56 400	343	8.56 429	344	11.43 571	9.99 971	54
7	8.56 743	341	8.56 773	341	11.43 227	9.99 970	53
8	8.57 084	337	8.57 114	338	11.42 886	9.99 970	52
9	8.57 421	336	8.57 452	336	11.42 548	9.99 969	51
10	8.57 757	332	8.57 788	333	11.42 212	9.99 969	50
11	8.58 089	330	8.58 121	330	11.41 879	9.99 968	49
12	8.58 419	328	8.58 451	328	11.41 549	9.99 968	48
13	8.58 747	325	8.58 779	326	11.41 221	9.99 967	47
14	8.59 072	323	8.59 105	323	11.40 895	9.99 967	46
15	8.59 395	320	8.59 428	321	11.40 572	9.99 967	45
16	8.59 715	318	8.59 749	319	11.40 251	9.99 966	44
17	8.60 033	316	8.60 068	316	11.39 932	9.99 966	43
18	8.60 349	313	8.60 384	314	11.39 616	9.99 965	42
19	8.60 662	311	8.60 698	311	11.39 302	9.99 964	41
20	8.60 973	309	8.61 009	310	11.38 991	9.99 964	40
21	8.61 282	307	8.61 319	307	11.38 681	9.99 963	39
22	8.61 589	305	8.61 626	305	11.38 374	9.99 963	38
23	8.61 894	302	8.61 931	303	11.38 069	9.99 962	37
24	8.62 196	301	8.62 234	301	11.37 766	9.99 962	36
25	8.62 497	298	8.62 535	299	11.37 465	9.99 961	35
26	8.62 795	296	8.62 834	297	11.37 166	9.99 961	34
27	8.63 091	294	8.63 131	295	11.36 869	9.99 960	33
28	8.63 385	293	8.63 426	292	11.36 574	9.99 960	32
29	8.63 678	290	8.63 718	291	11.36 282	9.99 959	31
30	8.63 968	288	8.64 009	289	11.35 991	9.99 959	30
31	8.64 256	287	8.64 298	287	11.35 702	9.99 958	29
32	8.64 543	284	8.64 585	285	11.35 415	9.99 958	28
33	8.64 827	283	8.64 870	284	11.35 130	9.99 957	27
34	8.65 110	281	8.65 154	281	11.34 846	9.99 956	26
35	8.65 391	279	8.65 435	280	11.34 565	9.99 956	25
36	8.65 670	277	8.65 715	278	11.34 285	9.99 955	24
37	8.65 947	276	8.65 993	276	11.34 007	9.99 955	23
38	8 66 223	274	8.66 269	274	11.33 731	9.99 954	22
39	8.66 497	272	8.66 543	273	11.33 457	9.99 954	21
40	8.66 769	270	8.66 816	271	11.33 184	9.99 953	20
41	8.67 039	269	8.67 087	269	11.32 913	9.99 952	19
42	8.67 308	267	8.67 356	268	11.32 644	9.99 952	18
43	8.67 575	266	8.67 624	266	11.32 376	9.99 951	17
44	8.67 841	263	8.67 890	264	11.32 110	9.99 951	16
45	8.68 104	263	8.68 154	263	11.31 846	9.99 950	15
46	8.68 367	260	8.68 417	261	11.31 583	9.99 949	14
47	8.68 627	259	8.68 678	260	11.31 322	9.99 949	13
48	8.68 886	258	8.68 938	258	11.31 062	9.99 948	12
49	8.69 144	256	8.69 196	257	11.30 804	9.99 948	11
50	8.69 400	254	8.69 453	255	11.30 547	9.99 947	10
51	8.69 654	253	8.69 708	254	11.30 292	9.99 946	9
52	8.69 907	252	8.69 962	252	11.30 038	9.99 946	8
53	8.70 159	250	8.70 214	251	11.29 786	9.99 945	7
54	8.70 409	249	8.70 465	249	11.29 535	9.99 944	6
55	8.70 658	247	8.70 714	248	11.29 286	9.99 944	5
56	8.70 905	246	8.70 962	246	11.29 038	9.99 943	4
57	8.71 151	244	8.71 208	245	11.28 792	9.99 942	3
58	8.71 395	243	8.71 453	244	11.28 547	9.99 942	2
59	8.71 638	242	8.71 697	243	11.28 303	9.99 941	1
60	8.71 880		8.71 940		11.28 060	9.99 940	0
	L Cos	d	L Cot	c d	L Tan	L Sin	′

If ordinary linear interpolation is not sufficiently accurate, use the special methods described on page 270 of the tables.

TABLE

3

87°

3°

′	L Sin	d	L Tan	c d	L Cot	L Cos		Prop. Pts.
0	8.71 880	240	8.71 940	241	11.28 060	9.99 940	60	
1	8.72 120	239	8.72 181	239	11.27 819	9.99 940	59	
2	8.72 359	238	8.72 420	239	11.27 580	9.99 939	58	
3	8.72 597	237	8.72 659	237	11.27 341	9.99 938	57	
4	8.72 834	235	8.72 896	236	11.27 104	9.99 938	56	
5	8.73 069	234	8.73 132	234	11.26 868	9.99 937	55	
6	8.73 303	232	8.73 366	234	11.26 634	9.99 936	54	
7	8.73 535	232	8.73 600	232	11.26 400	9.99 936	53	
8	8.73 767	230	8.73 832	231	11.26 168	9.99 935	52	
9	8.73 997	229	8.74 063	229	11.25 937	9.99 934	51	
10	8.74 226	228	8.74 292	229	11.25 708	9.99 934	50	
11	8.74 454	226	8.74 521	227	11.25 479	9.99 933	49	
12	8.74 680	226	8.74 748	226	11.25 252	9.99 932	48	
13	8.74 906	224	8.74 974	225	11.25 026	9.99 932	47	
14	8.75 130	223	8.75 199	224	11.24 801	9.99 931	46	
15	8.75 353	222	8.75 423	222	11.24 577	9.99 930	45	
16	8.75 575	220	8.75 645	222	11.24 355	9.99 929	44	
17	8.75 795	220	8.75 867	220	11.24 133	9.99 929	43	
18	8.76 015	219	8.76 087	219	11.23 913	9.99 928	42	
19	8.76 234	217	8.76 306	219	11.23 694	9.99 927	41	
20	8.76 451	216	8.76 525	217	11.23 475	9.99 926	40	
21	8.76 667	216	8.76 742	216	11.23 258	9.99 926	39	
22	8.76 883	214	8.76 958	215	11.23 042	9.99 925	38	
23	8.77 097	213	8.77 173	214	11.22 827	9.99 924	37	
24	8.77 310	212	8.77 387	213	11.22 613	9.99 923	36	
25	8.77 522	211	8.77 600	211	11.22 400	9.99 923	35	
26	8.77 733	210	8.77 811	211	11.22 189	9.99 922	34	
27	8.77 943	209	8.78 022	210	11.21 978	9.99 921	33	
28	8.78 152	208	8.78 232	209	11.21 768	9.99 920	32	
29	8.78 360	208	8.78 441	208	11.21 559	9.99 920	31	
30	8.78 568	206	8.78 649	206	11.21 351	9.99 919	30	
31	8.78 774	205	8.78 855	206	11.21 145	9.99 918	29	
32	8.78 979	204	8.79 061	205	11.20 939	9.99 917	28	
33	8.79 183	203	8.79 266	204	11.20 734	9.99 917	27	
34	8.79 386	202	8.79 470	203	11.20 530	9.99 916	26	
35	8.79 588	201	8.79 673	202	11.20 327	9.99 915	25	
36	8.79 789	201	8.79 875	201	11.20 125	9.99 914	24	
37	8.79 990	199	8.80 076	201	11.19 924	9.99 913	23	
38	8.80 189	199	8.80 277	199	11.19 723	9.99 913	22	
39	8.80 388	197	8.80 476	198	11.19 524	9.99 912	21	
40	8.80 585	197	8.80 674	198	11.19 326	9.99 911	20	
41	8.80 782	196	8.80 872	196	11.19 128	9.99 910	19	
42	8.80 978	195	8.81 068	196	11.18 932	9.99 909	18	
43	8.81 173	194	8.81 264	195	11.18 736	9.99 909	17	
44	8.81 367	193	8.81 459	194	11.18 541	9.99 908	16	
45	8.81 560	192	8.81 653	193	11.18 347	9.99 907	15	
46	8.81 752	192	8.81 846	192	11.18 154	9.99 906	14	
47	8.81 944	190	8.82 038	192	11.17 962	9.99 905	13	
48	8.82 134	190	8.82 230	190	11.17 770	9.99 904	12	
49	8.82 324	189	8.82 420	190	11.17 580	9.99 904	11	
50	8.82 513	188	8.82 610	189	11.17 390	9.99 903	10	
51	8.82 701	187	8.82 799	188	11.17 201	9.99 902	9	
52	8.82 888	187	8.82 987	188	11.17 013	9.99 901	8	
53	8.83 075	186	8.83 175	186	11.16 825	9.99 900	7	
54	8.83 261	185	8.83 361	186	11.16 639	9.99 899	6	
55	8.83 446	184	8.83 547	185	11.16 453	9.99 898	5	
56	8.83 630	183	8.83 732	184	11.16 268	9.99 898	4	
57	8.83 813	183	8.83 916	184	11.16 084	9.99 897	3	
58	8.83 996	181	8.84 100	182	11.15 900	9.99 896	2	
59	8.84 177	181	8.84 282	182	11.15 718	9.99 895	1	
60	8.84 358		8.84 464		11.15 536	9.99 894	0	
	L Cos	d	L Cot	c d	L Tan	L Sin	′	Prop. Pts.

Prop. Pts.

	239	237	235	234
2	47.8	47.4	47.0	46.8
3	71.7	71.1	70.5	70.2
4	95.6	94.8	94.0	93.6
5	119.5	118.5	117.5	117.0
6	143.4	142.2	141.0	140.4
7	167.3	165.9	164.5	163.8
8	191.2	189.6	188.0	187.2
9	215.1	213.3	211.5	210.6

	232	229	227	226
2	46.4	45.8	45.4	45.2
3	69.6	68.7	68.1	67.8
4	92.8	91.6	90.8	90.4
5	116.0	114.5	113.5	113.0
6	139.2	137.4	136.2	135.6
7	162.4	160.3	158.9	158.2
8	185.6	183.2	181.6	180.8
9	208.8	206.1	204.3	203.4

	224	222	220	219
2	44.8	44.4	44.0	43.8
3	67.2	66.6	66.0	65.7
4	89.6	88.8	88.0	87.6
5	112.0	111.0	110.0	109.5
6	134.4	133.2	132.0	131.4
7	156.8	155.4	154.0	153.3
8	179.2	177.6	176.0	175.2
9	201.6	199.8	198.0	197.1

	217	215	213	211
2	43.4	43.0	42.6	42.2
3	65.1	64.5	63.9	63.3
4	86.8	86.0	85.2	84.4
5	108.5	107.5	106.5	105.5
6	130.2	129.0	127.8	126.6
7	151.9	150.5	149.1	147.7
8	173.6	172.0	170.4	168.8
9	195.3	193.5	191.7	189.9

	208	206	203	201
2	41.6	41.2	40.6	40.2
3	62.4	61.8	60.9	60.3
4	83.2	82.4	81.2	80.4
5	104.0	103.0	101.5	100.5
6	124.8	123.6	121.8	120.6
7	145.6	144.2	142.1	140.7
8	166.4	164.8	162.4	160.8
9	187.2	185.4	182.7	180.9

	199	197	195	193
2	39.8	39.4	39.0	38.6
3	59.7	59.1	58.5	57.9
4	79.6	78.8	78.0	77.2
5	99.5	98.5	97.5	96.5
6	119.4	118.2	117.0	115.8
7	139.3	137.9	136.5	135.1
8	159.2	157.6	156.0	154.4
9	179.1	177.3	175.5	173.7

	192	190	188	186
2	38.4	38.0	37.6	37.2
3	57.6	57.0	56.4	55.8
4	76.8	76.0	75.2	74.4
5	96.0	95.0	94.0	93.0
6	115.2	114.0	112.8	111.6
7	134.4	133.0	131.6	130.2
8	153.6	152.0	150.4	148.8
9	172.8	171.0	169.2	167.4

	184	183	182	181
2	36.8	36.6	36.4	36.2
3	55.2	54.9	54.6	54.3
4	73.6	73.2	72.8	72.4
5	92.0	91.5	91.0	90.5
6	110.4	109.8	109.2	108.6
7	128.8	128.1	127.4	126.7
8	147.2	146.4	145.6	144.8
9	165.6	164.7	163.8	162.9

TABLE 3

86°

4°

′	L Sin	d	L Tan	c d	L Cot	L Cos	
0	8.84 358		8.84 464		11.15 536	9.99 894	60
1	8.84 539	181	8.84 646	182	11.15 354	9.99 893	59
2	8.84 718	179	8.84 826	180	11.15 174	9.99 892	58
3	8.84 897	179	8.85 006	180	11.14 994	9.99 891	57
4	8.85 075	178	8.85 185	179	11.14 815	9.99 891	56
5	8.85 252	177	8.85 363	178	11.14 637	9.99 890	55
6	8.85 429	177	8.85 540	177	11.14 460	9.99 889	54
7	8.85 605	176	8.85 717	177	11.14 283	9.99 888	53
8	8.85 780	175	8.85 893	176	11.14 107	9.99 887	52
9	8.85 955	175	8.86 069	176	11.13 931	9.99 886	51
10	8.86 128	173	8.86 243	174	11.13 757	9.99 885	50
11	8.86 301	173	8.86 417	174	11.13 583	9.99 884	49
12	8.86 474	173	8.86 591	172	11.13 409	9.99 883	48
13	8.86 645	171	8.86 763	172	11.13 237	9.99 882	47
14	8.86 816	171	8.86 935	171	11.13 065	9.99 881	46
15	8.86 987	171	8.87 106	171	11.12 894	9.99 881	45
16	8.87 156	169	8.87 277	171	11.12 723	9.99 879	44
17	8.87 325	169	8.87 447	170	11.12 553	9.99 879	43
18	8.87 494	169	8.87 616	169	11.12 384	9.99 878	42
19	8.87 661	167	8.87 785	169	11.12 215	9.99 877	41
20	8.87 829	168	8.87 953	168	11.12 047	9.99 876	40
21	8.87 995	166	8.88 120	167	11.11 880	9.99 875	39
22	8.88 161	166	8.88 287	167	11.11 713	9.99 874	38
23	8.88 326	165	8.88 453	166	11.11 547	9.99 873	37
24	8.88 490	164	8.88 618	165	11.11 382	9.99 872	36
25	8.88 654	164	8.88 783	165	11.11 217	9.99 871	35
26	8.88 817	163	8.88 948	165	11.11 052	9.99 870	34
27	8.88 980	162	8.89 111	163	11.10 889	9.99 869	33
28	8.89 142	162	8.89 274	163	11.10 726	9.99 868	32
29	8.89 304	160	8.89 437	163	11.10 563	9.99 867	31
30	8.89 464	161	8.89 598	162	11.10 402	9.99 866	30
31	8.89 625	159	8.89 760	160	11.10 240	9.99 865	29
32	8.89 784	159	8.89 920	160	11.10 080	9.99 864	28
33	8.89 943	159	8.90 080	160	11.09 920	9.99 863	27
34	8.90 102	158	8.90 240	159	11.09 760	9.99 862	26
35	8.90 260	157	8.90 399	158	11.09 601	9.99 861	25
36	8.90 417	157	8.90 557	158	11.09 443	9.99 860	24
37	8.90 574	156	8.90 715	157	11.09 285	9.99 859	23
38	8.90 730	155	8.90 872	157	11.09 128	9.99 858	22
39	8.90 885	155	8.91 029	156	11.08 971	9.99 857	21
40	8.91 040	155	8.91 185	155	11.08 815	9.99 856	20
41	8.91 195	154	8.91 340	155	11.08 660	9.99 855	19
42	8.91 349	153	8.91 495	155	11.08 505	9.99 854	18
43	8.91 502	153	8.91 650	153	11.08 350	9.99 853	17
44	8.91 655	152	8.91 803	154	11.08 197	9.99 852	16
45	8.91 807	152	8.91 957	153	11.08 043	9.99 851	15
46	8.91 959	151	8.92 110	152	11.07 890	9.99 850	14
47	8.92 110	151	8.92 262	152	11.07 738	9.99 848	13
48	8.92 261	150	8.92 414	151	11.07 586	9.99 847	12
49	8.92 411	150	8.92 565	151	11.07 435	9.99 846	11
50	8.92 561	149	8.92 716	150	11.07 284	9.99 845	10
51	8.92 710	149	8.92 866	150	11.07 134	9.99 844	9
52	8.92 859	148	8.93 016	149	11.06 984	9.99 843	8
53	8.93 007	147	8.93 165	148	11.06 835	9.99 842	7
54	8.93 154	147	8.93 313	149	11.06 687	9.99 841	6
55	8.93 301	147	8.93 462	147	11.06 538	9.99 840	5
56	8.93 448	146	8.93 609	147	11.06 391	9.99 839	4
57	8.93 594	146	8.93 756	147	11.06 244	9.99 838	3
58	8.93 740	145	8.93 903	146	11.06 097	9.99 837	2
59	8.93 885	145	8.94 049	146	11.05 951	9.99 836	1
60	8.94 030		8.94 195		11.05 805	9.99 834	0
	L Cos	d	L Cot	c d	L Tan	L Sin	′

Prop. Pts.

	182	181	180	179
2	36.4	36.2	36.0	35.8
3	54.6	54.3	54.0	53.7
4	72.8	72.4	72.0	71.6
5	91.0	90.5	90.0	89.5
6	109.2	108.6	108.0	107.4
7	127.4	126.7	126.0	125.3
8	145.6	144.8	144.0	143.2
9	163.8	162.9	162.0	161.1

	178	177	176	175
2	35.6	35.4	35.2	35.0
3	53.4	53.1	52.8	52.5
4	71.2	70.8	70.4	70.0
5	89.0	88.5	88.0	87.5
6	106.8	106.2	105.6	105.0
7	124.6	123.9	123.2	122.5
8	142.4	141.6	140.8	140.0
9	160.2	159.3	158.4	157.5

	174	173	172	171
2	34.8	34.6	34.4	34.2
3	52.2	51.9	51.6	51.3
4	69.6	69.2	68.8	68.4
5	87.0	86.5	86.0	85.5
6	104.4	103.8	103.2	102.6
7	121.8	121.1	120.4	119.7
8	139.2	138.4	137.6	136.8
9	156.6	155.7	154.8	153.9

	170	169	168	167
2	34.0	33.8	33.6	33.4
3	51.0	50.7	50.4	50.1
4	68.0	67.6	67.2	66.8
5	85.0	84.5	84.0	83.5
6	102.0	101.4	100.8	100.2
7	119.0	118.3	117.6	116.9
8	136.0	135.2	134.4	133.6
9	153.0	152.1	151.2	150.3

	166	165	164	163
2	33.2	33.0	32.8	32.6
3	49.8	49.5	49.2	48.9
4	66.4	66.0	65.6	65.2
5	83.0	82.5	82.0	81.5
6	99.6	99.0	98.4	97.8
7	116.2	115.5	114.8	114.1
8	132.8	132.0	131.2	130.4
9	149.4	148.5	147.6	146.7

	162	161	160	159
2	32.4	32.2	32.0	31.8
3	48.6	48.3	48.0	47.7
4	64.8	64.4	64.0	63.6
5	81.0	80.5	80.0	79.5
6	97.2	96.6	96.0	95.4
7	113.4	112.7	112.0	111.3
8	129.6	128.8	128.0	127.2
9	145.8	144.9	144.0	143.1

	158	157	156	155
2	31.6	31.4	31.2	31.0
3	47.4	47.1	46.8	46.5
4	63.2	62.8	62.4	62.0
5	79.0	78.5	78.0	77.5
6	94.8	94.2	93.6	93.0
7	110.6	109.9	109.2	108.5
8	126.4	125.6	124.8	124.0
9	142.2	141.3	140.4	139.5

	154	153	152
2	30.8	30.6	30.4
3	46.2	45.9	45.6
4	61.6	61.2	60.8
5	77.0	76.5	76.0
6	92.4	91.8	91.2
7	107.8	107.1	106.4
8	123.2	122.4	121.6
9	138.6	137.7	136.8

TABLE

3

5°

'	L Sin	d	L Tan	c d	L Cot	L Cos	'
0	8.94 030	144	8.94 195	145	11.05 805	9.99 834	60
1	8.94 174	143	8.94 340	145	11.05 660	9.99 833	59
2	8.94 317	144	8.94 485	145	11.05 515	9.99 832	58
3	8.94 461	142	8.94 630	143	11.05 370	9.99 831	57
4	8.94 603	143	8.94 773	144	11.05 227	9.99 830	56
5	8.94 746	141	8.94 917	143	11.05 083	9.99 829	55
6	8.94 887	142	8.95 060	142	11.04 940	9.99 828	54
7	8.95 029	141	8.95 202	142	11.04 798	9.99 827	53
8	8.95 170	140	8.95 344	142	11.04 656	9.99 825	52
9	8.95 310	140	8.95 486	141	11.04 514	9.99 824	51
10	8.95 450	139	8.95 627	140	11.04 373	9.99 823	50
11	8.95 589	139	8.95 767	141	11.04 233	9.99 822	49
12	8.95 728	139	8.95 908	139	11.04 092	9.99 821	48
13	8.95 867	138	8.96 047	140	11.03 953	9.99 820	47
14	8.96 005	138	8.96 187	138	11.03 813	9.99 819	46
15	8.96 143	137	8.96 325	139	11.03 675	9.99 817	45
16	8.96 280	137	8.96 464	138	11.03 536	9.99 816	44
17	8.96 417	136	8.96 602	137	11.03 398	9.99 815	43
18	8.96 553	136	8.96 739	138	11.03 261	9.99 814	42
19	8.96 689	136	8.96 877	136	11.03 123	9.99 813	41
20	8.96 825	135	8.97 013	137	11.02 987	9.99 812	40
21	8.96 960	135	8.97 150	135	11.02 850	9.99 810	39
22	8.97 095	134	8.97 285	136	11.02 715	9.99 809	38
23	8.97 229	134	8.97 421	135	11.02 579	9.99 808	37
24	8.97 363	133	8.97 556	135	11.02 444	9.99 807	36
25	8.97 496	133	8.97 691	134	11.02 309	9.99 806	35
26	8.97 629	133	8.97 825	134	11.02 175	9.99 804	34
27	8.97 762	132	8.97 959	133	11.02 041	9.99 803	33
28	8.97 894	132	8.98 092	133	11.01 908	9.99 802	32
29	8.98 026	131	8.98 225	133	11.01 775	9.99 801	31
30	8.98 157	131	8.98 358	132	11.01 642	9.99 800	30
31	8.98 288	131	8.98 490	131	11.01 510	9.99 798	29
32	8.98 419	130	8.98 622	131	11.01 378	9.99 797	28
33	8.98 549	130	8.98 753	131	11.01 247	9.99 796	27
34	8.98 679	129	8.98 884	131	11.01 116	9.99 795	26
35	8.98 808	129	8.99 015	130	11.00 985	9.99 793	25
36	8.98 937	129	8.99 145	130	11.00 855	9.99 792	24
37	8.99 066	128	8.99 275	130	11.00 725	9.99 791	23
38	8.99 194	128	8.99 405	129	11.00 595	9.99 790	22
39	8.99 322	128	8.99 534	128	11.00 466	9.99 788	21
40	8.99 450	127	8.99 662	129	11.00 338	9.99 787	20
41	8.99 577	127	8.99 791	128	11.00 209	9.99 786	19
42	8.99 704	126	8.99 919	127	11.00 081	9.99 785	18
43	8.99 830	126	9.00 046	128	10.99 954	9.99 783	17
44	8.99 956	126	9.00 174	127	10.99 826	9.99 782	16
45	9.00 082	125	9.00 301	126	10.99 699	9.99 781	15
46	9.00 207	125	9.00 427	126	10.99 573	9.99 780	14
47	9.00 332	124	9.00 553	126	10.99 447	9.99 778	13
48	9.00 456	125	9.00 679	126	10.99 321	9.99 777	12
49	9.00 581	123	9.00 805	125	10.99 195	9.99 776	11
50	9.00 704	124	9.00 930	125	10.99 070	9.99 775	10
51	9.00 828	123	9.01 055	124	10.98 945	9.99 773	9
52	9.00 951	123	9.01 179	124	10.98 821	9.99 772	8
53	9.01 074	122	9.01 303	124	10.98 697	9.99 771	7
54	9.01 196	122	9.01 427	123	10.98 573	9.99 769	6
55	9.01 318	122	9.01 550	123	10.98 450	9.99 768	5
56	9.01 440	121	9.01 673	123	10.98 327	9.99 767	4
57	9.01 561	121	9.01 796	122	10.98 204	9.99 765	3
58	9.01 682	121	9.01 918	122	10.98 082	9.99 764	2
59	9.01 803	120	9.02 040	122	10.97 960	9.99 763	1
60	9.01 923		9.02 162		10.97 838	9.99 761	0

| | L Cos | d | L Cot | c d | L Tan | L Sin | ' |

Prop. Pts.

	151	150	149	148
2	30.2	30.0	29.8	29.6
3	45.3	45.0	44.7	44.4
4	60.4	60.0	59.6	59.2
5	75.5	75.0	74.5	74.0
6	90.6	90.0	89.4	88.8
7	105.7	105.0	104.3	103.6
8	120.8	120.0	119.2	118.4
9	135.9	135.0	134.1	133.2

	147	146	145	144
2	29.4	29.2	29.0	28.8
3	44.1	43.8	43.5	43.2
4	58.8	58.4	58.0	57.6
5	73.5	73.0	72.5	72.0
6	88.2	87.6	87.0	86.4
7	102.9	102.2	101.5	100.8
8	117.6	116.8	116.0	115.2
9	132.3	131.4	130.5	129.6

	143	142	141	140
2	28.6	28.4	28.2	28.0
3	42.9	42.6	42.3	42.0
4	57.2	56.8	56.4	56.0
5	71.5	71.0	70.5	70.0
6	85.8	85.2	84.6	84.0
7	100.1	99.4	98.7	98.0
8	114.4	113.6	112.8	112.0
9	128.7	127.8	126.9	126.0

	139	138	137	136
2	27.8	27.6	27.4	27.2
3	41.7	41.4	41.1	40.8
4	55.6	55.2	54.8	54.4
5	69.5	69.0	68.5	68.0
6	83.4	82.8	82.2	81.6
7	97.3	96.6	95.9	95.2
8	111.2	110.4	109.6	108.8
9	125.1	124.2	123.3	122.4

	135	134	133	132
2	27.0	26.8	26.6	26.4
3	40.5	40.2	39.9	39.6
4	54.0	53.6	53.2	52.8
5	67.5	67.0	66.5	66.0
6	81.0	80.4	79.8	79.2
7	94.5	93.8	93.1	92.4
8	108.0	107.2	106.4	105.6
9	121.5	120.6	119.7	118.8

	131	130	129	128
2	26.2	26.0	25.8	25.6
3	39.3	39.0	38.7	38.4
4	52.4	52.0	51.6	51.2
5	65.5	65.0	64.5	64.0
6	78.6	78.0	77.4	76.8
7	91.7	91.0	90.3	89.6
8	104.8	104.0	103.2	102.4
9	117.9	117.0	116.1	115.2

	127	126	125	124
2	25.4	25.2	25.0	24.8
3	38.1	37.8	37.5	37.2
4	50.8	50.4	50.0	49.6
5	63.5	63.0	62.5	62.0
6	76.2	75.6	75.0	74.4
7	88.9	88.2	87.5	86.8
8	101.6	100.8	100.0	99.2
9	114.3	113.4	112.5	111.6

	123	122	121	120
2	24.6	24.4	24.2	24.0
3	36.9	36.6	36.3	36.0
4	49.2	48.8	48.4	48.0
5	61.5	61.0	60.5	60.0
6	73.8	73.2	72.6	72.0
7	86.1	85.4	84.7	84.0
8	98.4	97.6	96.8	96.0
9	110.7	109.8	108.9	108.0

TABLE 3

84°

6°

′	L Sin	d	L Tan	c d	L Cot	L Cos	′
0	9.01 923		9.02 162		10.97 838	9.99 761	60
1	9.02 043	120	9.02 283	121	10.97 717	9.99 760	59
2	9.02 163	120	9.02 404	121	10.97 596	9.99 759	58
3	9.02 283	120	9.02 525	121	10.97 475	9.99 757	57
4	9.02 402	119	9.02 645	120	10.97 355	9.99 756	56
5	9.02 520	118	9.02 766	121	10.97 234	9.99 755	55
6	9.02 639	119	9.02 885	119	10.97 115	9.99 753	54
7	9.02 757	118	9.03 005	120	10.96 995	9.99 752	53
8	9.02 874	117	9.03 124	119	10.96 876	9.99 751	52
9	9.02 992	118	9.03 242	118	10.96 758	9.99 749	51
10	9.03 109	117	9.03 361	119	10.96 639	9.99 748	50
11	9.03 226	117	9.03 479	118	10.96 521	9.99 747	49
12	9.03 342	116	9.03 597	118	10.96 403	9.99 745	48
13	9.03 458	116	9.03 714	117	10.96 286	9.99 744	47
14	9.03 574	116	9.03 832	118	10.96 168	9.99 742	46
15	9.03 690	116	9.03 948	116	10.96 052	9.99 741	45
16	9.03 805	115	9.04 065	117	10.95 935	9.99 740	44
17	9.03 920	115	9.04 181	116	10.95 819	9.99 738	43
18	9.04 034	114	9.04 297	116	10.95 703	9.99 737	42
19	9.04 149	115	9.04 413	116	10.95 587	9.99 736	41
20	9.04 262	113	9.04 528	115	10.95 472	9.99 734	40
21	9.04 376	114	9.04 643	115	10.95 357	9.99 733	39
22	9.04 490	114	9.04 758	115	10.95 242	9.99 731	38
23	9.04 603	113	9.04 873	115	10.95 127	9.99 730	37
24	9.04 715	112	9.04 987	114	10.95 013	9.99 728	36
25	9.04 828	113	9.05 101	114	10 94 899	9.99 727	35
26	9.04 940	112	9.05 214	113	10.94 786	9.99 726	34
27	9.05 052	112	9.05 328	114	10.94 672	9.99 724	33
28	9.05 164	112	9.05 441	113	10.94 559	9.99 723	32
29	9.05 275	111	9.05 553	112	10.94 447	9.99 721	31
30	9.05 386	111	9.05 666	113	10.94 334	9.99 720	30
31	9.05 497	111	9.05 778	112	10.94 222	9.99 718	29
32	9.05 607	110	9.05 890	112	10.94 110	9.99 717	28
33	9.05 717	110	9.06 002	112	10.93 998	9.99 716	27
34	9.05 827	110	9.06 113	111	10.93 887	9.99 714	26
35	9.05 937	110	9.06 224	111	10.93 776	9.99 713	25
36	9.06 046	109	9.06 335	111	10.93 665	9.99 711	24
37	9.06 155	109	9.06 445	110	10.93 555	9.99 710	23
38	9.06 264	109	9.06 556	111	10.93 444	9.99 708	22
39	9.06 372	108	9.06 666	110	10.93 334	9.99 707	21
40	9.06 481	109	9.06 775	109	10.93 225	9.99 705	20
41	9.06 589	108	9.06 885	110	10.93 115	9.99 704	19
42	9.06 696	107	9.06 994	109	10.93 006	9.99 702	18
43	9.06 804	108	9.07 103	109	10.92 897	9.99 701	17
44	9.06 911	107	9.07 211	108	10.92 789	9.99 699	16
45	9.07 018	107	9.07 320	109	10.92 680	9.99 698	15
46	9.07 124	106	9.07 428	108	10.92 572	9.99 696	14
47	9.07 231	107	9.07 536	108	10.92 464	9.99 695	13
48	9.07 337	106	9.07 643	107	10.92 357	9.99 693	12
49	9.07 442	105	9.07 751	108	10.92 249	9.99 692	11
50	9.07 548	106	9.07 858	107	10.92 142	9.99 690	10
51	9.07 653	105	9.07 964	106	10.92 036	9.99 689	9
52	9.07 758	105	9.08 071	107	10.91 929	9.99 687	8
53	9.07 863	105	9.08 177	106	10.91 823	9.99 686	7
54	9.07 968	105	9.08 283	106	10.91 717	9.99 684	6
55	9.08 072	104	9.08 389	106	10.91 611	9.99 683	5
56	9.08 176	104	9.08 495	105	10.91 505	9.99 681	4
57	9.08 280	104	9.08 600	105	10.91 400	9.99 680	3
58	9.08 383	103	9.08 705	105	10.91 295	9.99 678	2
59	9.08 486	103	9.08 810	104	10.91 190	9.99 677	1
60	9.08 589	103	9.08 914		10.91 086	9.99 675	0
	L Cos	d	L Cot	c d	L Tan	L Sin	′

Prop. Pts.

	121	120	119
1	12.1	12.0	11.9
2	24.2	24.0	23.8
3	36.3	36.0	35.7
4	48.4	48.0	47.6
5	60.5	60.0	59.5
6	72.6	72.0	71.4
7	84.7	84.0	83.3
8	96.8	96.0	95.2
9	108.9	108.0	107.1

	118	117	116
1	11.8	11.7	11.6
2	23.6	23.4	23.2
3	35.4	35.1	34.8
4	47.2	46.8	46.4
5	59.0	58.5	58.0
6	70.8	70.2	69.6
7	82.6	81.9	81.2
8	94.4	93.6	92.8
9	106.2	105.3	104.4

	115	114	113
1	11.5	11.4	11.3
2	23.0	22.8	22.6
3	34.5	34.2	33.9
4	46.0	45.6	45.2
5	57.5	57.0	56.5
6	69.0	68.4	67.8
7	80.5	79.8	79.1
8	92.0	91.2	90.4
9	103.5	102.6	101.7

	112	111	110
1	11.2	11.1	11.0
2	22.4	22.2	22.0
3	33.6	33.3	33.0
4	44.8	44.4	44.0
5	56.0	55.5	55.0
6	67.2	66.6	66.0
7	78.4	77.7	77.0
8	89.6	88.8	88.0
9	100.8	99.9	99.0

	109	108	107	106
1	10.9	10.8	10.7	10.6
2	21.8	21.6	21.4	21.2
3	32.7	32.4	32.1	31.8
4	43.6	43.2	42.8	42.4
5	54.5	54.0	53.5	53.0
6	65.4	64.8	64.2	63.6
7	76.3	75.6	74.9	74.2
8	87.2	86.4	85.6	84.8
9	98.1	97.2	96.3	95.4

Prop. Pts.

TABLE

3

7°

′	L Sin	d	L Tan	c d	L Cot	L Cos	′
0	9.08 589	103	9.08 914	105	10.91 086	9.99 675	60
1	9.08 692	103	9.09 019	104	10.90 981	9.99 674	59
2	9.08 795	102	9.09 123	104	10.90 877	9.99 672	58
3	9.08 897	102	9.09 227	103	10.90 773	9.99 670	57
4	9.08 999	102	9.09 330	104	10.90 670	9.99 669	56
5	9.09 101	101	9.09 434	103	10.90 566	9.99 667	55
6	9.09 202	102	9.09 537	103	10.90 463	9.99 666	54
7	9.09 304	101	9.09 640	102	10.90 360	9.99 664	53
8	9.09 405	101	9.09 742	103	10.90 258	9.99 663	52
9	9.09 506	100	9.09 845	102	10.90 155	9.99 661	51
10	9.09 606	101	9.09 947	102	10.90 053	9.99 659	50
11	9.09 707	100	9.10 049	101	10.89 951	9.99 658	49
12	9.09 807	100	9.10 150	102	10.89 850	9.99 656	48
13	9.09 907	99	9.10 252	101	10.89 748	9.99 655	47
14	9.10 006	100	9.10 353	101	10.89 647	9.99 653	46
15	9.10 106	99	9.10 454	101	10.89 546	9.99 651	45
16	9.10 205	99	9.10 555	101	10.89 445	9.99 650	44
17	9.10 304	98	9.10 656	100	10.89 344	9.99 648	43
18	9.10 402	99	9.10 756	100	10.89 244	9.99 647	42
19	9.10 501	98	9.10 856	100	10.89 144	9.99 645	41
20	9.10 599	98	9.10 956	100	10.89 044	9.99 643	40
21	9.10 697	98	9.11 056	99	10.88 944	9.99 642	39
22	9.10 795	98	9.11 155	99	10.88 845	9.99 640	38
23	9.10 893	97	9.11 254	99	10.88 746	9.99 638	37
24	9.10 990	97	9.11 353	99	10.88 647	9.99 637	36
25	9.11 087	97	9.11 452	99	10.88 548	9.99 635	35
26	9.11 184	97	9.11 551	98	10.88 449	9.99 633	34
27	9.11 281	96	9.11 649	98	10.88 351	9.99 632	33
28	9.11 377	97	9.11 747	98	10.88 253	9.99 630	32
29	9.11 474	96	9.11 845	98	10.88 155	9.99 629	31
30	9.11 570	96	9.11 943	97	10.88 057	9.99 627	30
31	9.11 666	95	9.12 040	98	10.87 960	9.99 625	29
32	9.11 761	96	9.12 138	97	10.87 862	9.99 624	28
33	9.11 857	95	9.12 235	97	10.87 765	9.99 622	27
34	9.11 952	95	9.12 332	96	10.87 668	9.99 620	26
35	9.12 047	95	9.12 428	97	10.87 572	9.99 618	25
36	9.12 142	94	9.12 525	96	10.87 475	9.99 617	24
37	9.12 236	95	9.12 621	96	10.87 379	9.99 615	23
38	9.12 331	94	9.12 717	96	10.87 283	9.99 613	22
39	9.12 425	94	9.12 813	96	10.87 187	9.99 612	21
40	9.12 519	93	9.12 909	95	10.87 091	9.99 610	20
41	9.12 612	94	9.13 004	95	10.86 996	9.99 608	19
42	9.12 706	93	9.13 099	95	10.86 901	9.99 607	18
43	9.12 799	93	9.13 194	95	10.86 806	9.99 605	17
44	9.12 892	93	9.13 289	95	10.86 711	9.99 603	16
45	9.12 985	93	9.13 384	94	10.86 616	9.99 601	15
46	9.13 078	93	9.13 478	95	10.86 522	9.99 600	14
47	9 13 171	92	9.13 573	94	10.86 427	9.99 598	13
48	9.13 263	92	9.13 667	94	10.86 333	9.99 596	12
49	9.13 355	92	9.13 761	93	10.86 239	9.99 595	11
50	9.13 447	92	9.13 854	94	10.86 146	9.99 593	10
51	9.13 539	91	9.13 948	93	10.86 052	9.99 591	9
52	9.13 630	92	9.14 041	93	10.85 959	9.99 589	8
53	9.13 722	91	9.14 134	93	10.85 866	9.99 588	7
54	9.13 813	91	9.14 227	93	10.85 773	9.99 586	6
55	9.13 904	90	9.14 320	92	10.85 680	9.99 584	5
56	9.13 994	91	9.14 412	92	10.85 588	9.99 582	4
57	9.14 085	90	9.14 504	93	10.85 496	9.99 581	3
58	9.14 175	91	9.14 597	91	10.85 403	9.99 579	2
59	9.14 266	90	9.14 688	92	10.85 312	9.99 577	1
60	9.14 356		9.14 780		10.85 220	9.99 575	0
	L Cos	d	L Cot	c d	L Tan	L Sin	′

82°

Prop. Pts.

	105	104	103
1	10.5	10.4	10.3
2	21.0	20.8	20.6
3	31.5	31.2	30.9
4	42.0	41.6	41.2
5	52.5	52.0	51.5
6	63.0	62.4	61.8
7	73.5	72.8	72.1
8	84.0	83.2	82.4
9	94.5	93.6	92.7

	102	101	99
1	10.2	10.1	9.9
2	20.4	20.2	19.8
3	30.6	30.3	29.7
4	40.8	40.4	39.6
5	51.0	50.5	49.5
6	61.2	60.6	59.4
7	71.4	70.7	69.3
8	81.6	80.8	79.2
9	91.8	90.9	89.1

	98	97	96
1	9.8	9.7	9.6
2	19.6	19.4	19.2
3	29.4	29.1	28.8
4	39.2	38.8	38.4
5	49.0	48.5	48.0
6	58.8	58.2	57.6
7	68.6	67.9	67.2
8	78.4	77.6	76.8
9	88.2	87.3	86.4

	95	94	93
1	9.5	9.4	9.3
2	19.0	18.8	18.6
3	28.5	28.2	27.9
4	38.0	37.6	37.2
5	47.5	47.0	46.5
6	57.0	56.4	55.8
7	66.5	65.8	65.1
8	76.0	75.2	74.4
9	85.5	84.6	83.7

	92	91	90
1	9.2	9.1	9.0
2	18.4	18.2	18.0
3	27.6	27.3	27.0
4	36.8	36.4	36.0
5	46.0	45.5	45.0
6	55.2	54.6	54.0
7	64.4	63.7	63.0
8	73.6	72.8	72.0
9	82.8	81.9	81.0

TABLE

3

8°

′	L Sin	d	L Tan	c d	L Cot	L Cos	′
0	9.14 356	89	9.14 780	92	10.85 220	9.99 575	60
1	9.14 445	90	9.14 872	91	10.85 128	9.99 574	59
2	9.14 535	89	9.14 963	91	10.85 037	9.99 572	58
3	9.14 624	90	9.15 054	91	10.84 946	9.99 570	57
4	9.14 714	89	9.15 145	91	10.84 855	9.99 568	56
5	9.14 803	88	9.15 236	91	10.84 764	9.99 566	55
6	9.14 891	89	9.15 327	90	10.84 673	9.99 565	54
7	9.14 980	89	9.15 417	91	10.84 583	9.99 563	53
8	9.15 069	88	9.15 508	90	10.84 492	9.99 561	52
9	9.15 157	88	9.15 598	90	10.84 402	9.99 559	51
10	9.15 245	88	9.15 688	89	10.84 312	9.99 557	50
11	9.15 333	88	9.15 777	90	10.84 223	9.99 556	49
12	9.15 421	87	9.15 867	89	10.84 133	9.99 554	48
13	9.15 508	88	9.15 956	90	10.84 044	9.99 552	47
14	9.15 596	87	9.16 046	89	10.83 954	9.99 550	46
15	9.15 683	87	9.16 135	89	10.83 865	9.99 548	45
16	9.15 770	87	9.16 224	88	10.83 776	9.99 546	44
17	9.15 857	87	9.16 312	89	10.83 688	9.99 545	43
18	9.15 944	86	9.16 401	88	10.83 599	9.99 543	42
19	9.16 030	86	9.16 489	88	10.83 511	9.99 541	41
20	9.16 116	87	9.16 577	88	10.83 423	9.99 539	40
21	9.16 203	86	9.16 665	88	10.83 335	9.99 537	39
22	9.16 289	85	9.16 753	88	10.83 247	9.99 535	38
23	9.16 374	86	9.16 841	87	10.83 159	9.99 533	37
24	9.16 460	85	9.16 928	88	10.83 072	9.99 532	36
25	9.16 545	86	9.17 016	87	10.82 984	9.99 530	35
26	9.16 631	85	9.17 103	87	10.82 897	9.99 528	34
27	9.16 716	85	9.17 190	87	10.82 810	9.99 526	33
28	9.16 801	85	9.17 277	86	10.82 723	9.99 524	32
29	9.16 886	84	9.17 363	87	10.82 637	9.99 522	31
30	9.16 970	85	9.17 450	86	10.82 550	9.99 520	30
31	9.17 055	84	9.17 536	86	10.82 464	9.99 518	29
32	9.17 139	84	9.17 622	86	10.82 378	9.99 517	28
33	9.17 223	84	9.17 708	86	10.82 292	9.99 515	27
34	9.17 307	84	9.17 794	86	10.82 206	9.99 513	26
35	9.17 391	83	9.17 880	85	10.82 120	9.99 511	25
36	9.17 474	84	9.17 965	86	10.82 035	9.99 509	24
37	9.17 558	83	9.18 051	85	10.81 949	9.99 507	23
38	9.17 641	83	9.18 136	85	10.81 864	9.99 505	22
39	9.17 724	83	9.18 221	85	10.81 779	9.99 503	21
40	9.17 807	83	9.18 306	85	10.81 694	9.99 501	20
41	9.17 890	83	9.18 391	84	10.81 609	9.99 499	19
42	9.17 973	82	9.18 475	85	10.81 525	9.99 497	18
43	9.18 055	82	9.18 560	84	10.81 440	9.99 495	17
44	9.18 137	83	9.18 644	84	10.81 356	9.99 494	16
45	9.18 220	82	9.18 728	84	10.81 272	9.99 492	15
46	9.18 302	81	9.18 812	84	10.81 188	9.99 490	14
47	9.18 383	82	9.18 896	83	10.81 104	9.99 488	13
48	9.18 465	82	9.18 979	84	10.81 021	9.99 486	12
49	9.18 547	81	9.19 063	83	10.80 937	9.99 484	11
50	9.18 628	81	9.19 146	83	10.80 854	9.99 482	10
51	9.18 709	81	9.19 229	83	10.80 771	9.99 480	9
52	9.18 790	81	9.19 312	83	10.80 688	9.99 478	8
53	9.18 871	81	9.19 395	83	10.80 605	9.99 476	7
54	9.18 952	81	9.19 478	83	10.80 522	9.99 474	6
55	9.19 033	80	9.19 561	82	10.80 439	9.99 472	5
56	9.19 113	80	9.19 643	82	10.80 357	9.99 470	4
57	9.19 193	80	9.19 725	82	10.80 275	9.99 468	3
58	9.19 273	80	9.19 807	82	10.80 193	9.99 466	2
59	9.19 353	80	9.19 889	82	10.80 111	9.99 464	1
60	9.19 433		9.19 971		10.80 029	9.99 462	0
	L Cos	d	L Cot	c d	L Tan	L Sin	′

Prop. Pts.

	92	91	90
1	9.2	9.1	9.0
2	18.4	18.2	18.0
3	27.6	27.3	27.0
4	36.8	36.4	36.0
5	46.0	45.5	45.0
6	55.2	54.6	54.0
7	64.4	63.7	63.0
8	73.6	72.8	72.0
9	82.8	81.9	81.0

	89	88	87
1	8.9	8.8	8.7
2	17.8	17.6	17.4
3	26.7	26.4	26.1
4	35.6	35.2	34.8
5	44.5	44.0	43.5
6	53.4	52.8	52.2
7	62.3	61.6	60.9
8	71.2	70.4	69.6
9	80.1	79.2	78.3

	86	85	84
1	8.6	8.5	8.4
2	17.2	17.0	16.8
3	25.8	25.5	25.2
4	34.4	34.0	33.6
5	43.0	42.5	42.0
6	51.6	51.0	50.4
7	60.2	59.5	58.8
8	68.8	68.0	67.2
9	77.4	76.5	75.6

	83	82	81	80
1	8.3	8.2	8.1	8.0
2	16.6	16.4	16.2	16.0
3	24.9	24.6	24.3	24.0
4	33.2	32.8	32.4	32.0
5	41.5	41.0	40.5	40.0
6	49.8	49.2	48.6	48.0
7	58.1	57.4	56.7	56.0
8	66.4	65.6	64.8	64.0
9	74.7	73.8	72.9	72.0

TABLE 3

81°

9°

′	L Sin	d	L Tan	c d	L Cot	L Cos		Prop. Pts.
0	9.19 433		9.19 971		10.80 029	9.99 462	60	
1	9.19 513	80	9.20 053	82	10.79 947	9.99 460	59	
2	9.19 592	79	9.20 134	81	10.79 866	9.99 458	58	
3	9.19 672	80	9.20 216	82	10.79 784	9.99 456	57	
4	9.19 751	79	9.20 297	81	10.79 703	9.99 454	56	**82 · 81 · 80**
5	9.19 830	79	9.20 378	81	10.79 622	9.99 452	55	1 8.2 8.1 8.0
6	9.19 909	79	9.20 459	81	10.79 541	9.99 450	54	2 16.4 16.2 16.0
7	9.19 988	79	9.20 540	81	10.79 460	9.99 448	53	3 24.6 24.3 24.0
8	9.20 067	79	9.20 621	81	10.79 379	9.99 446	52	4 32.8 32.4 32.0
9	9.20 145	78	9.20 701	80	10.79 299	9.99 444	51	5 41.0 40.5 40.0
10	9.20 223	78	9.20 782	81	10.79 218	9.99 442	50	6 49.2 48.6 48.0
11	9.20 302	79	9.20 862	80	10.79 138	9.99 440	49	7 57.4 56.7 56.0
12	9.20 380	78	9.20 942	80	10.79 058	9.99 438	48	8 65.6 64.8 64.0
13	9.20 458	78	9.21 022	80	10.78 978	9.99 436	47	9 73.8 72.9 72.0
14	9.20 535	77	9.21 102	80	10.78 898	9.99 434	46	
15	9.20 613	78	9.21 182	80	10.78 818	9.99 432	45	**79 · 78 · 77**
16	9.20 691	78	9.21 261	79	10.78 739	9.99 429	44	1 7.9 7.8 7.7
17	9.20 768	77	9.21 341	80	10.78 659	9.99 427	43	2 15.8 15.6 15.4
18	9.20 845	77	9.21 420	79	10.78 580	9.99 425	42	3 23.7 23.4 23.1
19	9.20 922	77	9.21 499	79	10.78 501	9.99 423	41	4 31.6 31.2 30.8
20	9.20 999	77	9.21 578	79	10.78 422	9.99 421	40	5 39.5 39.0 38.5
21	9.21 076	77	9.21 657	79	10.78 343	9.99 419	39	6 47.4 46.8 46.2
22	9.21 153	76	9.21 736	78	10.78 264	9.99 417	38	7 55.3 54.6 53.9
23	9.21 229	77	9.21 814	79	10.78 186	9.99 415	37	8 63.2 62.4 61.6
24	9.21 306	76	9.21 893	78	10.78 107	9.99 413	36	9 71.1 70.2 69.3
25	9.21 382	76	9.21 971	78	10.78 029	9.99 411	35	
26	9.21 458	76	9.22 049	78	10.77 951	9.99 409	34	**76 · 75 · 74**
27	9.21 534	76	9.22 127	78	10.77 873	9.99 407	33	1 7.6 7.5 7.4
28	9.21 610	75	9.22 205	78	10.77 795	9.99 404	32	2 15.2 15.0 14.8
29	9.21 685	76	9.22 283	78	10.77 717	9.99 402	31	3 22.8 22.5 22.2
30	9.21 761	75	9.22 361	77	10.77 639	9.99 400	30	4 30.4 30.0 29.6
31	9.21 836	76	9.22 438	78	10.77 562	9.99 398	29	5 38.0 37.5 37.0
32	9.21 912	75	9.22 516	77	10.77 484	9.99 396	28	6 45.6 45.0 44.4
33	9.21 987	75	9.22 593	77	10.77 407	9.99 394	27	7 53.2 52.5 51.8
34	9.22 062	75	9.22 670	77	10.77 330	9.99 392	26	8 60.8 60.0 59.2
35	9.22 137	74	9.22 747	77	10.77 253	9.99 390	25	9 68.4 67.5 66.6
36	9.22 211	75	9.22 824	77	10.77 176	9.99 388	24	
37	9.22 286	75	9.22 901	76	10.77 099	9.99 385	23	**73 · 72 · 71**
38	9.22 361	74	9.22 977	77	10.77 023	9.99 383	22	1 7.3 7.2 7.1
39	9.22 435	74	9.23 054	76	10.76 946	9.99 381	21	2 14.6 14.4 14.2
40	9.22 509	74	9.23 130	76	10.76 870	9.99 379	20	3 21.9 21.6 21.3
41	9.22 583	74	9.23 206	77	10.76 794	9.99 377	19	4 29.2 28.8 28.4
42	9.22 657	74	9.23 283	76	10.76 717	9.99 375	18	5 36.5 36.0 35.5
43	9.22 731	74	9.23 359	76	10.76 641	9.99 372	17	6 43.8 43.2 42.6
44	9.22 805	73	9.23 435	75	10.76 565	9.99 370	16	7 51.1 50.4 49.7
45	9.22 878	74	9.23 510	76	10.76 490	9.99 368	15	8 58.4 57.6 56.8
46	9.22 952	73	9.23 586	75	10.76 414	9.99 366	14	9 65.7 64.8 63.9
47	9.23 025	73	9.23 661	76	10.76 339	9.99 364	13	
48	9.23 098	73	9.23 737	75	10.76 263	9.99 362	12	
49	9.23 171	73	9.23 812	75	10.76 188	9.99 359	11	
50	9.23 244	73	9.23 887	75	10.76 113	9.99 357	10	**3 · 2**
51	9.23 317	73	9.23 962	75	10.76 038	9.99 355	9	1 0.3 0.2
52	9.23 390	72	9.24 037	75	10.75 963	9.99 353	8	2 0.6 0.4
53	9.23 462	73	9.24 112	74	10.75 888	9.99 351	7	3 0.9 0.6
54	9.23 535	72	9.24 186	75	10.75 814	9.99 348	6	4 1.2 0.8
55	9.23 607	72	9.24 261	74	10.75 739	9.99 346	5	5 1.5 1.0
56	9.23 679	73	9.24 335	75	10.75 665	9.99 344	4	6 1.8 1.2
57	9.23 752	71	9.24 410	74	10.75 590	9.99 342	3	7 2.1 1.4
58	9.23 823	72	9.24 484	74	10.75 516	9.99 340	2	8 2.4 1.6
59	9.23 895	72	9.24 558	74	10.75 442	9.99 337	1	9 2.7 1.8
60	9.23 967		9.24 632		10.75 368	9.99 335	0	
	L Cos	d	L Cot	c d	L Tan	L Sin	′	Prop. Pts.

80°

10°

′	L Sin	d	L Tan	c d	L Cot	L Cos	d	′
0	9.23 967	72	9.24 632	74	10.75 368	9.99 335	2	**60**
1	9.24 039	71	9.24 706	73	10.75 294	9.99 333	2	59
2	9.24 110	71	9.24 779	74	10.75 221	9.99 331	3	58
3	9.24 181	72	9.24 853	73	10.75 147	9.99 328	2	57
4	9.24 253	71	9.24 926	74	10.75 074	9.99 326	2	56
5	9.24 324	71	9.25 000	73	10.75 000	9.99 324	2	55
6	9.24 395	71	9.25 073	73	10.74 927	9.99 322	3	54
7	9.24 466	70	9.25 146	73	10.74 854	9.99 319	2	53
8	9.24 536	71	9.25 219	73	10.74 781	9.99 317	2	52
9	9.24 607	70	9.25 292	73	10.74 708	9.99 315	2	51
10	9.24 677	71	9.25 365	72	10.74 635	9.99 313	3	**50**
11	9.24 748	70	9.25 437	73	10.74 563	9.99 310	2	49
12	9.24 818	70	9.25 510	72	10.74 490	9.99 308	2	48
13	9.24 888	70	9.25 582	73	10.74 418	9.99 306	2	47
14	9.24 958	70	9.25 655	72	10.74 345	9.99 304	3	46
15	9.25 028	70	9.25 727	72	10.74 273	9.99 301	2	45
16	9.25 098	70	9.25 799	72	10.74 201	9.99 299	2	44
17	9.25 168	69	9.25 871	72	10.74 129	9.99 297	3	43
18	9.25 237	70	9.25 943	72	10.74 057	9.99 294	2	42
19	9.25 307	69	9.26 015	71	10.73 985	9.99 292	2	41
20	9.25 376	69	9.26 086	72	10.73 914	9.99 290	2	**40**
21	9.25 445	69	9.26 158	71	10.73 842	9.99 288	3	39
22	9.25 514	69	9.26 229	72	10.73 771	9.99 285	2	38
23	9.25 583	69	9.26 301	71	10.73 699	9.99 283	2	37
24	9.25 652	69	9.26 372	71	10.73 628	9.99 281	3	36
25	9.25 721	69	9.26 443	71	10.73 557	9.99 278	2	35
26	9.25 790	68	9.26 514	71	10.73 486	9.99 276	2	34
27	9.25 858	69	9.26 585	70	10.73 415	9.99 274	3	33
28	9.25 927	68	9.26 655	71	10.73 345	9.99 271	2	32
29	9.25 995	68	9.26 726	71	10.73 274	9.99 269	2	31
30	9.26 063	68	9.26 797	70	10.73 203	9.99 267	3	**30**
31	9.26 131	68	9.26 867	70	10.73 133	9.99 264	2	29
32	9.26 199	68	9.26 937	71	10.73 063	9.99 262	2	28
33	9.26 267	68	9.27 008	70	10.72 992	9.99 260	3	27
34	9.26 335	68	9.27 078	70	10.72 922	9.99 257	2	26
35	9.26 403	67	9.27 148	70	10.72 852	9.99 255	3	25
36	9.26 470	68	9.27 218	70	10.72 782	9.99 252	2	24
37	9.26 538	67	9.27 288	69	10.72 712	9.99 250	2	23
38	9.26 605	67	9.27 357	70	10.72 643	9.99 248	3	22
39	9.26 672	67	9.27 427	69	10.72 573	9.99 245	2	21
40	9.26 739	67	9.27 496	70	10.72 504	9.99 243	2	**20**
41	9.26 806	67	9.27 566	69	10.72 434	9.99 241	3	19
42	9.26 873	67	9.27 635	69	10.72 365	9.99 238	2	18
43	9.26 940	67	9.27 704	69	10.72 296	9.99 236	3	17
44	9.27 007	66	9.27 773	69	10.72 227	9.99 233	2	16
45	9.27 073	67	9.27 842	69	10.72 158	9.99 231	2	15
46	9.27 140	66	9.27 911	69	10.72 089	9.99 229	3	14
47	9.27 206	67	9.27 980	69	10.72 020	9.99 226	2	13
48	9.27 273	66	9.28 049	68	10.71 951	9.99 224	3	12
49	9.27 339	66	9.28 117	69	10.71 883	9.99 221	2	11
50	9.27 405	66	9.28 186	68	10.71 814	9.99 219	2	**10**
51	9.27 471	66	9.28 254	69	10.71 746	9.99 217	3	9
52	9.27 537	65	9.28 323	68	10.71 677	9.99 214	2	8
53	9.27 602	66	9.28 391	68	10.71 609	9.99 212	3	7
54	9.27 668	66	9.28 459	68	10.71 541	9.99 209	2	6
55	9.27 734	65	9.28 527	68	10.71 473	9.99 207	3	5
56	9.27 799	65	9.28 595	67	10.71 405	9.99 204	2	4
57	9.27 864	66	9.28 662	68	10.71 338	9.99 202	2	3
58	9.27 930	65	9.28 730	68	10.71 270	9.99 200	3	2
59	9.27 995	65	9.28 798	67	10.71 202	9.99 197	2	1
60	9.28 060		9.28 865		10.71 135	9.99 195		**0**
′	L Cos	d	L Cot	c d	L Tan	L Sin	d	′

Prop. Pts.

	74	73	72
1	7.4	7.3	7.2
2	14.8	14.6	14.4
3	22.2	21.9	21.6
4	29.6	29.2	28.8
5	37.0	36.5	36.0
6	44.4	43.8	43.2
7	51.8	51.1	50.4
8	59.2	58.4	57.6
9	66.6	65.7	64.8

	71	70	69
1	7.1	7.0	6.9
2	14.2	14.0	13.8
3	21.3	21.0	20.7
4	28.4	28.0	27.6
5	35.5	35.0	34.5
6	42.6	42.0	41.4
7	49.7	49.0	48.3
8	56.8	56.0	55.2
9	63.9	63.0	62.1

	68	67	66
1	6.8	6.7	6.6
2	13.6	13.4	13.2
3	20.4	20.1	19.8
4	27.2	26.8	26.4
5	34.0	33.5	33.0
6	40.8	40.2	39.6
7	47.6	46.9	46.2
8	54.4	53.6	52.8
9	61.2	60.3	59.4

	65	3	2
1	6.5	0.3	0.2
2	13.0	0.6	0.4
3	19.5	0.9	0.6
4	26.0	1.2	0.8
5	32.5	1.5	1.0
6	39.0	1.8	1.2
7	45.5	2.1	1.4
8	52.0	2.4	1.6
9	58.5	2.7	1.8

79°

TABLE

3

11°

′	L Sin	d	L Tan	c d	L Cot	L Cos	d	′	Prop. Pts.
0	9.28 060		9.28 865		10.71 135	9.99 195		60	
1	9.28 125	65	9.28 933	68	10.71 067	9.99 192	3	59	
2	9.28 190	65	9.29 000	67	10.71 000	9.99 190	2	58	
3	9.28 254	64	9.29 067	67	10.70 933	9.99 187	2	57	
4	9.28 319	65	9.29 134	67	10.70 866	9.99 185	3	56	
5	9.28 384	65	9.29 201	67	10.70 799	9.99 182	2	55	
6	9.28 448	64	9.29 268	67	10.70 732	9.99 180	3	54	
7	9.28 512	64	9.29 335	67	10.70 665	9.99 177	2	53	
8	9.28 577	65	9.29 402	67	10.70 598	9.99 175	3	52	
9	9.28 641	64	9.29 468	66	10.70 532	9.99 172	2	51	
10	9.28 705	64	9.29 535	67	10.70 465	9.99 170	3	50	
11	9.28 769	64	9.29 601	66	10.70 399	9.99 167	2	49	
12	9.28 833	64	9.29 668	67	10.70 332	9.99 165	3	48	
13	9.28 896	63	9.29 734	66	10.70 266	9.99 162	2	47	
14	9.28 960	64	9.29 800	66	10.70 200	9.99 160	3	46	
15	9.29 024	64	9.29 866	66	10.70 134	9.99 157	2	45	
16	9.29 087	63	9.29 932	66	10.70 068	9.99 155	2	44	
17	9.29 150	63	9.29 998	66	10.70 002	9.99 152	3	43	
18	9.29 214	64	9.30 064	66	10.69 936	9.99 150	2	42	
19	9.29 277	63	9.30 130	66	10.69 870	9.99 147	3	41	
20	9.29 340	63	9.30 195	65	10.69 805	9.99 145	2	40	
21	9.29 403	63	9.30 261	66	10.69 739	9.99 142	3	39	
22	9.29 466	63	9.30 326	65	10.69 674	9.99 140	2	38	
23	9.29 529	63	9.30 391	65	10.69 609	9.99 137	3	37	
24	9.29 591	62	9.30 457	66	10.69 543	9.99 135	2	36	
25	9.29 654	63	9.30 522	65	10.69 478	9.99 132	2	35	
26	9.29 716	62	9.30 587	65	10.69 413	9.99 130	3	34	
27	9.29 779	63	9.30 652	65	10.69 348	9.99 127	3	33	
28	9.29 841	62	9.30 717	65	10.69 283	9.99 124	2	32	
29	9.29 903	62	9.30 782	64	10.69 218	9.99 122	3	31	
30	9.29 966	63	9.30 846	65	10.69 154	9.99 119	2	30	
31	9.30 028	62	9.30 911	64	10.69 089	9.99 117	3	29	
32	9.30 090	62	9.30 975	65	10.69 025	9.99 114	2	28	
33	9.30 151	61	9.31 040	64	10.68 960	9.99 112	3	27	
34	9.30 213	62	9.31 104	64	10.68 896	9.99 109	3	26	
35	9.30 275	62	9.31 168	65	10.68 832	9.99 106	2	25	
36	9.30 336	61	9.31 233	64	10.68 767	9.99 104	3	24	
37	9.30 398	62	9.31 297	64	10.68 703	9.99 101	2	23	
38	9.30 459	61	9.31 361	64	10.68 639	9.99 099	3	22	
39	9.30 521	62	9.31 425	64	10.68 575	9.99 096	3	21	
40	9.30 582	61	9.31 489	64	10.68 511	9.99 093	2	20	
41	9.30 643	61	9.31 552	63	10.68 448	9.99 091	3	19	
42	9.30 704	61	9.31 616	64	10.68 384	9.99 088	2	18	
43	9.30 765	61	9.31 679	63	10.68 321	9.99 086	3	17	
44	9.30 826	61	9.31 743	64	10.68 257	9.99 083	3	16	
45	9.30 887	60	9.31 806	64	10.68 194	9.99 080	2	15	
46	9.30 947	61	9.31 870	63	10.68 130	9.99 078	3	14	
47	9.31 008	60	9.31 933	63	10.68 067	9.99 075	3	13	
48	9.31 068	61	9.31 996	63	10.68 004	9.99 072	2	12	
49	9.31 129	60	9.32 059	63	10.67 941	9.99 070	3	11	
50	9.31 189	61	9.32 122	63	10.67 878	9.99 067	3	10	
51	9.31 250	60	9.32 185	63	10.67 815	9.99 064	2	9	
52	9.31 310	60	9.32 248	63	10.67 752	9.99 062	3	8	
53	9.31 370	60	9.32 311	62	10.67 689	9.99 059	3	7	
54	9.31 430	60	9.32 373	63	10.67 627	9.99 056	2	6	
55	9.31 490	59	9.32 436	62	10.67 564	9.99 054	3	5	
56	9.31 549	60	9.32 498	63	10.67 502	9.99 051	3	4	
57	9.31 609	60	9.32 561	62	10.67 439	9.99 048	2	3	
58	9.31 669	59	9.32 623	62	10.67 377	9.99 046	3	2	
59	9.31 728	60	9.32 685	62	10.67 315	9.99 043	3	1	
60	9.31 788		9.32 747		10.67 253	9.99 040		0	
	L Cos	d	L Cot	c d	L Tan	L Sin	d	′	Prop. Pts.

Prop. Pts.

	68	67	66
1	6.8	6.7	6.6
2	13.6	13.4	13.2
3	20.4	20.1	19.8
4	27.2	26.8	26.4
5	34.0	33.5	33.0
6	40.8	40.2	39.6
7	47.6	46.9	46.2
8	54.4	53.6	52.8
9	61.2	60.3	59.4

	65	64	63
1	6.5	6.4	6.3
2	13.0	12.8	12.6
3	19.5	19.2	18.9
4	26.0	25.6	25.2
5	32.5	32.0	31.5
6	39.0	38.4	37.8
7	45.5	44.8	44.1
8	52.0	51.2	50.4
9	58.5	57.6	56.7

	62	61	60
1	6.2	6.1	6.0
2	12.4	12.2	12.0
3	18.6	18.3	18.0
4	24.8	24.4	24.0
5	31.0	30.5	30.0
6	37.2	36.6	36.0
7	43.4	42.7	42.0
8	49.6	48.8	48.0
9	55.8	54.9	54.0

	59	3	2
1	5.9	0.3	0.2
2	11.8	0.6	0.4
3	17.7	0.9	0.6
4	23.6	1.2	0.8
5	29.5	1.5	1.0
6	35.4	1.8	1.2
7	41.3	2.1	1.4
8	47.2	2.4	1.6
9	53.1	2.7	1.8

TABLE 3

78°

12°

′	L Sin	d	L Tan	c d	L Cot	L Cos	d		Prop. Pts.
0	9.31 788		9.32 747		10.67 253	9.99 040		60	
1	9.31 847	59	9.32 810	63	10.67 190	9.99 038	2	59	
2	9.31 907	60	9.32 872	62	10.67 128	9.99 035	3	58	
3	9.31 966	59	9.32 933	61	10.67 067	9.99 032	3	57	
4	9.32 025	59	9.32 995	62	10.67 005	9.99 030	2	56	
5	9.32 084	59	9.33 057	62	10.66 943	9.99 027	3	55	**63** **62** **61**
6	9.32 143	59	9.33 119	62	10.66 881	9.99 024	3	54	1 6.3 6.2 6.1
7	9.32 202	59	9.33 180	61	10.66 820	9.99 022	3	53	2 12.6 12.4 12.2
8	9.32 261	59	9.33 242	62	10.66 758	9.99 019	3	52	3 18.9 18.6 18.3
9	9.32 319	58	9.33 303	61	10.66 697	9.99 016	3	51	4 25.2 24.8 24.4
10	9.32 378	59	9.33 365	62	10.66 635	9.99 013	3	50	5 31.5 31.0 30.5
11	9.32 437	59	9.33 426	61	10.66 574	9.99 011	2	49	6 37.8 37.2 36.6
12	9.32 495	58	9.33 487	61	10.66 513	9.99 008	3	48	7 44.1 43.4 42.7
13	9.32 553	58	9.33 548	61	10.66 452	9.99 005	3	47	8 50.4 49.6 48.8
14	9.32 612	59	9.33 609	61	10.66 391	9.99 002	3	46	9 56.7 55.8 54.9
15	9.32 670	58	9.33 670	61	10.66 330	9.99 000	2	45	
16	9.32 728	58	9.33 731	61	10.66 269	9.98 997	3	44	
17	9.32 786	58	9.33 792	61	10.66 208	9.98 994	3	43	
18	9.32 844	58	9.33 853	61	10.66 147	9.98 991	3	42	
19	9.32 902	58	9.33 913	60	10.66 087	9.98 989	2	41	**60** **59** **58**
20	9.32 960	58	9.33 974	61	10.66 026	9.98 986	3	40	1 6.0 5.9 5.8
21	9.33 018	58	9.34 034	60	10.65 966	9.98 983	3	39	2 12.0 11.8 11.6
22	9.33 075	57	9.34 095	61	10.65 905	9.98 980	3	38	3 18.0 17.7 17.4
23	9.33 133	58	9.34 155	60	10.65 845	9.98 978	2	37	4 24.0 23.6 23.2
24	9.33 190	57	9.34 215	60	10.65 785	9.98 975	3	36	5 30.0 29.5 29.0
25	9.33 248	58	9.34 276	61	10.65 724	9.98 972	3	35	6 36.0 35.4 34.8
26	9.33 305	57	9.34 336	60	10.65 664	9.98 969	3	34	7 42.0 41.3 40.6
27	9.33 362	57	9.34 396	60	10.65 604	9.98 967	2	33	8 48.0 47.2 46.4
28	9.33 420	58	9.34 456	60	10.65 544	9.98 964	3	32	9 54.0 53.1 52.2
29	9.33 477	57	9.34 516	60	10.65 484	9.98 961	3	31	
30	9.33 534	57	9.34 576	60	10.65 424	9.98 958	3	30	
31	9.33 591	57	9.34 635	59	10.65 365	9.98 955	2	29	
32	9.33 647	56	9.34 695	60	10.65 305	9.98 953	3	28	
33	9.33 704	57	9.34 755	60	10.65 245	9.98 950	3	27	**57** **56** **55**
34	9.33 761	57	9.34 814	59	10.65 186	9.98 947	3	26	1 5.7 5.6 5.5
35	9.33 818	57	9.34 874	60	10.65 126	9.98 944	3	25	2 11.4 11.2 11.0
36	9.33 874	56	9.34 933	59	10.65 067	9.98 941	3	24	3 17.1 16.8 16.5
37	9.33 931	57	9.34 992	59	10.65 008	9.98 938	2	23	4 22.8 22.4 22.0
38	9.33 987	56	9.35 051	59	10.64 949	9.98 936	3	22	5 28.5 28.0 27.5
39	9.34 043	56	9.35 111	60	10.64 889	9.98 933	3	21	6 34.2 33.6 33.0
40	9.34 100	57	9.35 170	59	10.64 830	9.98 930	3	20	7 39.9 39.2 38.5
41	9.34 156	56	9.35 229	59	10.64 771	9.98 927	3	19	8 45.6 44.8 44.0
42	9.34 212	56	9.35 288	59	10.64 712	9.98 924	3	18	9 51.3 50.4 49.5
43	9.34 268	56	9.35 347	59	10.64 653	9.98 921	3	17	
44	9.34 324	56	9.35 405	58	10.64 595	9.98 919	2	16	
45	9.34 380	56	9.35 464	59	10.64 536	9.98 916	3	15	
46	9.34 436	56	9.35 523	59	10.64 477	9.98 913	3	14	
47	9.34 491	55	9.35 581	58	10.64 419	9.98 910	3	13	**3** **2**
48	9.34 547	56	9.35 640	59	10.64 360	9.98 907	3	12	1 0.3 0.2
49	9.34 602	55	9.35 698	58	10.64 302	9.98 904	3	11	2 0.6 0.4
50	9.34 658	56	9.35 757	59	10.64 243	9.98 901	3	10	3 0.9 0.6
51	9.34 713	55	9.35 815	58	10.64 185	9.98 898	2	9	4 1.2 0.8
52	9.34 769	56	9.35 873	58	10.64 127	9.98 896	3	8	5 1.5 1.0
53	9.34 824	55	9.35 931	58	10.64 069	9.98 893	3	7	6 1.8 1.2
54	9.34 879	55	9.35 989	58	10.64 011	9.98 890	3	6	7 2.1 1.4
55	9.34 934	55	9.36 047	58	10.63 953	9.98 887	3	5	8 2.4 1.6
56	9.34 989	55	9.36 105	58	10.63 895	9.98 884	3	4	9 2.7 1.8
57	9.35 044	55	9.36 163	58	10.63 837	9.98 881	3	3	
58	9.35 099	55	9.36 221	58	10.63 779	9.98 878	3	2	
59	9.35 154	55	9.36 279	57	10.63 721	9.98 875	3	1	
60	9.35 209		9.36 336		10.63 664	9.98 872		0	
	L Cos	d	L Cot	c d	L Tan	L Sin	d	′	Prop. Pts.

77°

TABLE

13°

′	L Sin	d	L Tan	c d	L Cot	L Cos	d	′
0	9.35 209	54	9.36 336	58	10.63 664	9.98 872	3	60
1	9.35 263	55	9.36 394	58	10.63 606	9.98 869	2	59
2	9.35 318	55	9.36 452	57	10.63 548	9.98 867	3	58
3	9.35 373	54	9.36 509	57	10.63 491	9.98 864	3	57
4	9.35 427	54	9.36 566	58	10.63 434	9.98 861	3	56
5	9.35 481	55	9.36 624	57	10.63 376	9.98 858	3	55
6	9.35 536	54	9.36 681	57	10.63 319	9.98 855	3	54
7	9.35 590	54	9.36 738	57	10.63 262	9.98 852	3	53
8	9.35 644	54	9.36 795	57	10.63 205	9.98 849	3	52
9	9.35 698	54	9.36 852	57	10.63 148	9.98 846	3	51
10	9 35 752	54	9.36 909	57	10.63 091	9.98 843	3	50
11	9.35 806	54	9.36 966	57	10.63 034	9.98 840	3	49
12	9.35 860	54	9.37 023	57	10.62 977	9.98 837	3	48
13	9.35 914	54	9.37 080	57	10.62 920	9.98 834	3	47
14	9.35 968	54	9.37 137	56	10.62 863	9.98 831	3	46
15	9.36 022	53	9.37 193	57	10.62 807	9.98 828	3	45
16	9.36 075	54	9.37 250	57	10.62 750	9.98 825	3	44
17	9.36 129	53	9.37 306	57	10.62 694	9.98 822	3	43
18	9.36 182	54	9.37 363	56	10.62 637	9.98 819	3	42
19	9.36 236	53	9.37 419	57	10.62 581	9.98 816	3	41
20	9.36 289	53	9.37 476	56	10.62 524	9.98 813	3	40
21	9.36 342	53	9.37 532	56	10.62 468	9.98 810	3	39
22	9.36 395	54	9.37 588	56	10.62 412	9.98 807	3	38
23	9.36 449	53	9.37 644	56	10.62 356	9.98 804	3	37
24	9.36 502	53	9.37 700	56	10.62 300	9.98 801	3	36
25	9.36 555	53	9.37 756	56	10.62 244	9.98 798	3	35
26	9.36 608	52	9.37 812	56	10.62 188	9.98 795	3	34
27	9.36 660	53	9.37 868	56	10.62 132	9.98 792	3	33
28	9.36 713	53	9.37 924	56	10.62 076	9.98 789	3	32
29	9.36 766	53	9.37 980	55	10.62 020	9.98 786	3	31
30	9.36 819	52	9.38 035	56	10.61 965	9.98 783	3	30
31	9.36 871	53	9.38 091	56	10.61 909	9.98 780	3	29
32	9.36 924	52	9.38 147	55	10.61 853	9.98 777	3	28
33	9.36 976	52	9.38 202	55	10.61 798	9.98 774	3	27
34	9.37 028	52	9.38 257	56	10.61 743	9.98 771	3	26
35	9.37 081	52	9.38 313	55	10.61 687	9.98 768	3	25
36	9.37 133	52	9.38 368	55	10.61 632	9.98 765	3	24
37	9.37 185	52	9.38 423	56	10.61 577	9.98 762	3	23
38	9.37 237	52	9.38 479	55	10.61 521	9.98 759	3	22
39	9.37 289	52	9.38 534	55	10.61 466	9.98 756	3	21
40	9.37 341	52	9.38 589	55	10.61 411	9.98 753	3	20
41	9.37 393	52	9.38 644	55	10.61 356	9.98 750	4	19
42	9.37 445	52	9.38 699	55	10.61 301	9.98 746	3	18
43	9.37 497	52	9.38 754	54	10.61 246	9.98 743	3	17
44	9.37 549	51	9.38 808	55	10.61 192	9.98 740	3	16
45	9.37 600	52	9.38 863	55	10.61 137	9.98 737	3	15
46	9.37 652	51	9.38 918	54	10.61 082	9.98 734	3	14
47	9.37 703	52	9.38 972	55	10.61 028	9.98 731	3	13
48	9.37 755	51	9.39 027	55	10.60 973	9.98 728	3	12
49	9.37 806	52	9.39 082	54	10.60 918	9.98 725	3	11
50	9.37 858	51	9.39 136	55	10.60 864	9.98 722	3	10
51	9.37 909	51	9.39 190	55	10.60 810	9.98 719	4	9
52	9.37 960	51	9.39 245	54	10.60 755	9.98 715	3	8
53	9.38 011	51	9.39 299	54	10.60 701	9.98 712	3	7
54	9.38 062	51	9.39 353	54	10.60 647	9.98 709	3	6
55	9.38 113	51	9.39 407	54	10.60 593	9.98 706	3	5
56	9.38 164	51	9.39 461	54	10.60 539	9.98 703	3	4
57	9.38 215	51	9.39 515	54	10.60 485	9.98 700	3	3
58	9.38 266	51	9.39 569	54	10.60 431	9.98 697	3	2
59	9.38 317	51	9.39 623	54	10.60 377	9.98 694	4	1
60	9.38 368		9.39 677		10.60 323	9.98 690		0
	L Cos	d	L Cot	c d	L Tan	L Sin	d	′

Prop. Pts.

	58	57	56
1	5.8	5.7	5.6
2	11.6	11.4	11.2
3	17.4	17.1	16.8
4	23.2	22.8	22.4
5	29.0	28.5	28.0
6	34.8	34.2	33.6
7	40.6	39.9	39.2
8	46.4	45.6	44.8
9	52.2	51.3	50.4

	55	54	53
1	5.5	5.4	5.3
2	11.0	10.8	10.6
3	16.5	16.2	15.9
4	22.0	21.6	21.2
5	27.5	27.0	26.5
6	33.0	32.4	31.8
7	38.5	37.8	37.1
8	44.0	43.2	42.4
9	49.5	48.6	47.7

	52	51
1	5.2	5.1
2	10.4	10.2
3	15.6	15.3
4	20.8	20.4
5	26.0	25.5
6	31.2	30.6
7	36.4	35.7
8	41.6	40.8
9	46.8	45.9

	4	3	2
1	0.4	0.3	0.2
2	0.8	0.6	0.4
3	1.2	0.9	0.6
4	1.6	1.2	0.8
5	2.0	1.5	1.0
6	2.4	1.8	1.2
7	2.8	2.1	1.4
8	3.2	2.4	1.6
9	3.6	2.7	1.8

TABLE

3

76°

14°

′	L Sin	d	L Tan	c d	L Cot	L Cos	d	′
0	9.38 368	50	9.39 677	54	10.60 323	9.98 690		60
1	9.38 418	51	9.39 731	54	10.60 269	9.98 687	3	59
2	9.38 469	50	9.39 785	53	10.60 215	9.98 684	3	58
3	9.38 519	51	9.39 838	54	10.60 162	9.98 681	3	57
4	9.38 570	50	9.39 892	53	10.60 108	9.98 678	3	56
5	9.38 620	50	9.39 945	54	10.60 055	9.98 675	3	55
6	9.38 670	51	9.39 999	53	10.60 001	9.98 671	4	54
7	9.38 721	50	9.40 052	54	10.59 948	9.98 668	3	53
8	9.38 771	50	9.40 106	53	10.59 894	9.98 665	3	52
9	9.38 821	50	9.40 159	53	10.59 841	9.98 662	3	51
10	9.38 871	50	9.40 212	54	10.59 788	9.98 659	3	50
11	9.38 921	50	9.40 266	53	10.59 734	9.98 656	4	49
12	9.38 971	50	9.40 319	53	10.59 681	9.98 652	3	48
13	9.39 021	50	9.40 372	53	10.59 628	9.98 649	3	47
14	9.39 071	50	9.40 425	53	10.59 575	9.98 646	3	46
15	9.39 121	49	9.40 478	53	10.59 522	9.98 643	3	45
16	9.39 170	50	9.40 531	53	10.59 469	9.98 640	4	44
17	9.39 220	50	9.40 584	52	10.59 416	9.98 636	3	43
18	9.39 270	49	9.40 636	53	10.59 364	9.98 633	3	42
19	9.39 319	50	9.40 689	53	10.59 311	9.98 630	3	41
20	9.39 369	49	9.40 742	53	10.59 258	9.98 627	4	40
21	9.39 418	49	9.40 795	52	10.59 205	9.98 623	3	39
22	9.39 467	50	9.40 847	53	10.59 153	9.98 620	3	38
23	9.39 517	49	9.40 900	52	10.59 100	9.98 617	3	37
24	9.39 566	49	9.40 952	53	10.59 048	9.98 614	4	36
25	9.39 615	49	9.41 005	52	10.58 995	9.98 610	3	35
26	9.39 664	49	9.41 057	52	10.58 943	9.98 607	3	34
27	9.39 713	49	9.41 109	52	10.58 891	9.98 604	3	33
28	9.39 762	49	9.41 161	53	10.58 839	9.98 601	4	32
29	9.39 811	49	9.41 214	52	10.58 786	9.98 597	3	31
30	9.39 860	49	9.41 266	52	10.58 734	9.98 594	3	30
31	9.39 909	49	9.41 318	52	10.58 682	9.98 591	3	29
32	9.93 958	48	9.41 370	52	10.58 630	9.98 588	4	28
33	9.40 006	49	9.41 422	52	10.58 578	9.98 584	3	27
34	9.40 055	48	9.41 474	52	10.58 526	9.98 581	3	26
35	9.40 103	49	9.41 526	52	10.58 474	9.98 578	4	25
36	9.40 152	48	9.41 578	51	10.58 422	9.98 574	3	24
37	9.40 200	49	9.41 629	52	10.58 371	9.98 571	3	23
38	9.40 249	48	9.41 681	52	10.58 319	9.98 568	3	22
39	9.40 297	49	9.41 733	51	10.58 267	9.98 565	4	21
40	9.40 346	48	9.41 784	52	10.58 216	9.98 561	3	20
41	9.40 394	48	9.41 836	51	10.58 164	9.98 558	3	19
42	9.40 442	48	9.41 887	52	10.58 113	9.98 555	4	18
43	9.40 490	48	9.41 939	51	10.58 061	9.98 551	3	17
44	9.40 538	48	9.41 990	51	10.58 010	9.98 548	3	16
45	9.40 586	48	9.42 041	52	10.57 959	9.98 545	4	15
46	9.40 634	48	9.42 093	51	10.57 907	9.98 541	3	14
47	9.40 682	48	9.42 144	51	10.57 856	9.98 538	3	13
48	9.40 730	48	9.42 195	51	10.57 805	9.98 535	4	12
49	9.40 778	47	9.42 246	51	10.57 754	9.98 531	3	11
50	9.40 825	48	9.42 297	51	10.57 703	9.98 528	3	10
51	9.40 873	48	9.42 348	51	10.57 652	9.98 525	4	9
52	9.40 921	47	9.42 399	51	10.57 601	9.98 521	3	8
53	9.40 968	48	9.42 450	51	10.57 550	9.98 518	3	7
54	9.41 016	47	9.42 501	51	10.57 499	9.98 515	4	6
55	9.41 063	48	9.42 552	51	10.57 448	9.98 511	3	5
56	9.41 111	47	9.42 603	50	10.57 397	9.98 508	3	4
57	9.41 158	47	9.42 653	51	10.57 347	9.98 505	4	3
58	9.41 205	47	9.42 704	51	10.57 296	9.98 501	4	2
59	9.41 252	48	9.42 755	50	10.57 245	9.98 498	4	1
60	9.41 300		9.42 805		10.57 195	9.98 494		0
	L Cos	d	L Cot	c d	L Tan	L Sin	d	′

75°

Prop. Pts.

	54	53	52
1	5.4	5.3	5.2
2	10.8	10.6	10.4
3	16.2	15.9	15.6
4	21.6	21.2	20.8
5	27.0	26.5	26.0
6	32.4	31.8	31.2
7	37.8	37.1	36.4
8	43.2	42.4	41.6
9	48.6	47.7	46.8

	51	50	49
1	5.1	5.0	4.9
2	10.2	10.0	9.8
3	15.3	15.0	14.7
4	20.4	20.0	19.6
5	25.5	25.0	24.5
6	30.6	30.0	29.4
7	35.7	35.0	34.3
8	40.8	40.0	39.2
9	45.9	45.0	44.1

	48	47
1	4.8	4.7
2	9.6	9.4
3	14.4	14.1
4	19.2	18.8
5	24.0	23.5
6	28.8	28.2
7	33.6	32.9
8	38.4	37.6
9	43.2	42.3

	4	3
1	0.4	0.3
2	0.8	0.6
3	1.2	0.9
4	1.6	1.2
5	2.0	1.5
6	2.4	1.8
7	2.8	2.1
8	3.2	2.4
9	3.6	2.7

TABLE

3

15°

TABLE 3

′	L Sin	d	L Tan	c d	L Cot	L Cos	d	′	Prop. Pts.
0	9.41 300	47	9.42 805	51	10.57 195	9.98 494		**60**	
1	9.41 347	47	9.42 856	50	10.57 144	9.98 491	3	59	
2	9.41 394	47	9.42 906	51	10.57 094	9.98 488	3	58	
3	9.41 441	47	9.42 957	50	10.57 043	9.98 484	4	57	
4	9.41 488	47	9.43 007	50	10.56 993	9.98 481	3	56	
5	9.41 535	47	9.43 057	51	10.56 943	9.98 477	4	55	**51 / 50 / 49**
6	9.41 582	46	9.43 108	50	10.56 892	9.98 474	3	54	1 · 5.1 · 5.0 · 4.9
7	9.41 628	47	9.43 158	50	10.56 842	9.98 471	3	53	2 · 10.2 · 10.0 · 9.8
8	9.41 675	47	9.43 208	50	10.56 792	9.98 467	4	52	3 · 15.3 · 15.0 · 14.7
9	9.41 722	46	9.43 258	50	10.56 742	9.98 464	3	51	4 · 20.4 · 20.0 · 19.6
10	9.41 768	47	9.43 308	50	10.56 692	9.98 460	4	**50**	5 · 25.5 · 25.0 · 24.5
11	9.41 815	46	9.43 358	50	10.56 642	9.98 457	3	49	6 · 30.6 · 30.0 · 29.4
12	9.41 861	47	9.43 408	50	10.56 592	9.98 453	4	48	7 · 35.7 · 35.0 · 34.3
13	9.41 908	46	9.43 458	50	10.56 542	9.98 450	3	47	8 · 40.8 · 40.0 · 39.2
14	9.41 954	47	9.43 508	50	10.56 492	9.98 447	4	46	9 · 45.9 · 45.0 · 44.1
15	9.42 001	46	9.43 558	49	10.56 442	9.98 443	3	45	
16	9.42 047	46	9.43 607	50	10.56 393	9.98 440	4	44	
17	9.42 093	47	9.43 657	50	10.56 343	9.98 436	3	43	
18	9.42 140	46	9.43 707	49	10.56 293	9.98 433	4	42	
19	9.42 186	46	9.43 756	50	10.56 244	9.98 429	3	41	**48 / 47 / 46**
20	9.42 232	46	9.43 806	49	10.56 194	9.98 426	4	**40**	1 · 4.8 · 4.7 · 4.6
21	9.42 278	46	9.43 855	50	10.56 145	9.98 422	4	39	2 · 9.6 · 9.4 · 9.2
22	9.42 324	46	9.43 905	49	10.56 095	9.98 419	3	38	3 · 14.4 · 14.1 · 13.8
23	9.42 370	46	9.43 954	50	10.56 046	9.98 415	4	37	4 · 19.2 · 18.8 · 18.4
24	9.42 416	45	9.44 004	49	10.55 996	9.98 412	3	36	5 · 24.0 · 23.5 · 23.0
25	9.42 461	46	9.44 053	49	10.55 947	9.98 409	4	35	6 · 28.8 · 28.2 · 27.6
26	9.42 507	46	9.44 102	49	10.55 898	9.98 405	3	34	7 · 33.6 · 32.9 · 32.2
27	9.42 553	46	9.44 151	50	10.55 849	9.98 402	4	33	8 · 38.4 · 37.6 · 36.8
28	9.42 599	45	9.44 201	49	10.55 799	9.98 398	3	32	9 · 43.2 · 42.3 · 41.4
29	9.42 644	46	9.44 250	49	10.55 750	9.98 395	4	31	
30	9.42 690	45	9.44 299	49	10.55 701	9.98 391	3	**30**	
31	9.42 735	46	9.44 348	49	10.55 652	9.98 388	4	29	
32	9.42 781	45	9.44 397	49	10.55 603	9.98 384	3	28	
33	9.42 826	46	9.44 446	49	10.55 554	9.98 381	4	27	**45 / 44**
34	9.42 872	45	9.44 495	49	10.55 505	9.98 377	3	26	1 · 4.5 · 4.4
35	9.42 917	45	9.44 544	48	10.55 456	9.98 373	4	25	2 · 9.0 · 8.8
36	9.42 962	46	9.44 592	49	10.55 408	9.98 370	3	24	3 · 13.5 · 13.2
37	9.43 008	45	9.44 641	49	10.55 359	9.98 366	3	23	4 · 18.0 · 17.6
38	9.43 053	45	9.44 690	48	10.55 310	9.98 363	4	22	5 · 22.5 · 22.0
39	9.43 098	45	9.44 738	49	10.55 262	9.98 359	3	21	6 · 27.0 · 26.4
40	9.43 143	45	9.44 787	49	10.55 213	9.98 356	4	**20**	7 · 31.5 · 30.8
41	9.43 188	45	9.44 836	48	10.55 164	9.98 352	3	19	8 · 36.0 · 35.2
42	9.43 233	45	9.44 884	49	10.55 116	9.98 349	4	18	9 · 40.5 · 39.6
43	9.43 278	45	9.44 933	48	10.55 067	9.98 345	3	17	
44	9.43 323	44	9.44 981	48	10.55 019	9.98 342	4	16	
45	9.43 367	45	9.45 029	49	10.54 971	9.98 338	4	15	
46	9.43 412	45	9.45 078	48	10.54 922	9.98 334	3	14	
47	9.43 457	45	9.45 126	48	10.54 874	9.98 331	4	13	**4 / 3**
48	9.43 502	44	9.45 174	48	10.54 826	9.98 327	3	12	1 · 0.4 · 0.3
49	9.43 546	45	9.45 222	49	10.54 778	9.98 324	4	11	2 · 0.8 · 0.6
50	9.43 591	44	9.45 271	48	10.54 729	9.98 320	3	**10**	3 · 1.2 · 0.9
51	9.43 635	45	9.45 319	48	10.54 681	9.98 317	4	9	4 · 1.6 · 1.2
52	9.43 680	44	9.45 367	48	10.54 633	9.98 313	4	8	5 · 2.0 · 1.5
53	9.43 724	45	9.45 415	48	10.54 585	9.98 309	3	7	6 · 2.4 · 1.8
54	9.43 769	44	9.45 463	48	10.54 537	9.98 306	4	6	7 · 2.8 · 2.1
55	9.43 813	44	9.45 511	48	10.54 489	9.98 302	3	5	8 · 3.2 · 2.4
56	9.43 857	44	9.45 559	47	10.54 441	9.98 299	4	4	9 · 3.6 · 2.7
57	9.43 901	45	9.45 606	48	10.54 394	9.98 295	4	3	
58	9.43 946	44	9.45 654	48	10.54 346	9.98 291	4	2	
59	9.43 990	44	9.45 702	48	10.54 298	9.98 288	4	1	
60	9.44 034		9.45 750		10.54 250	9.98 284		**0**	
	L Cos	d	L Cot	c d	L Tan	L Sin	d	′	Prop. Pts.

74°

16°

′	L Sin	d	L Tan	c d	L Cot	L Cos	d	′
0	9.44 034		9.45 750		10.54 250	9.98 284		60
1	9.44 078	44	9.45 797	47	10.54 203	9.98 281	3	59
2	9.44 122	44	9.45 845	48	10.54 155	9.98 277	4	58
3	9.44 166	44	9.45 892	47	10.54 108	9.98 273	4	57
4	9.44 210	44	9.45 940	48	10.54 060	9.98 270	4	56
5	9.44 253	43	9.45 987	47	10.54 013	9.98 266	3	55
6	9.44 297	44	9.46 035	48	10.53 965	9.98 262	4	54
7	9.44 341	44	9.46 082	47	10.53 918	9.98 259	3	53
8	9.44 385	44	9.46 130	48	10.53 870	9.98 255	4	52
9	9.44 428	43	9.46 177	47	10.53 823	9.98 251	4	51
10	9.44 472	44	9.46 224	47	10.53 776	9.98 248	3	50
11	9.44 516	44	9.46 271	47	10.53 729	9.98 244	4	49
12	9.44 559	43	9.46 319	48	10.53 681	9.98 240	4	48
13	9.44 602	43	9.46 366	47	10.53 634	9.98 237	3	47
14	9.44 646	44	9.46 413	47	10.53 587	9.98 233	4	46
15	9.44 689	43	9.46 460	47	10.53 540	9.98 229	4	45
16	9.44 733	44	9.46 507	47	10.53 493	9.98 226	3	44
17	9.44 776	43	9.46 554	47	10.53 446	9.98 222	4	43
18	9.44 819	43	9.46 601	47	10.53 399	9.98 218	4	42
19	9.44 862	43	9.46 648	46	10.53 352	9.98 215	3	41
20	9.44 905	43	9.46 694	47	10.53 306	9.98 211	4	40
21	9.44 948	44	9.46 741	47	10.53 259	9.98 207	3	39
22	9.44 992	43	9.46 788	47	10.53 212	9.98 204	4	38
23	9.45 035	42	9.46 835	46	10.53 165	9.98 200	4	37
24	9.45 077	43	9.46 881	47	10.53 119	9.98 196	4	36
25	9.45 120	43	9.46 928	47	10.53 072	9.98 192	4	35
26	9.45 163	43	9.46 975	46	10.53 025	9.98 189	3	34
27	9.45 206	43	9.47 021	47	10.52 979	9.98 185	4	33
28	9.45 249	43	9.47 068	46	10.52 932	9.98 181	4	32
29	9.45 292	42	9.47 114	46	10.52 886	9.98 177	4	31
30	9.45 334	43	9.47 160	47	10.52 840	9.98 174	3	30
31	9.45 377	42	9.47 207	46	10.52 793	9.98 170	4	29
32	9.45 419	43	9.47 253	46	10.52 747	9.98 166	4	28
33	9.45 462	42	9.47 299	47	10.52 701	9.98 162	4	27
34	9.45 504	43	9.47 346	46	10.52 654	9.98 159	3	26
35	9.45 547	42	9.47 392	46	10.52 608	9.98 155	4	25
36	9.45 589	43	9.47 438	46	10.52 562	9.98 151	4	24
37	9.45 632	42	9.47 484	46	10.52 516	9.98 147	3	23
38	9.45 674	42	9.47 530	46	10.52 470	9.98 144	4	22
39	9.45 716	42	9.47 576	46	10.52 424	9.98 140	4	21
40	9.45 758	43	9.47 622	46	10.52 378	9.98 136	4	20
41	9.45 801	42	9.47 668	46	10.52 332	9.98 132	4	19
42	9.45 843	42	9.47 714	46	10.52 286	9.98 129	4	18
43	9.45 885	42	9.47 760	46	10.52 240	9.98 125	4	17
44	9.45 927	42	9.47 806	46	10.52 194	9.98 121	4	16
45	9.45 969	42	9.47 852	45	10.52 148	9.98 117	4	15
46	9.46 011	42	9.47 897	46	10.52 103	9.98 113	3	14
47	9.46 053	42	9.47 943	46	10.52 057	9.98 110	4	13
48	9.46 095	41	9.47 989	46	10.52 011	9.98 106	4	12
49	9.46 136	42	9.48 035	45	10.51 965	9.98 102	4	11
50	9.46 178	42	9.48 080	46	10.51 920	9.98 098	4	10
51	9.46 220	42	9.48 126	45	10.51 874	9.98 094	4	9
52	9.46 262	41	9.48 171	46	10.51 829	9.98 090	4	8
53	9.46 303	42	9.48 217	45	10.51 783	9.98 087	4	7
54	9.46 345	41	9.48 262	45	10.51 738	9.98 083	4	6
55	9.46 386	42	9.48 307	46	10.51 693	9.98 079	4	5
56	9.46 428	41	9.48 353	45	10.51 647	9.98 075	4	4
57	9.46 469	42	9.48 398	45	10.51 602	9.98 071	4	3
58	9.46 511	41	9.48 443	46	10.51 557	9.98 067	4	2
59	9.46 552	42	9.48 489	45	10.51 511	9.98 063	3	1
60	9.46 594		9.48 534		10.51 466	9.98 060		0
	L Cos	d	L Cot	c d	L Tan	L Sin	d	′

73°

Prop. Pts.

	48	47	46
1	4.8	4.7	4.6
2	9.6	9.4	9.2
3	14.4	14.1	13.8
4	19.2	18.8	18.4
5	24.0	23.5	23.0
6	28.8	28.2	27.6
7	33.6	32.9	32.2
8	38.4	37.6	36.8
9	43.2	42.3	41.4

	45	44	43
1	4.5	4.4	4.3
2	9.0	8.8	8.6
3	13.5	13.2	12.9
4	18.0	17.6	17.2
5	22.5	22.0	21.5
6	27.0	26.4	25.8
7	31.5	30.8	30.1
8	36.0	35.2	34.4
9	40.5	39.6	38.7

	42	41
1	4.2	4.1
2	8.4	8.2
3	12.6	12.3
4	16.8	16.4
5	21.0	20.5
6	25.2	24.6
7	29.4	28.7
8	33.6	32.8
9	37.8	36.9

	4	3
1	0.4	0.3
2	0.8	0.6
3	1.2	0.9
4	1.6	1.2
5	2.0	1.5
6	2.4	1.8
7	2.8	2.1
8	3.2	2.4
9	3.6	2.7

TABLE 3

17°

′	L Sin	d	L Tan	c d	L Cot	L Cos	d	′
0	9.46 594	41	9.48 534	45	10.51 466	9.98 060	4	60
1	9.46 635	41	9.48 579	45	10.51 421	9.98 056	4	59
2	9.46 676	41	9.48 624	45	10.51 376	9.98 052	4	58
3	9.46 717	41	9.48 669	45	10.51 331	9.98 048	4	57
4	9.46 758	42	9.48 714	45	10.51 286	9.98 044	4	56
5	9.46 800	41	9.48 759	45	10.51 241	9.98 040	4	55
6	9.46 841	41	9.48 804	45	10.51 196	9.98 036	4	54
7	9.46 882	41	9.48 849	45	10.51 151	9.98 032	4	53
8	9.46 923	41	9.48 894	45	10.51 106	9.98 029	4	52
9	9.46 964	41	9.48 939	45	10.51 061	9.98 025	4	51
10	9.47 005	40	9.48 984	45	10.51 016	9.98 021	4	50
11	9.47 045	41	9.49 029	44	10.50 971	9.98 017	4	49
12	9.47 086	41	9.49 073	45	10.50 927	9.98 013	4	48
13	9.47 127	41	9.49 118	45	10.50 882	9.98 009	4	47
14	9.47 168	41	9.49 163	44	10.50 837	9.98 005	4	46
15	9.47 209	40	9.49 207	45	10.50 793	9.98 001	4	45
16	9.47 249	41	9.49 252	44	10.50 748	9.97 997	4	44
17	9.47 290	40	9.49 296	45	10.50 704	9.97 993	4	43
18	9.47 330	41	9.49 341	44	10.50 659	9.97 989	3	42
19	9.47 371	40	9.49 385	45	10.50 615	9.97 986	4	41
20	9.47 411	41	9.49 430	44	10.50 570	9.97 982	4	40
21	9.47 452	40	9.49 474	45	10.50 526	9.97 978	4	39
22	9.47 492	41	9.49 519	44	10.50 481	9.97 974	4	38
23	9.47 533	40	9.49 563	44	10.50 437	9.97 970	4	37
24	9.47 573	40	9.49 607	45	10.50 393	9.97 966	4	36
25	9.47 613	41	9.49 652	44	10.50 348	9.97 962	4	35
26	9.47 654	40	9.49 696	44	10.50 304	9.97 958	4	34
27	9.47 694	40	9.49 740	44	10.50 260	9.97 954	4	33
28	9.47 734	40	9.49 784	44	10.50 216	9.97 950	4	32
29	9.47 774	40	9.49 828	44	10.50 172	9.97 946	4	31
30	9.47 814	40	9.49 872	44	10.50 128	9.97 942	4	30
31	9.47 854	40	9.49 916	44	10.50 084	9.97 938	4	29
32	9.47 894	40	9.49 960	44	10.50 040	9.97 934	4	28
33	9.47 934	40	9.50 004	44	10.49 996	9.97 930	4	27
34	9.47 974	40	9.50 048	44	10.49 952	9.97 926	4	26
35	9.48 014	40	9.50 092	44	10.49 908	9.97 922	4	25
36	9.48 054	40	9.50 136	44	10.49 864	9.97 918	4	24
37	9.48 094	39	9.50 180	43	10.49 820	9.97 914	4	23
38	9.48 133	40	9.50 223	44	10.49 777	9.97 910	4	22
39	9.48 173	40	9.50 267	44	10.49 733	9.97 906	4	21
40	9.48 213	39	9.50 311	44	10.49 689	9.97 902	4	20
41	9.48 252	40	9.50 355	43	10.49 645	9.97 898	4	19
42	9.48 292	40	9.50 398	44	10.49 602	9.97 894	4	18
43	9.48 332	39	9.50 442	43	10.49 558	9.97 890	4	17
44	9.48 371	40	9.50 485	44	10.49 515	9.97 886	4	16
45	9.48 411	39	9.50 529	43	10.49 471	9.97 882	4	15
46	9.48 450	40	9.50 572	44	10.49 428	9.97 878	4	14
47	9.48 490	39	9.50 616	43	10.49 384	9.97 874	4	13
48	9.48 529	39	9.50 659	44	10.49 341	9.97 870	4	12
49	9.48 568	39	9.50 703	43	10.49 297	9.97 866	5	11
50	9.48 607	40	9.50 746	43	10.49 254	9.97 861	4	10
51	9.48 647	39	9.50 789	44	10.49 211	9.97 857	4	9
52	9.48 686	39	9.50 833	43	10.49 167	9.97 853	4	8
53	9.48 725	39	9.50 876	43	10.49 124	9.97 849	4	7
54	9.48 764	39	9.50 919	43	10.49 081	9.97 845	4	6
55	9.48 803	39	9.50 962	43	10.49 038	9.97 841	4	5
56	9.48 842	39	9.51 005	43	10.48 995	9.97 837	4	4
57	9.48 881	39	9.51 048	44	10.48 952	9.97 833	4	3
58	9.48 920	39	9.51 092	43	10.48 908	9.97 829	4	2
59	9.48 959	39	9.51 135	43	10.48 865	9.97 825	4	1
60	9.48 998		9.51 178		10.48 822	9.97 821		0
	L Cos	d	L Cot	c d	L Tan	L Sin	d	′

Prop. Pts.

	45	44	43
1	4.5	4.4	4.3
2	9.0	8.8	8.6
3	13.5	13.2	12.9
4	18.0	17.6	17.2
5	22.5	22.0	21.5
6	27.0	26.4	25.8
7	31.5	30.8	30.1
8	36.0	35.2	34.4
9	40.5	39.6	38.7

	42	41
1	4.2	4.1
2	8.4	8.2
3	12.6	12.3
4	16.8	16.4
5	21.0	20.5
6	25.2	24.6
7	29.4	28.7
8	33.6	32.8
9	37.8	36.9

	40	39
1	4.0	3.9
2	8.0	7.8
3	12.0	11.7
4	16.0	15.6
5	20.0	19.5
6	24.0	23.4
7	28.0	27.3
8	32.0	31.2
9	36.0	35.1

	5	4	3
1	0.5	0.4	0.3
2	1.0	0.8	0.6
3	1.5	1.2	0.9
4	2.0	1.6	1.2
5	2.5	2.0	1.5
6	3.0	2.4	1.8
7	3.5	2.8	2.1
8	4.0	3.2	2.4
9	4.5	3.6	2.7

TABLE 3

72°

18°

′	L Sin	d	L Tan	c d	L Cot	L Cos	d	′	Prop. Pts.
0	9.48 998		9.51 178		10.48 822	9.97 821		60	
1	9.49 037	39	9.51 221	43	10.48 779	9.97 817	4	59	
2	9.49 076	39	9.51 264	43	10.48 736	9.97 812	5	58	
3	9.49 115	39	9.51 306	42	10.48 694	9.97 808	4	57	
4	9.49 153	38	9.51 349	43	10.48 651	9.97 804	4	56	
5	9.49 192	39	9.51 392	43	10.48 608	9.97 800	4	55	
6	9.49 231	39	9.51 435	43	10.48 565	9.97 796	4	54	
7	9.49 269	38	9.51 478	43	10.48 522	9.97 792	4	53	
8	9.49 308	39	9.51 520	42	10.48 480	9.97 788	4	52	
9	9.49 347	39	9.51 563	43	10.48 437	9.97 784	4	51	**43 42 41**
10	9.49 385	38	9.51 606	43	10.48 394	9.97 779	5	50	1 4.3 4.2 4.1
11	9.49 424	39	9.51 648	42	10.48 352	9.97 775	4	49	2 8.6 8.4 8.2
12	9.49 462	38	9.51 691	43	10.48 309	9.97 771	4	48	3 12.9 12.6 12.3
13	9.49 500	38	9.51 734	43	10.48 266	9.97 767	4	47	4 17.2 16.8 16.4
14	9.49 539	38	9.51 776	42	10.48 224	9.97 763	4	46	5 21.5 21.0 20.5
15	9.49 577	38	9.51 819	43	10.48 181	9.97 759	4	45	6 25.8 25.2 24.6
16	9.49 615	39	9.51 861	42	10.48 139	9.97 754	5	44	7 30.1 29.4 28.7
17	9.49 654	38	9.51 903	43	10.48 097	9.97 750	4	43	8 34.4 33.6 32.8
18	9.49 692	38	9.51 946	42	10.48 054	9.97 746	4	42	9 38.7 37.8 36.9
19	9.49 730	38	9.51 988	43	10.48 012	9.97 742	4	41	
20	9.49 768	38	9.52 031	42	10.47 969	9.97 738	4	40	
21	9.49 806	38	9.52 073	42	10.47 927	9.97 734	5	39	
22	9.49 844	38	9.52 115	42	10.47 885	9.97 729	4	38	
23	9.49 882	38	9.52 157	43	10.47 843	9.97 725	4	37	
24	9.49 920	38	9.52 200	42	10.47 800	9.97 721	4	36	
25	9.49 958	38	9.52 242	42	10.47 758	9.97 717	4	35	
26	9.49 996	38	9.52 284	42	10.47 716	9.97 713	5	34	**39 38 37**
27	9.50 034	38	9.52 326	42	10.47 674	9.97 708	4	33	1 3.9 3.8 3.7
28	9.50 072	38	9.52 368	42	10.47 632	9.97 704	4	32	2 7.8 7.6 7.4
29	9.50 110	38	9.52 410	42	10.47 590	9.97 700	4	31	3 11.7 11.4 11.1
30	9.50 148	37	9.52 452	42	10.47 548	9.97 696	5	30	4 15.6 15.2 14.8
31	9.50 185	38	9.52 494	42	10.47 506	9.97 691	4	29	5 19.5 19.0 18.5
32	9.50 223	38	9.52 536	42	10.47 464	9.97 687	4	28	6 23.4 22.8 22.2
33	9.50 261	37	9.52 578	42	10.47 422	9.97 683	4	27	7 27.3 26.6 25.9
34	9.50 298	38	9.52 620	41	10.47 380	9.97 679	5	26	8 31.2 30.4 29.6
35	9.50 336	38	9.52 661	42	10.47 339	9.97 674	4	25	9 35.1 34.2 33.3
36	9.50 374	37	9.52 703	42	10.47 297	9.97 670	4	24	
37	9.50 411	38	9.52 745	42	10.47 255	9.97 666	4	23	
38	9.50 449	37	9.52 787	42	10.47 213	9.97 662	5	22	
39	9.50 486	37	9.52 829	41	10.47 171	9.97 657	4	21	
40	9.50 523	38	9.52 870	42	10.47 130	9.97 653	4	20	
41	9.50 561	37	9.52 912	41	10.47 088	9.97 649	4	19	
42	9.50 598	37	9.52 953	42	10.47 047	9.97 645	5	18	
43	9.50 635	38	9.52 995	42	10.47 005	9.97 640	4	17	**36 5 4**
44	9.50 673	37	9.53 037	41	10.46 963	9.97 636	4	16	1 3.6 0.5 0.4
45	9.50 710	37	9.53 078	42	10.46 922	9.97 632	4	15	2 7.2 1.0 0.8
46	9.50 747	37	9.53 120	41	10.46 880	9.97 628	5	14	3 10.8 1.5 1.2
47	9.50 784	37	9.53 161	41	10.46 839	9.97 623	4	13	4 14.4 2.0 1.6
48	9.50 821	37	9.53 202	42	10.46 798	9.97 619	4	12	5 18.0 2.5 2.0
49	9.50 858	38	9.53 244	41	10.46 756	9.97 615	5	11	6 21.6 3.0 2.4
50	9.50 896	37	9.53 285	42	10.46 715	9.97 610	4	10	7 25.2 3.5 2.8
51	9.50 933	37	9.53 327	41	10.46 673	9.97 606	4	9	8 28.8 4.0 3.2
52	9.50 970	37	9.53 368	41	10.46 632	9.97 602	5	8	9 32.4 4.5 3.6
53	9.51 007	36	9.53 409	41	10.46 591	9.97 597	4	7	
54	9.51 043	37	9.53 450	42	10.46 550	9.97 593	4	6	
55	9.51 080	37	9.53 492	41	10.46 508	9.97 589	5	5	
56	9.51 117	37	9.53 533	41	10.46 467	9.97 584	4	4	
57	9.51 154	37	9.53 574	41	10.46 426	9.97 580	4	3	
58	9.51 191	36	9.53 615	41	10.46 385	9.97 576	5	2	
59	9.51 227	37	9.53 656	41	10.46 344	9.97 571	4	1	
60	9.51 264		9.53 697		10.46 303	9.97 567		0	
	L Cos	d	L Cot	c d	L Tan	L Sin	d	′	Prop. Pts.

71°

TABLE

3

19°

′	L Sin	d	L Tan	c d	L Cot	L Cos	d	′
0	9.51 264		9.53 697		10.46 303	9.97 567		60
1	9.51 301	37	9.53 738	41	10.46 262	9.97 563	4	59
2	9.51 338	37	9.53 779	41	10.46 221	9.97 558	5	58
3	9.51 374	36	9.53 820	41	10.46 180	9.97 554	4	57
4	9.51 411	37	9.53 861	41	10.46 139	9.97 550	4	56
5	9.51 447	36	9.53 902	41	10.46 098	9.97 545	5	55
6	9.51 484	37	9.53 943	41	10.46 057	9.97 541	4	54
7	9.51 520	36	9.53 984	41	10.46 016	9.97 536	5	53
8	9.51 557	37	9.54 025	41	10.45 975	9.97 532	4	52
9	9.51 593	36	9.54 065	40	10.45 935	9.97 528	4	51
10	9.51 629	36	9.54 106	41	10.45 894	9.97 523	5	50
11	9.51 666	37	9.54 147	41	10.45 853	9.97 519	4	49
12	9.51 702	36	9.54 187	40	10.45 813	9.97 515	4	48
13	9.51 738	36	9.54 228	41	10.45 772	9.97 510	5	47
14	9.51 774	36	9.54 269	41	10.45 731	9.97 506	4	46
15	9.51 811	37	9.54 309	40	10.45 691	9.97 501	5	45
16	9.51 847	36	9.54 350	41	10.45 650	9.97 497	4	44
17	9.51 883	36	9.54 390	40	10.45 610	9.97 492	5	43
18	9.51 919	36	9.54 431	41	10.45 569	9.97 488	4	42
19	9.51 955	36	9.54 471	40	10.45 529	9.97 484	4	41
20	9.51 991	36	9.54 512	41	10.45 488	9.97 479	5	40
21	9.52 027	36	9.54 552	40	10.45 448	9.97 475	4	39
22	9.52 063	36	9.54 593	41	10.45 407	9.97 470	5	38
23	9.52 099	36	9.54 633	40	10.45 367	9.97 466	4	37
24	9.52 135	36	9.54 673	40	10.45 327	9.97 461	5	36
25	9.52 171	36	9.54 714	41	10.45 286	9.97 457	4	35
26	9.52 207	36	9.54 754	40	10.45 246	9.97 453	5	34
27	9.52 242	35	9.54 794	41	10.45 206	9.97 448	4	33
28	9.52 278	36	9.54 835	40	10.45 165	9.97 444	5	32
29	9.52 314	36	9.54 875	40	10.45 125	9.97 439	4	31
30	9.52 350	36	9.54 915	40	10.45 085	9.97 435	5	30
31	9.52 385	35	9.54 955	40	10.45 045	9.97 430	4	29
32	9.52 421	36	9.54 995	40	10.45 005	9.97 426	5	28
33	9.52 456	35	9.55 035	40	10.44 965	9.97 421	4	27
34	9.52 492	36	9.55 075	40	10.44 925	9.97 417	5	26
35	9.52 527	35	9.55 115	40	10.44 885	9.97 412	4	25
36	9.52 563	36	9.55 155	40	10.44 845	9.97 408	5	24
37	9.52 598	35	9.55 195	40	10.44 805	9.97 403	4	23
38	9.52 634	33	9.55 235	40	10.44 765	9.97 399	5	22
39	9.52 669	35	9.55 275	40	10.44 725	9.97 394	4	21
40	9.52 705	36	9.55 315	40	10.44 685	9.97 390	5	20
41	9.52 740	35	9.55 355	40	10.44 645	9.97 385	4	19
42	9.52 775	35	9.55 395	39	10.44 605	9.97 381	5	18
43	9.52 811	36	9.55 434	40	10.44 566	9.97 376	4	17
44	9.52 846	35	9.55 474	40	10.44 526	9.97 372	5	16
45	9.52 881	35	9.55 514	40	10.44 486	9.97 367	4	15
46	9.52 916	35	9.55 554	39	10.44 446	9.97 363	5	14
47	9.52 951	35	9.55 593	40	10.44 407	9.97 358	5	13
48	9.52 986	35	9.55 633	40	10.44 367	9.97 353	4	12
49	9.53 021	35	9.55 673	39	10.44 327	9.97 349	5	11
50	9.53 056	35	9.55 712	40	10.44 288	9.97 344	4	10
51	9.53 092	26	9.55 752	39	10.44 248	9.97 340	5	9
52	9.53 126	34	9.55 791	40	10.44 209	9.97 335	4	8
53	9.53 161	35	9.55 831	39	10.44 169	9.97 331	5	7
54	9.53 196	35	9.55 870	40	10.44 130	9.97 326	5	6
55	9.53 231	35	9.55 910	39	10.44 090	9.97 322	4	5
56	9.53 266	35	9.55 949	40	10.44 051	9.97 317	5	4
57	9.53 301	35	9.55 989	39	10.44 011	9.97 312	4	3
58	9.53 336	34	9.56 028	39	10.43 972	9.97 308	5	2
59	9.53 370	35	9.56 067	40	10.43 933	9.97 303	4	1
60	9.53 405		9.56 107		10.43 893	9.97 299		0
	L Cos	d	L Cot	c d	L Tan	L Sin	d	′

70°

Prop. Pts.

	41	40	39
1	4.1	4.0	3.9
2	8.2	8.0	7.8
3	12.3	12.0	11.7
4	16.4	16.0	15.6
5	20.5	20.0	19.5
6	24.6	24.0	23.4
7	28.7	28.0	27.3
8	32.8	32.0	31.2
9	36.9	36.0	35.1

	37	36	35
1	3.7	3.6	3.5
2	7.4	7.2	7.0
3	11.1	10.8	10.5
4	14.8	14.4	14.0
5	18.5	18.0	17.5
6	22.2	21.6	21.0
7	25.9	25.2	24.5
8	29.6	28.8	28.0
9	33.3	32.4	31.5

	34	5	4
1	3.4	0.5	0.4
2	6.8	1.0	0.8
3	10.2	1.5	1.2
4	13.6	2.0	1.6
5	17.0	2.5	2.0
6	20.4	3.0	2.4
7	23.8	3.5	2.8
8	27.2	4.0	3.2
9	30.6	4.5	3.6

TABLE

3

20°

′	L Sin	d	L Tan	c d	L Cot	L Cos	d	′
0	9.53 405		9.56 107		10.43 893	9.97 299		60
1	9.53 440	35	9.56 146	39	10.43 854	9.97 294	5	59
2	9.53 475	35	9.56 185	39	10.43 815	9.97 289	5	58
3	9.53 509	34	9.56 224	39	10.43 776	9.97 285	4	57
4	9.53 544	35	9.56 264	40	10.43 736	9.97 280	5	56
5	9.53 578	34	9.56 303	39	10.43 697	9.97 276	4	55
6	9.53 613	35	9.56 342	39	10.43 658	9.97 271	5	54
7	9.53 647	34	9.56 381	39	10.43 619	9.97 266	5	53
8	9.53 682	35	9.56 420	39	10.43 580	9.97 262	4	52
9	9.53 716	34	9.56 459	39	10.43 541	9.97 257	5	51
10	9.53 751	35	9.56 498	39	10.43 502	9.97 252		50
11	9.53 785	34	9.56 537	39	10.43 463	9.97 248	4	49
12	9.53 819	34	9.56 576	39	10.43 424	9.97 243	5	48
13	9.53 854	35	9.56 615	39	10.43 385	9.97 238	5	47
14	9.53 888	34	9.56 654	39	10.43 346	9.97 234	4	46
15	9.53 922	34	9.56 693	39	10.43 307	9.97 229	5	45
16	9.53 957	35	9.56 732	39	10.43 268	9.97 224	5	44
17	9.53 991	34	9.56 771	39	10.43 229	9.97 220	4	43
18	9.54 025	34	9.56 810	39	10.43 190	9.97 215	5	42
19	9.54 059	34	9.56 849	39	10.43 151	9.97 210	5	41
20	9.54 093	34	9.56 887	38	10.43 113	9.97 206	4	40
21	9.54 127	34	9.56 926	39	10.43 074	9.97 201	5	39
22	9.54 161	34	9.56 965	39	10.43 035	9.97 196	5	38
23	9.54 195	34	9.57 004	38	10.42 996	9.97 192	4	37
24	9.54 229	34	9.57 042	39	10.42 958	9.97 187	5	36
25	9.54 263	34	9.57 081	39	10.42 919	9.97 182	5	35
26	9.54 297	34	9.57 120	38	10.42 880	9.97 178	4	34
27	9.54 331	34	9.57 158	39	10.42 842	9.97 173	5	33
28	9.54 365	34	9.57 197	38	10.42 803	9.97 168	5	32
29	9.54 399	34	9.57 235	39	10.42 765	9.97 163	4	31
30	9.54 433	33	9.57 274	38	10.42 726	9.97 159	5	30
31	9.54 466	34	9.57 312	39	10.42 688	9.97 154	5	29
32	9.54 500	34	9.57 351	38	10.42 649	9.97 149	4	28
33	9.54 534	33	9.57 389	39	10.42 611	9.97 145	5	27
34	9.54 567	34	9.57 428	38	10.42 572	9.97 140	5	26
35	9.54 601	34	9.57 466	38	10.42 534	9.97 135	5	25
36	9.54 635	33	9.57 504	39	10.42 496	9.97 130	4	24
37	9.54 668	34	9.57 543	38	10.42 457	9.97 126	5	23
38	9.54 702	33	9.57 581	38	10.42 419	9.97 121	5	22
39	9.54 735	34	9.57 619	39	10.42 381	9.97 116	5	21
40	9.54 769	33	9.57 658	38	10.42 342	9.97 111	4	20
41	9.54 802	34	9.57 696	38	10.42 304	9.97 107	5	19
42	9.54 836	33	9.57 734	38	10.42 266	9.97 102	5	18
43	9.54 869	34	9.57 772	38	10.42 228	9.97 097	5	17
44	9.54 903	33	9.57 810	39	10.42 190	9.97 092	5	16
45	9.54 936	33	9.57 849	38	10.42 151	9.97 087	4	15
46	9.54 969	34	9.57 887	38	10.42 113	9.97 083	5	14
47	9.55 003	33	9.57 925	38	10.42 075	9.97 078	5	13
48	9.55 036	33	9.57 963	38	10.42 037	9.97 073	5	12
49	9.55 069	33	9.58 001	38	10.41 999	9.97 068	5	11
50	9.55 102	34	9.58 039	38	10.41 961	9.97 063	4	10
51	9.55 136	33	9.58 077	38	10.41 923	9.97 059	5	9
52	9.55 169	33	9.58 115	38	10.41 885	9.97 054	5	8
53	9.55 202	33	9.58 153	38	10.41 847	9.97 049	5	7
54	9.55 235	33	9.58 191	38	10.41 809	9.97 044	5	6
55	9.55 268	33	9.58 229	38	10.41 771	9.97 039	5	5
56	9.55 301	33	9.58 267	37	10.41 733	9.97 035	5	4
57	9.55 334	33	9.58 304	38	10.41 696	9.97 030	5	3
58	9.55 367	33	9.58 342	38	10.41 658	9.97 025	5	2
59	9.55 400	33	9.58 380	38	10.41 620	9.97 020	5	1
60	9.55 433		9.58 418		10.41 582	9.97 015		0
	L Cos	d	L Cot	c d	L Tan	L Sin	d	′

Prop. Pts.

	40	39	38
1	4.0	3.9	3.8
2	8.0	7.8	7.6
3	12.0	11.7	11.4
4	16.0	15.6	15.2
5	20.0	19.5	19.0
6	24.0	23.4	22.8
7	28.0	27.3	26.6
8	32.0	31.2	30.4
9	36.0	35.1	34.2

	37	35	34
1	3.7	3.5	3.4
2	7.4	7.0	6.8
3	11.1	10.5	10.2
4	14.8	14.0	13.6
5	18.5	17.5	17.0
6	22.2	21.0	20.4
7	25.9	24.5	23.8
8	29.6	28.0	27.2
9	33.3	31.5	30.6

	33	5	4
1	3.3	0.5	0.4
2	6.6	1.0	0.8
3	9.9	1.5	1.2
4	13.2	2.0	1.6
5	16.5	2.5	2.0
6	19.8	3.0	2.4
7	23.1	3.5	2.8
8	26.4	4.0	3.2
9	29.7	4.5	3.6

TABLE

3

69°

21°

′	L Sin	d	L Tan	c d	L Cot	L Cos	d	′
0	9.55 433	33	9.58 418	37	10.41 582	9.97 015	5	**60**
1	9.55 466	33	9.58 455	38	10.41 545	9.97 010	5	59
2	9.55 499	33	9.58 493	38	10.41 507	9.97 005	4	58
3	9.55 532	32	9.58 531	38	10.41 469	9.97 001	5	57
4	9.55 564	33	9.58 569	37	10.41 431	9.96 996	5	56
5	9.55 597	33	9.58 606	38	10.41 394	9.96 991	5	55
6	9.55 630	33	9.58 644	37	10.41 356	9.96 986	5	54
7	9.55 663	32	9.58 681	38	10.41 319	9.96 981	5	53
8	9.55 695	33	9.58 719	38	10.41 281	9.96 976	5	52
9	9.55 728	33	9.58 757	37	10.41 243	9.96 971	5	51
10	9.55 761	32	9.58 794	38	10.41 206	9.96 966	4	**50**
11	9.55 793	33	9.58 832	37	10.41 168	9.96 962	5	49
12	9.55 826	32	9.58 869	38	10.41 131	9.96 957	5	48
13	9.55 858	33	9.58 907	37	10.41 093	9.96 952	5	47
14	9.55 891	32	9.58 944	37	10.41 056	9.96 947	5	46
15	9.55 923	33	9.58 981	38	10.41 019	9.96 942	5	45
16	9.55 956	32	9.59 019	37	10.40 981	9.96 937	5	44
17	9.55 988	33	9.59 056	38	10.40 944	9.96 932	5	43
18	9.56 021	32	9.59 094	37	10.40 906	9.96 927	5	42
19	9.56 053	32	9.59 131	37	10.40 869	9.96 922	5	41
20	9.56 085	33	9.59 168	37	10.40 832	9.96 917	5	**40**
21	9.56 118	32	9.59 205	38	10.40 795	9.96 912	5	39
22	9.56 150	32	9.59 243	37	10.40 757	9.96 907	4	38
23	9.56 182	33	9.59 280	37	10.40 720	9.96 903	5	37
24	9.56 215	32	9.59 317	37	10.40 683	9.96 898	5	36
25	9.56 247	32	9.59 354	37	10.40 646	9.96 893	5	35
26	9.56 279	32	9.59 391	38	10.40 609	9.96 888	5	34
27	9.56 311	32	9.59 429	37	10.40 571	9.96 883	5	33
28	9.56 343	32	9.59 466	37	10.40 534	9.96 878	5	32
29	9.56 375	33	9.59 503	37	10.40 497	9.96 873	5	31
30	9.56 408	32	9.59 540	37	10.40 460	9.96 868	5	**30**
31	9.56 440	32	9.59 577	37	10.40 423	9.96 863	5	29
32	9.56 472	32	9.59 614	37	10.40 386	9.96 858	5	28
33	9.56 504	32	9.59 651	37	10.40 349	9.96 853	5	27
34	9.56 536	32	9.59 688	37	10.40 312	9.96 848	5	26
35	9.56 568	31	9.59 725	37	10.40 275	9.96 843	5	25
36	9.56 599	32	9.59 762	37	10.40 238	9.96 838	5	24
37	9.56 631	32	9.59 799	36	10.40 201	9.96 833	5	23
38	9.56 663	32	9.59 835	37	10.40 165	9.96 828	5	22
39	9.56 695	32	9.59 872	37	10.40 128	9.96 823	5	21
40	9.56 727	32	9.59 909	37	10.40 091	9.96 818	5	**20**
41	9.56 759	31	9.59 946	37	10.40 054	9.96 813	5	19
42	9.56 790	32	9.59 983	36	10.40 017	9.96 808	5	18
43	9.56 822	32	9.60 019	37	10.39 981	9.96 803	5	17
44	9.56 854	32	9.60 056	37	10.39 944	9.96 798	5	16
45	9.56 886	31	9.60 093	37	10.39 907	9.96 793	5	15
46	9.56 917	32	9.60 130	36	10.39 870	9.96 788	5	14
47	9.56 949	31	9.60 166	37	10.39 834	9.96 783	5	13
48	9.56 980	32	9.60 203	37	10.39 797	9.96 778	6	12
49	9.57 012	32	9.60 240	36	10.39 760	9.96 772	5	11
50	9.57 044	31	9.60 276	37	10.39 724	9.96 767	5	**10**
51	9.57 075	32	9.60 313	36	10.39 687	9.96 762	5	9
52	9.57 107	31	9.60 349	37	10.39 651	9.96 757	5	8
53	9.57 138	31	9.60 386	36	10.39 614	9.96 752	5	7
54	9.57 169	32	9.60 422	37	10.39 578	9.96 747	5	6
55	9.57 201	31	9.60 459	36	10.39 541	9.96 742	5	5
56	9.57 232	32	9.60 495	37	10.39 505	9.96 737	5	4
57	9.57 264	31	9.60 532	36	10.39 468	9.96 732	5	3
58	9.57 295	31	9.60 568	37	10.39 432	9.96 727	5	2
59	9.57 326	32	9.60 605	36	10.39 395	9.96 722	5	1
60	9.57 358		9.60 641		10.39 359	9.96 717		**0**
	L Cos	d	L Cot	c d	L Tan	L Sin	d	′

Prop. Pts.

	38	37	36
1	3.8	3.7	3.6
2	7.6	7.4	7.2
3	11.4	11.1	10.8
4	15.2	14.8	14.4
5	19.0	18.5	18.0
6	22.8	22.2	21.6
7	26.6	25.9	25.2
8	30.4	29.6	28.8
9	34.2	33.3	32.4

	33	32	31
1	3.3	3.2	3.1
2	6.6	6.4	6.2
3	9.9	9.6	9.3
4	13.2	12.8	12.4
5	16.5	16.0	15.5
6	19.8	19.2	18.6
7	23.1	22.4	21.7
8	26.4	25.6	24.8
9	29.7	28.8	27.9

	6	5	4
1	0.6	0.5	0.4
2	1.2	1.0	0.8
3	1.8	1.5	1.2
4	2.4	2.0	1.6
5	3.0	2.5	2.0
6	3.6	3.0	2.4
7	4.2	3.5	2.8
8	4.8	4.0	3.2
9	5.4	4.5	3.6

TABLE 3

68°

22°

′	L Sin	d	L Tan	c d	L Cot	L Cos	d	′
0	9.57 358		9.60 641		10.39 359	9.96 717		60
1	9.57 389	31	9.60 677	36	10.39 323	9.96 711	6	59
2	9.57 420	31	9.60 714	37	10.39 286	9.96 706	5	58
3	9.57 451	31	9.60 750	36	10.39 250	9.96 701	5	57
4	9.57 482	31	9.60 786	36	10.39 214	9.96 696	5	56
5	9.57 514	32	9.60 823	37	10.39 177	9.96 691	5	55
6	9.57 545	31	9.60 859	36	10.39 141	9.96 686	5	54
7	9.57 576	31	9.60 895	36	10.39 105	9.96 681	5	53
8	9.57 607	31	9.60 931	36	10.39 069	9.96 676	5	52
9	9.57 638	31	9.60 967	36	10.39 033	9.96 670	6	51
10	9.57 669	31	9.61 004	37	10.38 996	9.96 665	5	50
11	9.57 700	31	9.61 040	36	10.38 960	9.96 660	5	49
12	9.57 731	31	9.61 076	36	10.38 924	9.96 655	5	48
13	9.57 762	31	9.61 112	36	10.38 888	9.96 650	5	47
14	9.57 793	31	9.61 148	36	10.38 852	9.96 645	5	46
15	9.57 824	31	9.61 184	36	10.38 816	9.96 640	6	45
16	9.57 855	30	9.61 220	36	10.38 780	9.96 634	5	44
17	9.57 885	31	9.61 256	36	10.38 744	9.96 629	5	43
18	9.57 916	31	9.61 292	36	10.38 708	9.96 624	5	42
19	9.57 947	31	9.61 328	36	10.38 672	9.96 619	5	41
20	9.57 978	30	9.61 364	36	10.38 636	9.96 614	6	40
21	9.58 008	31	9.61 400	36	10.38 600	9.96 608	5	39
22	9.58 039	31	9.61 436	36	10.38 564	9.96 603	5	38
23	9.58 070	31	9.61 472	36	10.38 528	9.96 598	5	37
24	9.58 101	30	9.61 508	36	10.38 492	9.96 593	5	36
25	9.58 131	31	9.61 544	35	10.38 456	9.96 588	6	35
26	9.58 162	30	9.61 579	36	10.38 421	9.96 582	5	34
27	9.58 192	31	9.61 615	36	10.38 385	9.96 577	5	33
28	9.58 223	30	9.61 651	36	10.38 349	9.96 572	5	32
29	9.58 253	31	9.61 687	35	10.38 313	9.96 567	5	31
30	9.58 284	30	9.61 722	36	10.38 278	9.96 562	6	30
31	9.58 314	31	9.61 758	36	10.38 242	9.96 556	5	29
32	9.58 345	30	9.61 794	36	10.38 206	9.96 551	5	28
33	9.58 375	31	9.61 830	35	10.38 170	9.96 546	5	27
34	9.58 406	30	9.61 865	36	10.38 135	9.96 541	6	26
35	9.58 436	31	9.61 901	35	10.38 099	9.96 535	5	25
36	9.58 467	30	9.61 936	36	10.38 064	9.96 530	5	24
37	9.58 497	30	9.61 972	36	10.38 028	9.96 525	5	23
38	9.58 527	30	9.62 008	35	10.37 992	9.96 520	6	22
39	9.58 557	31	9.62 043	36	10.37 957	9.96 514	5	21
40	9.58 588	30	9.62 079	35	10.37 921	9.96 509	5	20
41	9.58 618	30	9.62 114	36	10.37 886	9.96 504	6	19
42	9.58 648	30	9.62 150	35	10.37 850	9.96 498	5	18
43	9.58 678	31	9.62 185	36	10.37 815	9.96 493	5	17
44	9.58 709	30	9.62 221	35	10.37 779	9.96 488	5	16
45	9.58 739	30	9.62 256	36	10.37 744	9.96 483	6	15
46	9.58 769	30	9.62 292	35	10.37 708	9.96 477	5	14
47	9.58 799	30	9.62 327	35	10.37 673	9.96 472	5	13
48	9.58 829	30	9.62 362	36	10.37 638	9.96 467	6	12
49	9.58 859	30	9.62 398	35	10.37 602	9.96 461	5	11
50	9.58 889	30	9.62 433	35	10.37 567	9.96 456	5	10
51	9.58 919	30	9.62 468	36	10.37 532	9.96 451	6	9
52	9.58 949	30	9.62 504	35	10.37 496	9.96 445	5	8
53	9.58 979	30	9.62 539	35	10.37 461	9.96 440	5	7
54	9.59 009	30	9.62 574	35	10.37 426	9.96 435	6	6
55	9.59 039	30	9.62 609	36	10.37 391	9.96 429	5	5
56	9.59 069	29	9.62 645	35	10.37 355	9.96 424	5	4
57	9.59 098	30	9.62 680	35	10.37 320	9.96 419	6	3
58	9.59 128	30	9.62 715	35	10.37 285	9.96 413	5	2
59	9.59 158	30	9.62 750	35	10.37 250	9.96 408	5	1
60	9.59 188		9.62 785	35	10.37 215	9.96 403		0
	L Cos	d	L Cot	c d	L Tan	L Sin	d	′

Prop. Pts.

	37	36	35
1	3.7	3.6	3.5
2	7.4	7.2	7.0
3	11.1	10.8	10.5
4	14.8	14.4	14.0
5	18.5	18.0	17.5
6	22.2	21.6	21.0
7	25.9	25.2	24.5
8	29.6	28.8	28.0
9	33.3	32.4	31.5

	32	31	30
1	3.2	3.1	3.0
2	6.4	6.2	6.0
3	9.6	9.3	9.0
4	12.8	12.4	12.0
5	16.0	15.5	15.0
6	19.2	18.6	18.0
7	22.4	21.7	21.0
8	25.6	24.8	24.0
9	28.8	27.9	27.0

	29	6	5
1	2.9	0.6	0.5
2	5.8	1.2	1.0
3	8.7	1.8	1.5
4	11.6	2.4	2.0
5	14.5	3.0	2.5
6	17.4	3.6	3.0
7	20.3	4.2	3.5
8	23.2	4.8	4.0
9	26.1	5.4	4.5

TABLE 3

67°

23°

′	L Sin	d	L Tan	c d	L Cot	L Cos	d	′		Prop. Pts.
0	9.59 188	30	9.62 785	35	10.37 215	9.96 403	6	60		
1	9.59 218	29	9.62 820	35	10.37 180	9.96 397	5	59		
2	9.59 247	30	9.62 855	35	10.37 145	9.96 392	5	58		
3	9.59 277	30	9.62 890	36	10.37 110	9.96 387	6	57		
4	9.59 307	29	9.62 926	35	10.37 074	9.96 381	5	56		
5	9.59 336	30	9.62 961	35	10.37 039	9.96 376	6	55		
6	9.59 366	30	9.62 996	35	10.37 004	9.96 370	5	54		
7	9.59 396	29	9.63 031	35	10.36 969	9.96 365	5	53		
8	9.59 425	30	9.63 066	35	10.36 934	9.96 360	6	52		
9	9.59 455	29	9.63 101	34	10.36 899	9.96 354	5	51		
10	9.59 484	30	9.63 135	35	10.36 865	9.96 349	6	50		
11	9.59 514	29	9.63 170	35	10.36 830	9.96 343	5	49		**36** · **35** · **34**
12	9.59 543	30	9.63 205	35	10.36 795	9.96 338	5	48		1 · 3.6 · 3.5 · 3.4
13	9.59 573	29	9.63 240	35	10.36 760	9.96 333	6	47		2 · 7.2 · 7.0 · 6.8
14	9.59 602	30	9.63 275	35	10.36 725	9.96 327	5	46		3 · 10.8 · 10.5 · 10.2
15	9.59 632	29	9.63 310	35	10.36 690	9.96 322	6	45		4 · 14.4 · 14.0 · 13.6
16	9.59 661	29	9.63 345	34	10.36 655	9.96 316	5	44		5 · 18.0 · 17.5 · 17.0
17	9.59 690	30	9.63 379	35	10.36 621	9.96 311	6	43		6 · 21.6 · 21.0 · 20.4
18	9.59 720	29	9.63 414	35	10.36 586	9.96 305	5	42		7 · 25.2 · 24.5 · 23.8
19	9.59 749	29	9.63 449	35	10.36 551	9.96 300	6	41		8 · 28.8 · 28.0 · 27.2
20	9.59 778	30	9.63 484	35	10.36 516	9.96 294	5	40		9 · 32.4 · 31.5 · 30.6
21	9.59 808	29	9.63 519	34	10.36 481	9.96 289	5	39		
22	9.59 839	29	9.63 553	35	10.36 447	9.96 284	6	38		
23	9.59 866	29	9.63 588	35	10.36 412	9.96 278	5	37		
24	9.59 895	29	9.63 623	34	10.36 377	9.96 273	6	36		
25	9.59 924	30	9.63 657	35	10.36 343	9.96 267	5	35		
26	9.59 954	29	9.63 692	34	10.36 308	9.96 262	6	34		**30** · **29** · **28**
27	9.59 983	29	9.63 726	35	10.36 274	9.96 256	5	33		1 · 3.0 · 2.9 · 2.8
28	9.60 012	29	9.63 761	35	10.36 239	9.96 251	6	32		2 · 6.0 · 5.8 · 5.6
29	9.60 041	29	9.63 796	34	10.36 204	9.96 245	5	31		3 · 9.0 · 8.7 · 8.4
30	9.60 070	29	9.63 830	35	10.36 170	9.96 240	6	30		4 · 12.0 · 11.6 · 11.2
31	9.60 099	29	9.63 865	34	10.36 135	9.96 234	5	29		5 · 15.0 · 14.5 · 14.0
32	9.60 128	29	9.63 899	35	10.36 101	9.96 229	6	28		6 · 18.0 · 17.4 · 16.8
33	9.60 157	29	9.63 934	34	10.36 066	9.96 223	5	27		7 · 21.0 · 20.3 · 19.6
34	9.60 186	29	9.63 968	35	10.36 032	9.96 218	6	26		8 · 24.0 · 23.2 · 22.4
35	9.60 215	29	9.64 003	34	10.35 997	9.96 212	5	25		9 · 27.0 · 26.1 · 25.2
36	9.60 244	29	9.64 037	35	10.35 963	9.96 207	6	24		
37	9.60 273	29	9.64 072	34	10.35 928	9.96 201	5	23		
38	9.60 302	29	9.64 106	34	10.35 894	9.96 196	6	22		
39	9.60 331	28	9.64 140	35	10.35 860	9.96 190	5	21		
40	9.60 359	29	9.64 175	34	10.35 825	9.96 185	6	20		
41	9.60 388	29	9.64 209	34	10.35 791	9.96 179	5	19		
42	9.60 417	29	9.64 243	35	10.35 757	9.96 174	6	18		
43	9.60 446	28	9.64 278	34	10.35 722	9.96 168	6	17		**6** · **5**
44	9.60 474	29	9.64 312	34	10.35 688	9.96 162	5	16		1 · 0.6 · 0.5
45	9.60 503	29	9.64 346	35	10.35 654	9.96 157	6	15		2 · 1.2 · 1.0
46	9.60 532	29	9.64 381	34	10.35 619	9.96 151	5	14		3 · 1.8 · 1.5
47	9.60 561	28	9.64 415	34	10.35 585	9.96 146	6	13		4 · 2.4 · 2.0
48	9.60 589	29	9.64 449	34	10.35 551	9.96 140	5	12		5 · 3.0 · 2.5
49	9.60 618	28	9.64 483	34	10.35 517	9.96 135	6	11		6 · 3.6 · 3.0
50	9.60 646	29	9.64 517	35	10.35 483	9.96 129	6	10		7 · 4.2 · 3.5
51	9.60 675	29	9.64 552	34	10.35 448	9.96 123	5	9		8 · 4.8 · 4.0
52	9.60 704	28	9.64 586	34	10.35 414	9.96 118	6	8		9 · 5.4 · 4.5
53	9.60 732	29	9.64 620	34	10.35 380	9.96 112	5	7		
54	9.60 761	28	9.64 654	34	10.35 346	9.96 107	6	6		
55	9.60 789	29	9.64 688	34	10.35 312	9.96 101	6	5		
56	9.60 818	28	9.64 722	34	10.35 278	9.96 095	5	4		
57	9.60 846	29	9.64 756	34	10.35 244	9.96 090	6	3		
58	9.60 875	28	9.64 790	34	10.35 210	9.96 084	5	2		
59	9.60 903	28	9.64 824	34	10.35 176	9.96 079	6	1		
60	9.60 931		9.64 858		10.35 142	9.96 073		0		
′	L Cos	d	L Cot	c d	L Tan	L Sin	d	′		Prop. Pts.

66°

TABLE 3

24°

′	L Sin	d	L Tan	c d	L Cot	L Cos	d	′
0	9.60 931		9.64 858		10.35 142	9.96 073		60
1	9.60 960	29	9.64 892	34	10.35 108	9.96 067	6	59
2	9.60 988	28	9.64 926	34	10.35 074	9.96 062	5	58
3	9.61 016	28	9.64 960	34	10.35 040	9.96 056	6	57
4	9.61 045	29	9.64 994	34	10.35 006	9.96 050	6	56
5	9.61 073	28	9.65 028	34	10.34 972	9.96 045	5	55
6	9.61 101	28	9.65 062	34	10.34 938	9.96 039	6	54
7	9.61 129	28	9.65 096	34	10.34 904	9.96 034	5	53
8	9.61 158	29	9.65 130	34	10.34 870	9.96 028	6	52
9	9.61 186	28	9.65 164	34	10.34 836	9.96 022	6	51
10	9.61 214	28	9.65 197	33	10.34 803	9.96 017	5	50
11	9.61 242	28	9.65 231	34	10.34 769	9.96 011	6	49
12	9.61 270	28	9.65 265	34	10.34 735	9.96 005	6	48
13	9.61 298	28	9.65 299	34	10.34 701	9.96 000	5	47
14	9.61 326	28	9.65 333	33	10.34 667	9.95 994	6	46
15	9.61 354	28	9.65 366	34	10.34 634	9.95 988	6	45
16	9.61 382	29	9.65 400	34	10.34 600	9.95 982	5	44
17	9.61 411	27	9.65 434	33	10.34 566	9.95 977	6	43
18	9.61 438	28	9.65 467	34	10.34 533	9.95 971	6	42
19	9.61 466	28	9.65 501	34	10.34 499	9.95 965	5	41
20	9.61 494	28	9.65 535	33	10.34 465	9.95 960	6	40
21	9.61 522	28	9.65 568	34	10.34 432	9.95 954	6	39
22	9.61 550	28	9.65 602	34	10.34 398	9.95 948	6	38
23	9.61 578	28	9.65 636	33	10.34 364	9.95 942	5	37
24	9.61 606	28	9.65 669	34	10.34 331	9.95 937	6	36
25	9.61 634	28	9.65 703	33	10.34 297	9.95 931	6	35
26	9.61 662	27	9.65 736	34	10.34 264	9.95 925	5	34
27	9.61 689	28	9.65 770	33	10.34 230	9.95 920	6	33
28	9.61 717	28	9.65 803	34	10.34 197	9.95 914	6	32
29	9.61 745	28	9.65 837	33	10.34 163	9.95 908	6	31
30	9.61 773	27	9.65 870	34	10.34 130	9.95 902	5	30
31	9.61 800	28	9.65 904	33	10.34 096	9.95 897	6	29
32	9.61 828	28	9.65 937	34	10.34 063	9.95 891	6	28
33	9.61 856	27	9.65 971	33	10.34 029	9.95 885	6	27
34	9.61 883	28	9.66 004	34	10.33 996	9.95 879	6	26
35	9.61 911	28	9.66 038	33	10.33 962	9.95 873	5	25
36	9.61 939	27	9.66 071	33	10.33 929	9.95 868	6	24
37	9.61 966	28	9.66 104	34	10.33 896	9.95 862	6	23
38	9.61 994	27	9.66 138	33	10.33 862	9.95 856	6	22
39	9.62 021	28	9.66 171	33	10.33 829	9.95 850	6	21
40	9.62 049	27	9.66 204	34	10.33 796	9.95 844	5	20
41	9.62 076	28	9.66 238	33	10.33 762	9.95 839	6	19
42	9.62 104	27	9.66 271	33	10.33 729	9.95 833	6	18
43	9.62 131	28	9.66 304	33	10.33 696	9.95 827	6	17
44	9.62 159	27	9.66 337	34	10.33 663	9.95 821	6	16
45	9.62 186	28	9.66 371	33	10.33 629	9.95 815	5	15
46	9.62 214	27	9.66 404	33	10.33 596	9.95 810	6	14
47	9.62 241	27	9.66 437	33	10.33 563	9.95 804	6	13
48	9.62 268	28	9.66 470	33	10.33 530	9.95 798	6	12
49	9.62 296	27	9.66 503	34	10.33 497	9.95 792	6	11
50	9.62 323	27	9.66 537	33	10.33 463	9.95 786	6	10
51	9.62 350	27	9.66 570	33	10.33 430	9.95 780	5	9
52	9.62 377	28	9.66 603	33	10.33 397	9.95 775	6	8
53	9.62 405	27	9.66 636	33	10.33 364	9.95 769	6	7
54	9.62 432	27	9.66 669	33	10.33 331	9.95 763	6	6
55	9.62 459	27	9.66 702	33	10.33 298	9.95 757	6	5
56	9.62 486	27	9.66 735	33	10.33 265	9.95 751	6	4
57	9.62 513	28	9.66 768	33	10.33 232	9.95 745	6	3
58	9.62 541	27	9.66 801	33	10.33 199	9.95 739	6	2
59	9.62 568	27	9.66 834	33	10.33 166	9.95 733	5	1
60	9.62 595		9.66 867		10.33 133	9.95 728		0
	L Cos	d	L Cot	c d	L Tan	L Sin	d	′

Prop. Pts.

	34	33
1	3.4	3.3
2	6.8	6.6
3	10.2	9.9
4	13.6	13.2
5	17.0	16.5
6	20.4	19.8
7	23.8	23.1
8	27.2	26.4
9	30.6	29.7

	29	28	27
1	2.9	2.8	2.7
2	5.8	5.6	5.4
3	8.7	8.4	8.1
4	11.6	11.2	10.8
5	14.5	14.0	13.5
6	17.4	16.8	16.2
7	20.3	19.6	18.9
8	23.2	22.4	21.6
9	26.1	25.2	24.3

	6	5
1	0.6	0.5
2	1.2	1.0
3	1.8	1.5
4	2.4	2.0
5	3.0	2.5
6	3.6	3.0
7	4.2	3.5
8	4.8	4.0
9	5.4	4.5

65°

TABLE
3

25°

′	L Sin	d	L Tan	c d	L Cot	L Cos	d	′	Prop. Pts.
0	9.62 595		9.66 867		10.33 133	9.95 728		**60**	
1	9.62 622	27	9.66 900	33	10.33 100	9.95 722	6	59	
2	9.62 649	27	9.66 933	33	10.33 067	9.95 716	6	58	
3	9.62 676	27	9.66 966	33	10.33 034	9.95 710	6	57	
4	9.62 703	27	9.66 999	33	10.33 001	9.95 704	6	56	
5	9.62 730	27	9.67 032	33	10.32 968	9.95 698	6	55	
6	9.62 757	27	9.67 065	33	10.32 935	9.95 692	6	54	
7	9.62 784	27	9.67 098	33	10.32 902	9.95 686	6	53	
8	9.62 811	27	9.67 131	33	10.32 869	9.95 680	6	52	
9	9.62 838	27	9.67 163	32	10.32 837	9.95 674	6	51	
10	9.62 865	27	9.67 196	33	10.32 804	9.95 668	5	**50**	
11	9.62 892	27	9.67 229	33	10.32 771	9.95 663	6	49	
12	9.62 918	26	9.67 262	33	10.32 738	9.95 657	6	48	
13	9.62 945	27	9.67 295	33	10.32 705	9.95 651	6	47	
14	9.62 972	27	9.67 327	32	10.32 673	9.95 645	6	46	
15	9.62 999	27	9.67 360	33	10.32 640	9.95 639	6	45	
16	9.63 026	27	9.67 393	33	10.32 607	9.95 633	6	44	
17	9.63 052	26	9.67 426	33	10.32 574	9.95 627	6	43	
18	9.63 079	27	9.67 458	32	10.32 542	9.95 621	6	42	
19	9.63 106	27	9.67 491	33	10.32 509	9.95 615	6	41	
20	9.63 133	27	9.67 524	33	10.32 476	9.95 609	6	**40**	
21	9.63 159	26	9.67 556	32	10.32 444	9.95 603	6	39	
22	9.63 186	27	9.67 589	33	10.32 411	9.95 597	6	38	
23	9.63 213	27	9.67 622	33	10.32 378	9.95 591	6	37	
24	9.63 239	26	9.67 654	32	10.32 346	9.95 585	6	36	
25	9.63 266	27	9.67 687	33	10.32 313	9.95 579	6	35	
26	9.63 292	26	9.67 719	32	10.32 281	9.95 573	6	34	
27	9.63 319	27	9.67 752	33	10.32 248	9.95 567	6	33	
28	9.63 345	26	9.67 785	33	10.32 215	9.95 561	6	32	
29	9.63 372	27	9.67 817	32	10.32 183	9.95 555	6	31	
30	9.63 398	26	9.67 850	33	10.32 150	9.95 549	6	**30**	
31	9.63 425	27	9.67 882	32	10.32 118	9.95 543	6	29	
32	9.63 451	26	9.67 915	33	10.32 085	9.95 537	6	28	
33	9.63 478	27	9.67 947	32	10.32 053	9.95 531	6	27	
34	9.63 504	26	9.67 980	33	10.32 020	9.95 525	6	26	
35	9.63 531	27	9.68 012	32	10.31 988	9.95 519	6	25	
36	9.63 557	26	9.68 044	32	10.31 956	9.95 513	6	24	
37	9.63 583	26	9.68 077	33	10.31 923	9.95 507	7	23	
38	9.63 610	27	9.68 109	32	10.31 891	9.95 500	6	22	
39	9.63 636	26	9.68 142	33	10.31 858	9.95 494	6	21	
40	9.63 662	26	9.68 174	32	10.31 826	9.95 488	6	**20**	
41	9.63 689	27	9.68 206	32	10.31 794	9.95 482	6	19	
42	9.63 715	26	9.68 239	33	10.31 761	9.95 476	6	18	
43	9.63 741	26	9.68 271	32	10.31 729	9.95 470	6	17	
44	9.63 767	27	9.68 303	33	10.31 697	9.95 464	6	16	
45	9.63 794	26	9.68 336	32	10.31 664	9.95 458	6	15	
46	9.63 820	26	9.68 368	32	10.31 632	9.95 452	6	14	
47	9.63 846	26	9.68 400	32	10.31 600	9.95 446	6	13	
48	9.63 872	26	9.68 432	33	10.31 568	9.95 440	6	12	
49	9.63 898	26	9.68 465	32	10.31 535	9.95 434	7	11	
50	9.63 924	26	9.68 497	32	10.31 503	9.95 427	6	**10**	
51	9.63 950	26	9.68 529	32	10.31 471	9.95 421	6	9	
52	9.63 976	26	9.68 561	32	10.31 439	9.95 415	6	8	
53	9.64 002	26	9.68 593	33	10.31 407	9.95 409	6	7	
54	9.64 028	26	9.68 626	32	10.31 374	9.95 403	6	6	
55	9.64 054	26	9.68 658	32	10.31 342	9.95 397	6	5	
56	9.64 080	26	9.68 690	32	10.31 310	9.95 391	7	4	
57	9.64 106	26	9.68 722	32	10.31 278	9.95 384	6	3	
58	9.64 132	26	9.68 754	32	10.31 246	9.95 378	6	2	
59	9.64 158	26	9.68 786	32	10.31 214	9.95 372	6	1	
60	9.64 184		9.68 818		10.31 182	9.95 366		**0**	
	L Cos	d	L Cot	c d	L Tan	L Sin	d	′	Prop. Pts.

Prop. Pts.

	33	32
1	3.3	3.2
2	6.6	6.4
3	9.9	9.6
4	13.2	12.8
5	16.5	16.0
6	19.8	19.2
7	23.1	22.4
8	26.4	25.6
9	29.7	28.8

	27	26
1	2.7	2.6
2	5.4	5.2
3	8.1	7.8
4	10.8	10.4
5	13.5	13.0
6	16.2	15.6
7	18.9	18.2
8	21.6	20.8
9	24.3	23.4

	7	6	5
1	0.7	0.6	0.5
2	1.4	1.2	1.0
3	2.1	1.8	1.5
4	2.8	2.4	2.0
5	3.5	3.0	2.5
6	4.2	3.6	3.0
7	4.9	4.2	3.5
8	5.6	4.8	4.0
9	6.3	5.4	4.5

TABLE

3

26°

′	L Sin	d	L Tan	c d	L Cot	L Cos	d	′	Prop. Pts.
0	9.64 184		9.68 818		10.31 182	9.95 366		60	
1	9.64 210	26	9.68 850	32	10.31 150	9.95 360	6	59	
2	9.64 236	26	9.68 882	32	10.31 118	9.95 354	6	58	
3	9.64 262	26	9.68 914	32	10.31 086	9.95 348	6	57	
4	9.64 288	26	9.68 946	32	10.31 054	9.95 341	7	56	
5	9.64 313	25	9.68 978	32	10.31 022	9.95 335	6	55	
6	9.64 339	26	9.69 010	32	10.30 990	9.95 329	6	54	
7	9.64 365	26	9.69 042	32	10.30 958	9.95 323	6	53	
8	9.64 391	26	9.69 074	32	10.30 926	9.95 317	6	52	
9	9.64 417	26	9.69 106	32	10.30 894	9.95 310	7	51	
10	9.64 442	25	9.69 138	32	10.30 862	9.95 304	6	50	
11	9.64 468	26	9.69 170	32	10.30 830	9.95 298	6	49	**32** / **31**
12	9.64 494	26	9.69 202	32	10.30 798	9.95 292	6	48	1 3.2 3.1
13	9.64 519	25	9.69 234	32	10.30 766	9.95 286	6	47	2 6.4 6.2
14	9.64 545	26	9.69 266	32	10.30 734	9.95 279	7	46	3 9.6 9.3
15	9.64 571	26	9.69 298	32	10.30 702	9.95 273	6	45	4 12.8 12.4
16	9.64 596	25	9.69 329	31	10.30 671	9.95 267	6	44	5 16.0 15.5
17	9.64 622	26	9.69 361	32	10.30 639	9.95 261	6	43	6 19.2 18.6
18	9.64 647	25	9.69 393	32	10.30 607	9.95 254	7	42	7 22.4 21.7
19	9.64 673	26	9.69 425	32	10.30 575	9.95 248	6	41	8 25.6 24.8
20	9.64 698	25	9.69 457	32	10.30 543	9.95 242	6	40	9 28.8 27.9
21	9.64 724	26	9.69 488	31	10.30 512	9.95 236	6	39	
22	9.64 749	25	9.69 520	32	10.30 480	9.95 229	7	38	
23	9.64 775	26	9.69 552	32	10.30 448	9.95 223	6	37	
24	9.64 800	25	9.69 584	32	10.30 416	9.95 217	6	36	
25	9.64 826	26	9.69 615	31	10.30 385	9.95 211	6	35	
26	9.64 851	25	9.69 647	32	10.30 353	9.95 204	7	34	**26** / **25** / **24**
27	9.64 877	26	9.69 679	32	10.30 321	9.95 198	6	33	1 2.6 2.5 2.4
28	9.64 902	25	9.69 710	31	10.30 290	9.95 192	6	32	2 5.2 5.0 4.8
29	9.64 927	25	9.69 742	32	10.30 258	9.95 185	7	31	3 7.8 7.5 7.2
30	9.64 953	26	9.69 774	32	10.30 226	9.95 179	6	30	4 10.4 10.0 9.6
31	9.64 978	25	9.69 805	31	10.30 195	9.95 173	6	29	5 13.0 12.5 12.0
32	9.65 003	25	9.69 837	32	10.30 163	9.95 167	6	28	6 15.6 15.0 14.4
33	9.65 029	26	9.69 868	31	10.30 132	9.95 160	7	27	7 18.2 17.5 16.8
34	9.65 054	25	9.69 900	32	10.30 100	9.95 154	6	26	8 20.8 20.0 19.2
35	9.65 079	25	9.69 932	32	10.30 068	9.95 148	7	25	9 23.4 22.5 21.6
36	9.65 104	25	9.69 963	31	10.30 037	9.95 141	6	24	
37	9.65 130	26	9.69 995	32	10.30 005	9.95 135	6	23	
38	9.65 155	25	9.70 026	31	10.29 974	9.95 129	6	22	
39	9.65 180	25	9.70 058	32	10.29 942	9.95 122	7	21	
40	9.65 205	25	9.70 089	31	10.29 911	9.95 116	6	20	
41	9.65 230	25	9.70 121	32	10.29 879	9.95 110	6	19	
42	9.65 255	25	9.70 152	31	10.29 848	9.95 103	7	18	
43	9.65 281	26	9.70 184	32	10.29 816	9.95 097	6	17	**7** / **6**
44	9.65 306	25	9.70 215	31	10.29 785	9.95 090	7	16	1 0.7 0.6
45	9.65 331	25	9.70 247	32	10.29 753	9.95 084	7	15	2 1.4 1.2
46	9.65 356	25	9.70 278	31	10.29 722	9.95 078	6	14	3 2.1 1.8
47	9.65 381	25	9.70 309	31	10.29 691	9.95 071	7	13	4 2.8 2.4
48	9.65 406	25	9.70 341	32	10.29 659	9.95 065	6	12	5 3.5 3.0
49	9.65 431	25	9.70 372	31	10.29 628	9.95 059	6	11	6 4.2 3.6
50	9.65 456	25	9.70 404	32	10.29 596	9.95 052	7	10	7 4.9 4.2
51	9.65 481	25	9.70 435	31	10.29 565	9.95 046	6	9	8 5.6 4.8
52	9.65 506	25	9.70 466	31	10.29 534	9.95 039	7	8	9 6.3 5.4
53	9.65 531	25	9.70 498	32	10.29 502	9.95 033	6	7	
54	9.65 556	25	9.70 529	31	10.29 471	9.95 027	6	6	
55	9.65 580	24	9.70 560	31	10.29 440	9.95 020	7	5	
56	9.65 605	25	9.70 592	32	10.29 408	9.95 014	6	4	
57	9.65 630	25	9.70 623	31	10.29 377	9.95 007	7	3	
58	9.65 655	25	9.70 654	31	10.29 346	9.95 001	6	2	
59	9.65 680	25	9.70 685	31	10.29 315	9.94 995	7	1	
60	9.65 705	25	9.70 717	32	10.29 283	9.94 988		0	
	L Cos	d	L Cot	c d	L Tan	L Sin	d	′	Prop. Pts.

63°

TABLE 3

27°

'	L Sin	d	L Tan	c d	L Cot	L Cos	d	'
0	9.65 705	24	9.70 717	31	10.29 283	9.94 988	6	60
1	9.65 729	25	9.70 748	31	10.29 252	9.94 982	7	59
2	9.65 754	25	9.70 779	31	10.29 221	9.94 975	6	58
3	9.65 779	25	9.70 810	31	10.29 190	9.94 969	7	57
4	9.65 804	24	9.70 841	32	10.29 159	9.94 962	6	56
5	9.65 828	25	9.70 873	31	10.29 127	9.94 956	7	55
6	9.65 853	25	9.70 904	31	10.29 096	9.94 949	6	54
7	9.65 878	24	9.70 935	31	10.29 065	9.94 943	7	53
8	9.65 902	25	9.70 966	31	10.29 034	9.94 936	6	52
9	9.65 927	25	9.70 997	31	10.29 003	9.94 930	7	51
10	9.65 952	24	9.71 028	31	10.28 972	9.94 923	6	50
11	9.65 976	25	9.71 059	31	10.28 941	9.94 917	6	49
12	9.66 001	24	9.71 090	31	10.28 910	9.94 911	7	48
13	9.66 025	25	9.71 121	32	10.28 879	9.94 904	6	47
14	9.66 050	25	9.71 153	31	10.28 847	9.94 898	7	46
15	9.66 075	24	9.71 184	31	10.28 816	9.94 891	6	45
16	9.66 099	25	9.71 215	31	10.28 785	9.94 885	7	44
17	9.66 124	24	9.71 246	31	10.28 754	9.94 878	7	43
18	9.66 148	25	9.71 277	31	10.28 723	9.94 871	6	42
19	9.66 173	24	9.71 308	31	10.28 692	9.94 865	7	41
20	9.66 197	24	9.71 339	31	10.28 661	9.94 858	6	40
21	9.66 221	25	9.71 370	31	10.28 630	9.94 852	7	39
22	9.66 246	24	9.71 401	30	10.28 599	9.94 845	6	38
23	9.66 270	25	9.71 431	31	10.28 569	9.94 839	7	37
24	9.66 295	24	9.71 462	31	10.28 538	9.94 832	6	36
25	9.66 319	24	9.71 493	31	10.28 507	9.94 826	7	35
26	9.66 343	25	9.71 524	31	10.28 476	9.94 819	6	34
27	9.66 368	24	9.71 555	31	10.28 445	9.94 813	7	33
28	9.66 392	24	9.71 586	31	10.28 414	9.94 806	7	32
29	9.66 416	25	9.71 617	31	10.28 383	9.94 799	6	31
30	9.66 441	24	9.71 648	31	10.28 352	9.94 793	7	30
31	9.66 465	24	9.71 679	30	10.28 321	9.94 786	6	29
32	9.66 489	24	9.71 709	31	10.28 291	9.94 780	7	28
33	9.66 513	24	9.71 740	31	10.28 260	9.94 773	6	27
34	9.66 537	25	9.71 771	31	10.28 229	9.94 767	7	26
35	9.66 562	24	9.71 802	31	10.28 198	9.94 760	7	25
36	9.66 586	24	9.71 833	30	10.28 167	9.94 753	6	24
37	9.66 610	24	9.71 863	31	10.28 137	9.94 747	7	23
38	9.66 634	24	9.71 894	31	10.28 106	9.94 740	6	22
39	9.66 658	24	9.71 925	30	10.28 075	9.94 734	7	21
40	9.66 682	24	9.71 955	31	10.28 045	9.94 727	7	20
41	9.66 706	25	9.71 986	31	10.28 014	9.94 720	6	19
42	9.66 731	24	9.72 017	31	10.27 983	9.94 714	7	18
43	9.66 755	24	9.72 048	30	10.27 952	9.94 707	7	17
44	9.66 779	24	9.72 078	31	10.27 922	9.94 700	6	16
45	9.66 803	24	9.72 109	31	10.27 891	9.94 694	7	15
46	9.66 827	24	9.72 140	30	10.27 860	9.94 687	7	14
47	9.66 851	24	9.72 170	31	10.27 830	9.94 680	6	13
48	9.66 875	24	9.72 201	30	10.27 799	9.94 674	7	12
49	9.66 899	23	9.72 231	31	10.27 769	9.94 667	7	11
50	9.66 922	24	9.72 262	31	10.27 738	9.94 660	6	10
51	9.66 946	24	9.72 293	30	10.27 707	9.94 654	7	9
52	9.66 970	24	9.72 323	31	10.27 677	9.94 647	7	8
53	9.66 994	24	9.72 354	30	10.27 646	9.94 640	6	7
54	9.67 018	24	9.72 384	31	10.27 616	9.94 634	7	6
55	9.67 042	24	9.72 415	30	10.27 585	9.94 627	7	5
56	9.67 066	24	9.72 445	31	10.27 555	9.94 620	6	4
57	9.67 090	23	9.72 476	30	10.27 524	9.94 614	7	3
58	9.67 113	24	9.72 506	31	10.27 494	9.94 607	7	2
59	9.67 137	24	9.72 537	30	10.27 463	9.94 600	7	1
60	9.67 161		9.72 567		10.27 433	9.94 593		0
	L Cos	d	L Cot	c d	L Tan	L Sin	d	'

62°

Prop. Pts.

	32	31	30
1	3.2	3.1	3.0
2	6.4	6.2	6.0
3	9.6	9.3	9.0
4	12.8	12.4	12.0
5	16.0	15.5	15.0
6	19.2	18.6	18.0
7	22.4	21.7	21.0
8	25.6	24.8	24.0
9	28.8	27.9	27.0

	25	24	23
1	2.5	2.4	2.3
2	5.0	4.8	4.6
3	7.5	7.2	6.9
4	10.0	9.6	9.2
5	12.5	12.0	11.5
6	15.0	14.4	13.8
7	17.5	16.8	16.1
8	20.0	19.2	18.4
9	22.5	21.6	20.7

	7	6
1	0.7	0.6
2	1.4	1.2
3	2.1	1.8
4	2.8	2.4
5	3.5	3.0
6	4.2	3.6
7	4.9	4.2
8	5.6	4.8
9	6.3	5.4

TABLE
3

28°

′	L Sin	d	L Tan	c d	L Cot	L Cos	d	′
0	9.67 161		9.72 567		10.27 433	9.94 593		60
1	9.67 185	24	9.72 598	31	10.27 402	9.94 587	6	59
2	9.67 208	23	9.72 628	30	10.27 372	9.94 580	7	58
3	9.67 232	24	9.72 659	31	10.27 341	9.94 573	7	57
4	9.67 256	24	9.72 689	30	10.27 311	9.94 567	6	56
5	9.67 280	24	9.72 720	31	10.27 280	9.94 560	7	55
6	9.67 303	23	9.72 750	30	10.27 250	9.94 553	7	54
7	9.67 327	24	9.72 780	30	10.27 220	9.94 546	7	53
8	9.67 350	23	9.72 811	31	10.27 189	9.94 540	6	52
9	9.67 374	24	9.72 841	30	10.27 159	9.94 533	7	51
10	9.67 398	24	9.72 872	31	10.27 128	9.94 526	7	50
11	9.67 421	23	9.72 902	30	10.27 098	9.94 519	6	49
12	9.67 445	24	9.72 932	30	10.27 068	9.94 513	7	48
13	9.67 468	23	9.72 963	31	10.27 037	9.94 506	7	47
14	9.67 492	24	9.72 993	30	10.27 007	9.94 499	7	46
15	9.67 515	24	9.73 023	31	10.26 977	9.94 492	7	45
16	9.67 539	23	9.73 054	30	10.26 946	9.94 485	6	44
17	9.67 562	24	9.73 084	30	10.26 916	9.94 479	7	43
18	9.67 586	23	9.73 114	30	10.26 886	9.94 472	7	42
19	9.67 609	24	9.73 144	31	10.26 856	9.94 465	7	41
20	9.67 633	23	9.73 175	30	10.26 825	9.94 458	7	40
21	9.67 656	24	9.73 205	30	10.26 795	9.94 451	6	39
22	9.67 680	23	9.73 235	30	10.26 765	9.94 445	7	38
23	9.67 703	23	9.73 265	30	10.26 735	9.94 438	7	37
24	9.67 726	24	9.73 295	31	10.26 705	9.94 431	7	36
25	9.67 750	23	9.73 326	30	10.26 674	9.94 424	7	35
26	9.67 773	23	9.73 356	30	10.26 644	9.94 417	7	34
27	9.67 796	24	9.73 386	30	10.26 614	9.94 410	6	33
28	9.67 820	23	9.73 416	30	10.26 584	9.94 404	7	32
29	9.67 843	23	9.73 446	30	10.26 554	9.94 397	7	31
30	9.67 866	24	9.73 476	31	10.26 524	9.94 390	7	30
31	9.67 890	23	9.73 507	30	10.26 493	9.94 383	7	29
32	9.67 913	23	9.73 537	30	10.26 463	9.94 376	7	28
33	9.67 936	23	9.73 567	30	10.26 433	9.94 369	7	27
34	9.67 959	23	9.73 597	30	10.26 403	9.94 362	7	26
35	9.67 982	24	9.73 627	30	10.26 373	9.94 355	6	25
36	9.68 006	23	9.73 657	30	10.26 343	9.94 349	7	24
37	9.68 029	23	9.73 687	30	10.26 313	9.94 342	7	23
38	9.68 052	23	9.73 717	30	10.26 283	9.94 335	7	22
39	9.68 075	23	9.73 747	30	10.26 253	9.94 328	7	21
40	9.68 098	23	9.73 777	30	10.26 223	9.94 321	7	20
41	9.68 121	23	9.73 807	30	10.26 193	9.94 314	7	19
42	9.68 144	23	9.73 837	30	10.26 163	9.94 307	7	18
43	9.68 167	23	9.73 867	30	10.26 133	9.94 300	7	17
44	9.68 190	23	9.73 897	30	10.26 103	9.94 293	7	16
45	9.68 213	24	9.73 927	30	10.26 073	9.94 286	7	15
46	9.68 237	23	9.73 957	30	10.26 043	9.94 279	6	14
47	9.68 260	23	9.73 987	30	10.26 013	9.94 273	7	13
48	9.68 283	22	9.74 017	30	10.25 983	9.94 266	7	12
49	9.68 305	23	9.74 047	30	10.25 953	9.94 259	7	11
50	9.68 328	23	9.74 077	30	10.25 923	9.94 252	7	10
51	9.68 351	23	9.74 107	30	10.25 893	9.94 245	7	9
52	9.68 374	23	9.74 137	29	10.25 863	9.94 238	7	8
53	9.68 397	23	9.74 166	30	10.25 834	9.94 231	7	7
54	9.68 420	23	9.74 196	30	10.25 804	9.94 224	7	6
55	9.68 443	23	9.74 226	30	10.25 774	9.94 217	7	5
56	9.68 466	23	9.74 256	30	10.25 744	9.94 210	7	4
57	9.68 489	23	9.74 286	30	10.25 714	9.94 203	7	3
58	9.68 512	22	9.74 316	29	10.25 684	9.94 196	7	2
59	9.68 534	23	9.74 345	30	10.25 655	9.94 189	7	1
60	9.68 557		9.74 375		10.25 625	9.94 182		0
	L Cos	d	L Cot	c d	L Tan	L Sin	d	′

Prop. Pts.

	31	30	29
1	3.1	3.0	2.9
2	6.2	6.0	5.8
3	9.3	9.0	8.7
4	12.4	12.0	11.6
5	15.5	15.0	14.5
6	18.6	18.0	17.4
7	21.7	21.0	20.3
8	24.8	24.0	23.2
9	27.9	27.0	26.1

	24	23	22
1	2.4	2.3	2.2
2	4.8	4.6	4.4
3	7.2	6.9	6.6
4	9.6	9.2	8.8
5	12.0	11.5	11.0
6	14.4	13.8	13.2
7	16.8	16.1	15.4
8	19.2	18.4	17.6
9	21.6	20.7	19.8

	7	6
1	0.7	0.6
2	1.4	1.2
3	2.1	1.8
4	2.8	2.4
5	3.5	3.0
6	4.2	3.6
7	4.9	4.2
8	5.6	4.8
9	6.3	5.4

TABLE

3

61°

29°

′	L Sin	d	L Tan	c d	L Cot	L Cos	d		Prop. Pts.
0	9.68 557		9.74 375		10.25 625	9.94 182		60	
1	9.68 580	23	9.74 405	30	10.25 595	9.94 175	7	59	
2	9.68 603	23	9.74 435	30	10.25 565	9.94 168	7	58	
3	9.68 625	22	9.74 465	30	10.25 535	9.94 161	7	57	
4	9.68 648	23	9.74 494	29	10.25 506	9.94 154	7	56	
5	9.68 671	23	9.74 524	30	10.25 476	9.94 147	7	55	
6	9.68 694	23	9.74 554	30	10.25 446	9.94 140	7	54	
7	9.68 716	22	9.74 583	29	10.25 417	9.94 133	7	53	
8	9.68 739	23	9.74 613	30	10.25 387	9.94 126	7	52	
9	9.68 762	23	9.74 643	30	10.25 357	9.94 119	7	51	
10	9.68 784	22	9.74 673	30	10.25 327	9.94 112	7	50	
11	9.68 807	23	9.74 702	29	10.25 298	9.94 105	7	49	
12	9.68 829	22	9.74 732	30	10.25 268	9.94 098	7	48	
13	9.68 852	23	9.74 762	30	10.25 238	9.94 090	8	47	
14	9.68 875	23	9.74 791	29	10.25 209	9.94 083	7	46	
15	9.68 897	22	9.74 821	30	10.25 179	9.94 076	7	45	
16	9.68 920	23	9.74 851	30	10.25 149	9.94 069	7	44	
17	9.68 942	22	9.74 880	29	10.25 120	9.94 062	7	43	
18	9.68 965	23	9.74 910	30	10.25 090	9.94 055	7	42	
19	9.68 987	22	9.74 939	29	10.25 061	9.94 048	7	41	
20	9.69 010	23	9.74 969	30	10.25 031	9.94 041	7	40	
21	9.69 032	22	9.74 998	30	10.25 002	9.94 034	7	39	
22	9.69 055	23	9.75 028	30	10.24 972	9.94 027	7	38	
23	9.69 077	22	9.75 058	29	10.24 942	9.94 020	8	37	
24	9.69 100	23	9.75 087	30	10.24 913	9.94 012	7	36	
25	9.69 122	22	9.75 117	29	10.24 883	9.94 005	7	35	
26	9.69 144	22	9.75 146	30	10.24 854	9.93 998	7	34	
27	9.69 167	23	9.75 176	30	10.24 824	9.93 991	7	33	
28	9.69 189	22	9.75 205	29	10.24 795	9.93 984	7	32	
29	9.69 212	23	9.75 235	29	10.24 765	9.93 977	7	31	
30	9.69 234	22	9.75 264	30	10.24 736	9.93 970	7	30	
31	9.69 256	22	9.75 294	29	10.24 706	9.93 963	8	29	
32	9.69 279	23	9.75 323	30	10.24 677	9.93 955	7	28	
33	9.69 301	22	9.75 353	29	10.24 647	9.93 948	7	27	
34	9.69 323	22	9.75 382	29	10.24 618	9.93 941	7	26	
35	9.69 345	23	9.75 411	30	10.24 589	9.93 934	7	25	
36	9.69 368	22	9.75 441	29	10.24 559	9.93 927	7	24	
37	9.69 390	22	9.75 470	30	10.24 530	9.93 920	8	23	
38	9.69 412	22	9.75 500	29	10.24 500	9.93 912	7	22	
39	9.69 434	22	9.75 529	29	10.24 471	9.93 905	7	21	
40	9.69 456	23	9.75 558	30	10.24 442	9.93 898	7	20	
41	9.69 479	22	9.75 588	29	10.24 412	9.93 891	7	19	
42	9.69 501	22	9.75 617	30	10.24 383	9.93 884	8	18	
43	9.69 523	22	9.75 647	29	10.24 353	9.93 876	7	17	
44	9.69 545	22	9.75 676	29	10.24 324	9.93 869	7	16	
45	9.69 567	22	9.75 705	30	10.24 295	9.93 862	7	15	
46	9.69 589	22	9.75 735	29	10.24 265	9.93 855	8	14	
47	9.69 611	22	9.75 764	29	10.24 236	9.93 847	7	13	
48	9.69 633	22	9.75 793	29	10.24 207	9.93 840	7	12	
49	9.69 655	22	9.75 822	30	10.24 178	9.93 833	7	11	
50	9.69 677	22	9.75 852	29	10.24 148	9.93 826	7	10	
51	9.69 699	22	9.75 881	29	10.24 119	9.93 819	8	9	
52	9.69 721	22	9.75 910	29	10.24 090	9.93 811	7	8	
53	9.69 743	22	9.75 939	30	10.24 061	9.93 804	7	7	
54	9.69 765	22	9.75 969	29	10.24 031	9.93 797	8	6	
55	9.69 787	22	9.75 998	29	10.24 002	9.93 789	7	5	
56	9.69 809	22	9.76 027	29	10.23 973	9.93 782	7	4	
57	9.69 831	22	9.76 056	30	10.23 944	9.93 775	7	3	
58	9.69 853	22	9.76 086	29	10.23 914	9.93 768	8	2	
59	9.69 875	22	9.76 115	29	10.23 885	9.93 760	7	1	
60	9.69 897		9.76 144		10.23 856	9.93 753		0	
	L Cos	d	L Cot	c d	L Tan	L Sin	d	′	Prop. Pts.

TABLE 3

Prop. Pts.

	30	29
1	3.0	2.9
2	6.0	5.8
3	9.0	8.7
4	12.0	11.6
5	15.0	14.5
6	18.0	17.4
7	21.0	20.3
8	24.0	23.2
9	27.0	26.1

	23	22
1	2.3	2.2
2	4.6	4.4
3	6.9	6.6
4	9.2	8.8
5	11.5	11.0
6	13.8	13.2
7	16.1	15.4
8	18.4	17.6
9	20.7	19.8

	8	7
1	0.8	0.7
2	1.6	1.4
3	2.4	2.1
4	3.2	2.8
5	4.0	3.5
6	4.8	4.2
7	5.6	4.9
8	6.4	5.6
9	7.2	6.3

60°

30°

'	L Sin	d	L Tan	c d	L Cot	L Cos	d	'
0	9.69 897	22	9.76 144	29	10.23 856	9.93 753	7	60
1	9.69 919	22	9.76 173	29	10.23 827	9.93 746	8	59
2	9.69 941	22	9.76 202	29	10.23 798	9.93 738	7	58
3	9.69 963	21	9.76 231	30	10.23 769	9.93 731	7	57
4	9.69 984	22	9.76 261	29	10.23 739	9.93 724	7	56
5	9.70 006	22	9.76 290	29	10.23 710	9.93 717	8	55
6	9.70 028	22	9.76 319	29	10.23 681	9.93 709	7	54
7	9.70 050	22	9.76 348	29	10.23 652	9.93 702	7	53
8	9.70 072	21	9.76 377	29	10.23 623	9.93 695	8	52
9	9.70 093	22	9.76 406	29	10.23 594	9.93 687	7	51
10	9.70 115	22	9.76 435	29	10.23 565	9.93 680	7	50
11	9.70 137	22	9.76 464	29	10.23 536	9.93 673	8	49
12	9.70 159	21	9.76 493	29	10.23 507	9.93 665	7	48
13	9.70 180	22	9.76 522	29	10.23 478	9.93 658	8	47
14	9.70 202	22	9.76 551	29	10.23 449	9.93 650	7	46
15	9.70 224	21	9.76 580	29	10.23 420	9.93 643	7	45
16	9.70 245	22	9.76 609	30	10.23 391	9.93 636	8	44
17	9.70 267	21	9.76 639	29	10.23 361	9.93 628	7	43
18	9.70 288	22	9.76 668	29	10.23 332	9.93 621	7	42
19	9.70 310	22	9.76 697	28	10.23 303	9.93 614	8	41
20	9.70 332	21	9.76 725	29	10.23 275	9.93 606	7	40
21	9.70 353	22	9.76 754	29	10.23 246	9.93 599	8	39
22	9.70 375	21	9.76 783	29	10.23 217	9.93 591	7	38
23	9.70 396	22	9.76 812	29	10.23 188	9.93 584	7	37
24	9.70 418	21	9.76 841	29	10.23 159	9.93 577	8	36
25	9.70 439	22	9.76 870	29	10.23 130	9.93 569	7	35
26	9.70 461	21	9.76 899	29	10.23 101	9.93 562	8	34
27	9.70 482	22	9.76 928	29	10.23 072	9.93 554	7	33
28	9.70 504	21	9.76 957	29	10.23 043	9.93 547	8	32
29	9.70 525	22	9.76 986	29	10.23 014	9.93 539	7	31
30	9.70 547	21	9.77 015	29	10.22 985	9.93 532	7	30
31	9.70 568	22	9.77 044	29	10.22 956	9.93 525	8	29
32	9.70 590	21	9.77 073	28	10.22 927	9.93 517	7	28
33	9.70 611	22	9.77 101	29	10.22 899	9.93 510	8	27
34	9.70 633	21	9.77 130	29	10.22 870	9.93 502	7	26
35	9.70 654	21	9.77 159	29	10.22 841	9.93 495	8	25
36	9.70 675	22	9.77 188	29	10.22 812	9.93 487	7	24
37	9.70 796	21	9.77 217	29	10.22 783	9.93 480	8	23
38	9.70 718	21	9.77 246	28	10.22 754	9.93 472	7	22
39	9.70 739	22	9.77 274	29	10.22 726	9.93 465	8	21
40	9.70 761	21	9.77 303	29	10.22 697	9.93 457	7	20
41	9.70 782	21	9.77 332	29	10.22 668	9.93 450	8	19
42	9.70 803	21	9.77 361	29	10.22 639	9.93 442	7	18
43	9.70 824	22	9.77 390	28	10.22 610	9.93 435	8	17
44	9.70 846	21	9.77 418	29	10.22 582	9.93 427	7	16
45	9.70 867	21	9.77 447	29	10.22 553	9.93 420	8	15
46	9.70 888	21	9.77 476	29	10.22 524	9.93 412	7	14
47	9.70 909	22	9.77 505	28	10.22 495	9.93 405	8	13
48	9.70 931	21	9.77 533	29	10.22 467	9.93 397	7	12
49	9.70 952	21	9.77 562	29	10.22 438	9.93 390	8	11
50	9.70 973	21	9.77 591	28	10.22 409	9.93 382	7	10
51	9.70 994	21	9.77 619	29	10.22 381	9.93 375	8	9
52	9.71 015	21	9.77 648	29	10.22 352	9.93 367	7	8
53	9.71 036	22	9.77 677	29	10.22 323	9.93 360	8	7
54	9.71 058	21	9.77 706	28	10.22 294	9.93 352	8	6
55	9.71 079	21	9.77 734	29	10.22 266	9.93 344	7	5
56	9.71 100	21	9.77 763	28	10.22 237	9.93 337	8	4
57	9.71 121	21	9.77 791	29	10.22 209	9.93 329	7	3
58	9.71 142	21	9.77 820	29	10.22 180	9.93 322	8	2
59	9.71 163	21	9.77 849	28	10.22 151	9.93 314	7	1
60	9.71 184		9.77 877		10.22 123	9.93 307		0
	L Cos	d	L Cot	c d	L Tan	L Sin	d	'

Prop. Pts.

	30	29	28
1	3.0	2.9	2.8
2	6.0	5.8	5.6
3	9.0	8.7	8.4
4	12.0	11.6	11.2
5	15.0	14.5	14.0
6	18.0	17.4	16.8
7	21.0	20.3	19.6
8	24.0	23.2	22.4
9	27.0	26.1	25.2

	22	21
1	2.2	2.1
2	4.4	4.2
3	6.6	6.3
4	8.8	8.4
5	11.0	10.5
6	13.2	12.6
7	15.4	14.7
8	17.6	16.8
9	19.8	18.9

	8	7
1	0.8	0.7
2	1.6	1.4
3	2.4	2.1
4	3.2	2.8
5	4.0	3.5
6	4.8	4.2
7	5.6	4.9
8	6.4	5.6
9	7.2	6.3

59°

TABLE 3

31°

′	L Sin	d	L Tan	c d	L Cot	L Cos	d	′	Prop. Pts.
0	9.71 184	21	9.77 877	29	10.22 123	9.93 307	8	60	
1	9.71 205	21	9.77 906	29	10.22 094	9.93 299	8	59	
2	9.71 226	21	9.77 935	28	10.22 065	9.93 291	7	58	
3	9.71 247	21	9.77 963	29	10.22 037	9.93 284	8	57	
4	9.71 268	21	9.77 992	28	10.22 008	9.93 276	7	56	
5	9.71 289	21	9.78 020	29	10.21 980	9.93 269	8	55	
6	9.71 310	21	9.78 049	28	10.21 951	9.93 261	8	54	
7	9.71 331	21	9.78 077	29	10.21 923	9.93 253	7	53	
8	9.71 352	21	9.78 106	29	10.21 894	9.93 246	8	52	
9	9.71 373	20	9.78 135	28	10.21 865	9.93 238	8	51	**29** **28**
10	9.71 393	21	9.78 163	29	10.21 837	9.93 230	7	50	1 2.9 2.8
11	9.71 414	21	9.78 192	28	10.21 808	9.93 223	8	49	2 5.8 5.6
12	9.71 435	21	9.78 220	29	10.21 780	9.93 215	8	48	3 8.7 8.4
13	9.71 456	21	9.78 249	28	10.21 751	9.93 207	7	47	4 11.6 11.2
14	9.71 477	21	9.78 277	29	10.21 723	9.93 200	8	46	5 14.5 14.0
15	9.71 498	21	9.78 306	28	10.21 694	9.93 192	8	45	6 17.4 16.8
16	9.71 519	20	9.78 334	29	10.21 666	9.93 184	7	44	7 20.3 19.6
17	9.71 539	21	9.78 363	28	10.21 637	9.93 177	8	43	8 23.2 22.4
18	9.71 560	21	9.78 391	28	10.21 609	9.93 169	8	42	9 26.1 25.2
19	9.71 581	21	9.78 419	29	10.21 581	9.93 161	7	41	
20	9.71 602	20	9.78 448	28	10.21 552	9.93 154	8	40	
21	9.71 622	21	9.78 476	29	10.21 524	9.93 146	8	39	
22	9.71 643	21	9.78 505	28	10.21 495	9.93 138	7	38	
23	9.71 664	21	9.78 533	29	10.21 467	9.93 131	8	37	
24	9.71 685	20	9.78 562	28	10.21 438	9.93 123	8	36	
25	9.71 705	21	9.78 590	28	10.21 410	9.93 115	7	35	
26	9.71 726	21	9.78 618	29	10.21 382	9.93 108	8	34	**21** **20**
27	9.71 747	20	9.78 647	28	10.21 353	9.93 100	8	33	1 2.1 2.0
28	9.71 767	21	9.78 675	29	10.21 325	9.93 092	8	32	2 4.2 4.0
29	9.71 788	21	9.78 704	28	10.21 296	9.93 084	7	31	3 6.3 6.0
30	9.71 809	20	9.78 732	28	10.21 268	9.93 077	8	30	4 8.4 8.0
31	9.71 829	21	9.78 760	29	10.21 240	9.93 069	8	29	5 10.5 10.0
32	9.71 850	20	9.78 789	28	10.21 211	9.93 061	8	28	6 12.6 12.0
33	9.71 870	21	9.78 817	28	10.21 183	9.93 053	7	27	7 14.7 14.0
34	9.71 891	20	9.78 845	29	10.21 155	9.93 046	8	26	8 16.8 16.0
35	9.71 911	21	9.78 874	28	10.21 126	9.93 038	8	25	9 18.9 18.0
36	9.71 932	20	9.78 902	28	10.21 098	9.93 030	8	24	
37	9.71 952	21	9.78 930	29	10.21 070	9.93 022	8	23	
38	9.71 973	21	9.78 959	28	10.21 041	9.93 014	7	22	
39	9.71 994	20	9.78 987	28	10.21 013	9.93 007	8	21	
40	9.72 014	20	9.79 015	28	10.20 985	9.92 999	8	20	
41	9.72 034	21	9.79 043	29	10.20 957	9.92 991	8	19	
42	9.72 055	20	9.79 072	28	10.20 928	9.92 983	7	18	
43	9.72 075	21	9.79 100	28	10.20 900	9.92 976	8	17	**8** **7**
44	9.72 096	20	9.79 128	28	10.20 872	9.92 968	8	16	1 0.8 0.7
45	9.72 116	21	9.79 156	29	10.20 844	9.92 960	8	15	2 1.6 1.4
46	9.72 137	20	9.79 185	28	10.20 815	9.92 952	8	14	3 2.4 2.1
47	9.72 157	20	9.79 213	28	10.20 787	9.92 944	8	13	4 3.2 2.8
48	9.72 177	21	9.79 241	28	10.20 759	9.92 936	7	12	5 4.0 3.5
49	9.72 198	20	9.79 269	28	10.20 731	9.92 929	8	11	6 4.8 4.2
50	9.72 218	20	9.79 297	29	10.20 703	9.92 921	8	10	7 5.6 4.9
51	9.72 238	21	9.79 326	28	10.20 674	9.92 913	8	9	8 6.4 5.6
52	9.72 259	20	9.79 354	28	10.20 646	9.92 905	8	8	9 7.2 6.3
53	9.72 279	20	9 79 382	28	10.20 618	9.92 897	8	7	
54	9.72 299	21	9 79 410	28	10.20 590	9.92 889	8	6	
55	9.72 320	20	9 79 438	28	10.20 562	9.92 881	7	5	
56	9.72 340	20	9.79 466	29	10.20 534	9.92 874	8	4	
57	9.72 360	21	9.79 495	28	10.20 505	9.92 866	8	3	
58	9.72 381	20	9.79 523	28	10.20 477	9.92 858	8	2	
59	9.72 401	20	9.79 551	28	10.20 449	9.92 850	8	1	
60	9.72 421		9.79 579		10.20 421	9.92 842		0	
	L Cos	d	L Cot	c d	L Tan	L Sin	d	′	Prop. Pts.

58°

TABLE 3

32°

′	L Sin	d	L Tan	c d	L Cot	L Cos	d	′
0	9.72 421		9.79 579		10.20 421	9.92 842		60
1	9.72 441	20	9.79 607	28	10.20 393	9.92 834	8	59
2	9.72 461	20	9.79 635	28	10.20 365	9.92 826	8	58
3	9.72 482	21	9.79 663	28	10.20 337	9.92 818		57
4	9.72 502	20	9.79 691	28	10.20 309	9.92 810	8	56
5	9.72 522	20	9.79 719	28	10.20 281	9.92 803	7	55
6	9.72 542	20	9.79 747	28	10.20 253	9.92 795	8	54
7	9.72 562	20	9.79 776	29	10.20 224	9.92 787	8	53
8	9.72 582	20	9.79 804	28	10.20 196	9.92 779	8	52
9	9.72 602	20	9.79 832	28	10.20 168	9.92 771	8	51
10	9.72 622	20	9.79 860	28	10.20 140	9.92 763	8	50
11	9.72 643	21	9.79 888	28	10.20 112	9.92 755	8	49
12	9.72 663	20	9.79 916	28	10.20 084	9.92 747	8	48
13	9.72 683	20	9.79 944	28	10.20 056	9.92 739	8	47
14	9.72 703	20	9.79 972	28	10.20 028	9.92 731	8	46
15	9.72 723	20	9.80 000	28	10.20 000	9.92 723	8	45
16	9.72 743	20	9.80 028	28	10.19 972	9.92 715	8	44
17	9.72 763	20	9.80 056	28	10.19 944	9.92 707	8	43
18	9.72 783	20	9.80 084	28	10.19 916	9.92 699	8	42
19	9.72 803	20	9.80 112	28	10.19 888	9.92 691	8	41
20	9.72 823	20	9.80 140	28	10.19 860	9.92 683	8	40
21	9.72 843	20	9.80 168	28	10.19 832	9.92 675	8	39
22	9.72 863	20	9.80 195	27	10.19 805	9.92 667	8	38
23	9.72 883	20	9.80 223	28	10.19 777	9.92 659	8	37
24	9.72 902	19	9.80 251	28	10.19 749	9.92 651	8	36
25	9.72 922	20	9.80 279	28	10.19 721	9.92 643	8	35
26	9.72 942	20	9.80 307	28	10.19 693	9.92 635	8	34
27	9.72 962	20	9.80 335	28	10.19 665	9.92 627	8	33
28	9.72 982	20	9.80 363	28	10.19 637	9.92 619	8	32
29	9.73 002	20	9.80 391	28	10.19 609	9.92 611	8	31
30	9.73 022	19	9.80 419	28	10.19 581	9.92 603	8	30
31	9.73 041	20	9.80 447	27	10.19 553	9.92 595	8	29
32	9.73 061	20	9.80 474	28	10.19 526	9.92 587	8	28
33	9.73 081	20	9.80 502	28	10.19 498	9.92 579	8	27
34	9.73 101	20	9.80 530	28	10.19 470	9.92 571	8	26
35	9.73 121	19	9.80 558	28	10.19 442	9.92 563	8	25
36	9.73 140	20	9.80 586	28	10.19 414	9.92 555	9	24
37	9.73 160	20	9.80 614	28	10.19 386	9.92 546	8	23
38	9.73 180	20	9.80 642	27	10.19 358	9.92 538	8	22
39	9.73 200	19	9.80 669	28	10.19 331	9.92 530	8	21
40	9.73 219	20	9.80 697	28	10.19 303	9.92 522	8	20
41	9.73 239	20	9.80 725	28	10.19 275	9.92 514	8	19
42	9.73 259	19	9.80 753	28	10.19 247	9.92 506	8	18
43	9.73 278	20	9.80 781	27	10.19 219	9.92 498	8	17
44	9.73 298	20	9.80 808	28	10.19 192	9.92 490	8	16
45	9.73 318	19	9.80 836	28	10.19 164	9.92 482	9	15
46	9.73 337	20	9.80 864	28	10.19 136	9.92 473	8	14
47	9.73 357	20	9.80 892	27	10.19 108	9.92 465	8	13
48	9.73 377	19	9.80 919	28	10.19 081	9.92 457	8	12
49	9.73 396	20	9.80 947	28	10.19 053	9.92 449	8	11
50	9.73 416	19	9.80 975	28	10.19 025	9.92 441	8	10
51	9.73 435	20	9.81 003	27	10.18 997	9.92 433	8	9
52	9.73 455	19	9.81 030	28	10.18 970	9.92 425	9	8
53	9.73 474	20	9.81 058	28	10.18 942	9.92 416	8	7
54	9.73 494	19	9.81 086	27	10.18 914	9.92 408	8	6
55	9.73 513	20	9.81 113	28	10.18 887	9.92 400	8	5
56	9.73 533	19	9.81 141	28	10.18 859	9.92 392	8	4
57	9.73 552	20	9.81 169	27	10.18 831	9.92 384	9	3
58	9.73 572	19	9.81 196	28	10.18 804	9.92 376	9	2
59	9.73 591	20	9.81 224	28	10.18 776	9.92 367	8	1
60	9.73 611		9.81 252		10.18 748	9.92 359		0
	L Cos	d	L Cot	c d	L Tan	L Sin	d	′

Prop. Pts.

	29	28	27
1	2.9	2.8	2.7
2	5.8	5.6	5.4
3	8.7	8.4	8.1
4	11.6	11.2	10.8
5	14.5	14.0	13.5
6	17.4	16.8	16.2
7	20.3	19.6	18.9
8	23.2	22.4	21.6
9	26.1	25.2	24.3

	21	20	19
1	2.1	2.0	1.9
2	4.2	4.0	3.8
3	6.3	6.0	5.7
4	8.4	8.0	7.6
5	10.5	10.0	9.5
6	12.6	12.0	11.4
7	14.7	14.0	13.3
8	16.8	16.0	15.2
9	18.9	18.0	17.1

	9	8	7
1	0.9	0.8	0.7
2	1.8	1.6	1.4
3	2.7	2.4	2.1
4	3.6	3.2	2.8
5	4.5	4.0	3.5
6	5.4	4.8	4.2
7	6.3	5.6	4.9
8	7.2	6.4	5.6
9	8.1	7.2	6.3

TABLE 3

57°

33°

′	L Sin	d	L Tan	c d	L Cot	L Cos	d	′	Prop. Pts.
0	9.73 611	19	9.81 252	27	10.18 748	9.92 359	8	60	
1	9.73 630	20	9.81 279	28	10.18 721	9.92 351	8	59	
2	9.73 650	19	9.81 307	28	10.18 693	9.92 343	8	58	
3	9.73 669	20	9.81 335	27	10.18 665	9.92 335	9	57	
4	9.73 689	19	9.81 362	28	10.18 638	9.92 326	8	56	
5	9.73 708	19	9.81 390	28	10.18 610	9.92 318	8	55	
6	9.73 727	20	9.81 418	27	10.18 582	9.92 310	8	54	
7	9.73 747	19	9.81 445	28	10.18 555	9.92 302	9	53	
8	9.73 766	19	9.81 473	27	10.18 527	9.92 293	8	52	
9	9.73 785	20	9.81 500	28	10.18 500	9.92 285	8	51	
10	9.73 805	19	9.81 528	28	10.18 472	9.92 277	8	**50**	**28 27**
11	9.73 824	19	9.81 556	27	10.18 444	9.92 269	9	49	1 2.8 2.7
12	9.73 843	20	9.81 583	28	10.18 417	9.92 260	8	48	2 5.6 5.4
13	9.73 863	19	9.81 611	27	10.18 389	9.92 252	8	47	3 8.4 8.1
14	9.73 882	19	9.81 638	28	10.18 362	9.92 244	9	46	4 11.2 10.8
15	9.73 901	20	9.81 666	27	10.18 334	9.92 235	8	45	5 14.0 13.5
16	9.73 921	19	9.81 693	28	10.18 307	9.92 227	8	44	6 16.8 16.2
17	9.73 940	19	9.81 721	27	10.18 279	9.92 219	8	43	7 19.6 18.9
18	9.73 959	19	9.81 748	28	10.18 252	9.92 211	9	42	8 22.4 21.6
19	9.73 978	19	9.81 776	27	10.18 224	9.92 202	8	41	9 25.2 24.3
20	9.73 997	20	9.81 803	28	10.18 197	9.92 194	8	**40**	
21	9.74 017	19	9.81 831	27	10.18 169	9.92 186	9	39	
22	9.74 036	19	9.81 858	28	10.18 142	9.92 177	8	38	
23	9.74 055	19	9.81 886	27	10.18 114	9.92 169	8	37	
24	9.74 074	19	9.81 913	28	10.18 087	9.92 161	9	36	
25	9.74 093	20	9.81 941	27	10.18 059	9.92 152	8	35	
26	9.74 113	19	9.81 968	28	10.18 032	9.92 144	8	34	
27	9.74 132	19	9.81 996	27	10.18 004	9.92 136	9	33	**20 19 18**
28	9.74 151	19	9.82 023	28	10.17 977	9.92 127	8	32	1 2.0 1.9 1.8
29	9.74 170	19	9.82 051	27	10.17 949	9.92 119	8	31	2 4.0 3.8 3.6
30	9.74 189	19	9.82 078	28	10.17 922	9.92 111	9	**30**	3 6.0 5.7 5.4
31	9.74 208	19	9.82 106	27	10.17 894	9.92 102	8	29	4 8.0 7.6 7.2
32	9.74 227	19	9.82 133	28	10.17 867	9.92 094	8	28	5 10.0 9.5 9.0
33	9.74 246	19	9.82 161	27	10.17 839	9.92 086	9	27	6 12.0 11.4 10.8
34	9.74 265	19	9.82 188	27	10.17 812	9.92 077	8	26	7 14.0 13.3 12.6
35	9.74 284	19	9.82 215	28	10.17 785	9.92 069	9	25	8 16.0 15.2 14.4
36	9.74 303	19	9.82 243	27	10.17 757	9.92 060	8	24	9 18.0 17.1 16.2
37	9.74 322	19	9.82 270	28	10.17 730	9.92 052	8	23	
38	9.74 341	19	9.82 298	27	10.17 702	9.92 044	9	22	
39	9.74 360	19	9.82 325	27	10.17 675	9.92 035	8	21	
40	9.74 379	19	9.82 352	28	10.17 648	9.92 027	9	**20**	
41	9.74 398	19	9.82 380	27	10.17 620	9.92 018	8	19	
42	9.74 417	19	9.82 407	28	10.17 593	9.92 010	8	18	
43	9.74 436	19	9.82 435	27	10.17 565	9.92 002	9	17	
44	9.74 455	19	9.82 462	27	10.17 538	9.91 993	8	16	**9 8**
45	9.74 474	19	9.82 489	28	10.17 511	9.91 985	9	15	1 0.9 0.8
46	9.74 493	19	9.82 517	27	10.17 483	9.91 976	8	14	2 1.8 1.6
47	9.74 512	19	9.82 544	27	10.17 456	9.91 968	9	13	3 2.7 2.4
48	9.74 531	18	9.82 571	28	10.17 429	9.91 959	8	12	4 3.6 3.2
49	9.74 549	19	9.82 599	27	10.17 401	9.91 951	9	11	5 4.5 4.0
50	9.74 568	19	9.82 626	27	10.17 374	9.91 942	8	**10**	6 5.4 4.8
51	9.74 587	19	9.82 653	28	10.17 347	9.91 934	9	9	7 6.3 5.6
52	9.74 606	19	9.82 681	27	10.17 319	9.91 925	8	8	8 7.2 6.4
53	9.74 625	19	9.82 708	27	10.17 292	9.91 917	9	7	9 8.1 7.2
54	9.74 644	18	9.82 735	27	10.17 265	9.91 908	8	6	
55	9.74 662	19	9.82 762	28	10.17 238	9.91 900	9	5	
56	9.74 681	19	9.82 790	27	10.17 210	9.91 891	8	4	
57	9.74 700	19	9.82 817	27	10.17 183	9.91 883	9	3	
58	9.74 719	18	9.82 844	27	10.17 156	9.91 874	9	2	
59	9.74 737	19	9.82 871	28	10.17 129	9.91 866	9	1	
60	9.74 756		9.82 899		10.17 101	9.91 857		**0**	
	L Cos	d	L Cot	c d	L Tan	L Sin	d	′	Prop. Pts.

56°

TABLE 3

34°

′	L Sin	d	L Tan	c d	L Cot	L Cos	d		Prop. Pts.
0	9.74 756		9.82 899		10.17 101	9.91 857		60	
1	9.74 775	19	9.82 926	27	10.17 074	9.91 849	8	59	
2	9.74 794	19	9.82 953	27	10.17 047	9.91 840	9	58	
3	9.74 812	18	9.82 980	27	10.17 020	9.91 832	8	57	
4	9.74 831	19	9.83 008	28	10.16 992	9.91 823	9	56	
5	9.74 850	19	9.83 035	27	10.16 965	9.91 815	8	55	
6	9.74 868	18	9.83 062	27	10.16 938	9.91 806	9	54	
7	9.74 887	19	9.83 089	27	10.16 911	9.91 798	8	53	
8	9.74 906	19	9.83 117	28	10.16 883	9.91 789	9	52	
9	9.74 924	18	9.83 144	27	10.16 856	9.91 781	8	51	
10	9.74 943	19	9.83 171	27	10.16 829	9.91 772	9	50	
11	9.74 961	18	9.83 198	27	10.16 802	9.91 763	9	49	
12	9.74 980	19	9.83 225	27	10.16 775	9.91 755	8	48	
13	9.74 999	19	9.83 252	27	10.16 748	9.91 746	9	47	
14	9.75 017	18	9.83 280	28	10.16 720	9.91 738	8	46	
15	9.75 036	19	9.83 307	27	10.16 693	9.91 729	9	45	
16	9.75 054	18	9.83 334	27	10.16 666	9.91 720	9	44	
17	9.75 073	19	9.83 361	27	10.16 639	9.91 712	8	43	
18	9.75 091	18	9.83 388	27	10.16 612	9.91 703	9	42	
19	9.75 110	19	9.83 415	27	10.16 585	9.91 695	8	41	
20	9.75 128	18	9.83 442	27	10.16 558	9.91 686	9	40	
21	9.75 147	19	9.83 470	28	10.16 530	9.91 677	9	39	
22	9.75 165	18	9.83 497	27	10.16 503	9.91 669	8	38	
23	9.75 184	19	9.83 524	27	10.16 476	9.91 660	9	37	
24	9.75 202	18	9.83 551	27	10.16 449	9.91 651	8	36	
25	9.75 221	19	9.83 578	27	10.16 422	9.91 643	9	35	
26	9.75 239	18	9.83 605	27	10.16 395	9.91 634	9	34	
27	9.75 258	19	9.83 632	27	10.16 368	9.91 625	8	33	
28	9.75 276	18	9.83 659	27	10.16 341	9.91 617	9	32	
29	9.75 294	19	9.83 686	27	10.16 314	9.91 608	9	31	
30	9.75 313	18	9.83 713	27	10.16 287	9.91 599	8	30	
31	9.75 331	19	9.83 740	28	10.16 260	9.91 591	9	29	
32	9.75 350	18	9.83 768	27	10.16 232	9.91 582	9	28	
33	9.75 368	18	9.83 795	27	10.16 205	9.91 573	8	27	
34	9.75 386	19	9.83 822	27	10.16 178	9.91 565	9	26	
35	9.75 405	18	9.83 849	27	10.16 151	9.91 556	9	25	
36	9.75 423	18	9.83 876	27	10.16 124	9.91 547	9	24	
37	9.75 441	18	9.83 903	27	10.16 097	9.91 538	8	23	
38	9.75 459	19	9.83 930	27	10.16 070	9.91 530	9	22	
39	9.75 478	18	9.83 957	27	10.16 043	9.91 521	9	21	
40	9.75 496	18	9.83 984	27	10.16 016	9.91 512	8	20	
41	9.75 514	19	9.84 011	27	10.15 989	9.91 504	9	19	
42	9.75 533	18	9.84 038	27	10.15 962	9.91 495	9	18	
43	9.75 551	18	9.84 065	27	10.15 935	9.91 486	9	17	
44	9.75 569	18	9.84 092	27	10.15 908	9.91 477	8	16	
45	9.75 587	18	9.84 119	27	10.15 881	9.91 469	9	15	
46	9.75 605	19	9.84 146	27	10.15 854	9.91 460	9	14	
47	9.75 624	18	9.84 173	27	10.15 827	9.91 451	9	13	
48	9.75 642	18	9.84 200	27	10.15 800	9.91 442	9	12	
49	9.75 660	18	9.84 227	27	10.15 773	9.91 433	8	11	
50	9.75 678	18	9.84 254	26	10.15 746	9.91 425	9	10	
51	9.75 696	18	9.84 280	27	10.15 720	9.91 416	9	9	
52	9.75 714	19	9.84 307	27	10.15 693	9.91 407	9	8	
53	9.75 733	18	9.84 334	27	10.15 666	9.91 398	9	7	
54	9.75 751	18	9.84 361	27	10.15 639	9.91 389	8	6	
55	9.75 769	18	9.84 388	27	10.15 612	9.91 381	9	5	
56	9.75 787	18	9.84 415	27	10.15 585	9.91 372	9	4	
57	9.75 805	18	9.84 442	27	10.15 558	9.91 363	9	3	
58	9.75 823	18	9.84 469	27	10.15 531	9.91 354	9	2	
59	9.75 841	18	9.84 496	27	10.15 504	9.91 345	9	1	
60	9.75 859		9.84 523		10.15 477	9.91 336		0	
	L Cos	d	L Cot	c d	L Tan	L Sin	d	′	Prop. Pts.

Prop. Pts.

	28	27	26
1	2.8	2.7	2.6
2	5.6	5.4	5.2
3	8.4	8.1	7.8
4	11.2	10.8	10.4
5	14.0	13.5	13.0
6	16.8	16.2	15.6
7	19.6	18.9	18.2
8	22.4	21.6	20.8
9	25.2	24.3	23.4

	19	18
1	1.9	1.8
2	3.8	3.6
3	5.7	5.4
4	7.6	7.2
5	9.5	9.0
6	11.4	10.8
7	13.3	12.6
8	15.2	14.4
9	17.1	16.2

	9	8
1	0.9	0.8
2	1.8	1.6
3	2.7	2.4
4	3.6	3.2
5	4.5	4.0
6	5.4	4.8
7	6.3	5.6
8	7.2	6.4
9	8.1	7.2

TABLE

3

55°

35°

'	L Sin	d	L Tan	c d	L Cot	L Cos	d	'	Prop. Pts.
0	9.75 859		9.84 523		10.15 477	9.91 336		60	
1	9.75 877	18	9.84 550	27	10.15 450	9.91 328	8	59	
2	9.75 895	18	9.84 576	26	10.15 424	9.91 319	9	58	
3	9.75 913	18	9.84 603	27	10.15 397	9.91 310	9	57	
4	9.75 931	18	9.84 630	27	10.15 370	9.91 301	9	56	
5	9.75 949	18	9.84 657	27	10.15 343	9.91 292	9	55	
6	9.75 967	18	9.84 684	27	10.15 316	9.91 283	9	54	
7	9.75 985	18	9.84 711	27	10.15 289	9.91 274	8	53	
8	9.76 003	18	9.84 738	27	10.15 262	9.91 266	9	52	
9	9.76 021	18	9.84 764	26	10.15 236	9.91 257	9	51	
10	9.76 039	18	9.84 791	27	10.15 209	9.91 248	9	50	
11	9.76 057	18	9.84 818	27	10.15 182	9.91 239	9	49	
12	9.76 075	18	9.84 845	27	10.15 155	9.91 230	9	48	
13	9.76 093	18	9.84 872	27	10.15 128	9.91 221	9	47	
14	9.76 111	18	9.84 899	26	10.15 101	9.91 212	9	46	
15	9.76 129	17	9.84 925	27	10.15 075	9.91 203	9	45	
16	9.76 146	18	9.84 952	27	10.15 048	9.91 194	9	44	
17	9.76 164	18	9.84 979	27	10.15 021	9.91 185	9	43	
18	9.76 182	18	9.85 006	27	10.14 994	9.91 176	9	42	
19	9.76 200	18	9.85 033	26	10.14 967	9.91 167	9	41	
20	9.76 218	18	9.85 059	27	10.14 941	9.91 158	9	40	
21	9.76 236	17	9.85 086	27	10.14 914	9.91 149	8	39	
22	9.76 253	18	9.85 113	27	19.14 887	9.91 141	9	38	
23	9.76 271	18	9.85 140	26	10.14 860	9.91 132	9	37	
24	9.76 289	18	9.85 166	27	10.14 834	9.91 123	9	36	
25	9.76 307	17	9.85 193	27	10.14 807	9.91 114	9	35	
26	9.76 324	18	9.85 220	27	10.14 780	9.91 105	9	34	
27	9.76 342	18	9.85 247	26	10.14 753	9.91 096	9	33	
28	9.76 360	18	9.85 273	27	10.14 727	9.91 087	9	32	
29	9.76 378	17	9.85 300	27	10.14 700	9.91 078	9	31	
30	9.76 395	18	9.85 327	27	10.14 673	9.91 069	9	30	
31	9.76 413	18	9.85 354	26	10.14 646	9.91 060	9	29	
32	9.76 431	17	9.85 380	27	10.14 620	9.91 051	9	28	
33	9.76 448	18	9.85 407	27	10.14 593	9.91 042	9	27	
34	9.76 466	18	9.85 434	26	10.14 566	9.91 033	10	26	
35	9.76 484	17	9.85 460	27	10.14 540	9.91 023	9	25	
36	9.76 501	18	9.85 487	27	10.14 513	9.91 014	9	24	
37	9.76 519	18	9.85 514	26	10.14 486	9.91 005	9	23	
38	9.76 537	17	9.85 540	27	10.14 460	9.90 996	9	22	
39	9.76 554	18	9.85 567	27	10.14 433	9.90 987	9	21	
40	9.76 572	18	9.85 594	26	10.14 406	9.90 978	9	20	
41	9.76 590	17	9.85 620	27	10.14 380	9.90 969	9	19	
42	9.76 607	18	9.85 647	27	10.14 353	9.90 960	9	18	
43	9.76 625	17	9.85 674	26	10.14 326	9.90 951	9	17	
44	9.76 642	18	9.85 700	27	10.14 300	9.90 942	9	16	
45	9.76 660	17	9.85 727	27	10.14 273	9.90 933	9	15	
46	9.76 677	18	9.85 754	26	10.14 246	9.90 924	9	14	
47	9.76 695	17	9.85 780	27	10.14 220	9.90 915	9	13	
48	9.76 712	18	9.85 807	27	10.14 193	9.90 906	10	12	
49	9.76 730	17	9.85 834	26	10.14 166	9.90 896	9	11	
50	9.76 747	18	9.85 860	27	10.14 140	9.90 887	9	10	
51	9.76 765	17	9.85 887	26	10.14 113	9.90 878	9	9	
52	9.76 782	18	9.85 913	27	10.14 087	9.90 869	9	8	
53	9.76 800	17	9.85 940	27	10.14 060	9.90 860	9	7	
54	9.76 817	18	9.85 967	26	10.14 033	9.90 851	9	6	
55	9.76 835	17	9.85 993	27	10.14 007	9.90 842	10	5	
56	9.76 852	18	9.86 020	26	10.13 980	9.90 832	9	4	
57	9.76 870	17	9.86 046	27	10.13 954	9.90 823	9	3	
58	9.76 887	17	9.86 073	27	10.13 927	9.90 814	9	2	
59	9.76 904	18	9.86 100	26	10.13 900	9.90 805	9	1	
60	9.76 922		9.86 126		10.13 874	9.90 796		0	
	L Cos	d	L Cot	c d	L Tan	L Sin	d	'	Prop. Pts.

Prop. Pts.

	27	26
1	2.7	2.6
2	5.4	5.2
3	8.1	7.8
4	10.8	10.4
5	13.5	13.0
6	16.2	15.6
7	18.9	18.2
8	21.6	20.8
9	24.3	23.4

	18	17
1	1.8	1.7
2	3.6	3.4
3	5.4	5.1
4	7.2	6.8
5	9.0	8.5
6	10.8	10.2
7	12.6	11.9
8	14.4	13.6
9	16.2	15.3

	10	9	8
1	1.0	0.9	0.8
2	2.0	1.8	1.6
3	3.0	2.7	2.4
4	4.0	3.6	3.2
5	5.0	4.5	4.0
6	6.0	5.4	4.8
7	7.0	6.3	5.6
8	8.0	7.2	6.4
9	9.0	8.1	7.2

TABLE 3

54°

36°

′	L Sin	d	L Tan	c d	L Cot	L Cos	d	′	Prop. Pts.
0	9.76 922		9.86 126		10.13 874	9.90 796		60	
1	9.76 939	17	9.86 153	27	10.13 847	9.90 787	9	59	
2	9.76 957	18	9.86 179	26	10.13 821	9.90 777	10	58	
3	9.76 974	17	9.86 206	27	10.13 794	9.90 768	9	57	
4	9.76 991	17	9.86 232	26	10.13 768	9.90 759	9	56	
5	9.77 009	18	9.86 259	27	10.13 741	9.90 750	9	55	
6	9.77 026	17	9.86 285	26	10.13 715	9.90 741	9	54	
7	9.77 043	17	9.86 312	27	10.13 688	9.90 731	10	53	
8	9.77 061	18	9.86 338	26	10.13 662	9.90 722	9	52	
9	9.77 078	17	9.86 365	27	10.13 635	9.90 713	9	51	
10	9.77 095	17	9.86 392	27	10.13 608	9.90 704	9	50	**27 26**
11	9.77 112	17	9.86 418	26	10.13 582	9.90 694	10	49	1 2.7 2.6
12	9.77 130	18	9.86 445	27	10.13 555	9.90 685	9	48	2 5.4 5.2
13	9.77 147	17	9.86 471	26	10.13 529	9.90 676	9	47	3 8.1 7.8
14	9.77 164	17	9.86 498	27	10.13 502	9.90 667	9	46	4 10.8 10.4
15	9.77 181	17	9.86 524	26	10.13 476	9.90 657	10	45	5 13.5 13.0
16	9.77 199	18	9.86 551	27	10.13 449	9.90 648	9	44	6 16.2 15.6
17	9.77 216	17	9.86 577	26	10.13 423	9.90 639	9	43	7 18.9 18.2
18	9.77 233	17	9.86 603	26	10.13 397	9.90 630	9	42	8 21.6 20.8
19	9.77 250	17	9.86 630	27	10.13 370	9.90 620	10	41	9 24.3 23.4
20	9.77 268	18	9.86 656	26	10.13 344	9.90 611	9	40	
21	9.77 285	17	9.86 683	27	10.13 317	9.90 602	9	39	
22	9.77 302	17	9.86 709	26	10.13 291	9.90 592	10	38	
23	9.77 319	17	9.86 736	27	10.13 264	9.90 583	9	37	
24	9.77 336	17	9.86 762	26	10.13 238	9.90 574	9	36	
25	9.77 353	17	9.86 789	27	10.13 211	9.90 565	9	35	
26	9.77 370	17	9.86 815	26	10.13 185	9.90 555	10	34	
27	9.77 387	17	9.86 842	27	10.13 158	9.90 546	9	33	**18 17 16**
28	9.77 405	18	9.86 868	26	10.13 132	9.90 537	9	32	1 1.8 1.7 1.6
29	9.77 422	17	9.86 894	26	10.13 106	9.90 527	10	31	2 3.6 3.4 3.2
30	9.77 439	17	9.86 921	27	10.13 079	9.90 518	9	30	3 5.4 5.1 4.8
31	9.77 456	17	9.86 947	26	10.13 053	9.90 509	9	29	4 7.2 6.8 6.4
32	9.77 473	17	9.86 974	27	10.13 026	9.90 499	10	28	5 9.0 8.5 8.0
33	9.77 490	17	9.87 000	26	10.13 000	9.90 490	10	27	6 10.8 10.2 9.6
34	9.77 507	17	9.87 027	27	10.12 973	9.90 480	9	26	7 12.6 11.9 11.2
35	9.77 524	17	9.87 053	26	10.12 947	9.90 471	9	25	8 14.4 13.6 12.8
36	9.77 541	17	9.87 079	27	10.12 921	9.90 462	10	24	9 16.2 15.3 14.4
37	9.77 558	17	9.87 106	26	10.12 894	9.90 452	9	23	
38	9.77 575	17	9.87 132	26	10.12 868	9.90 443	9	22	
39	9.77 592	17	9.87 158	27	10.12 842	9.90 434	10	21	
40	9.77 609	17	9.87 185	26	10.12 815	9.90 424	9	20	
41	9.77 626	17	9.87 211	27	10.12 789	9.90 415	10	19	
42	9.77 643	17	9.87 238	26	10.12 762	9.90 405	9	18	
43	9.77 660	17	9.87 264	26	10.12 736	9.90 396	10	17	
44	9.77 677	17	9.87 290	27	10.12 710	9.90 386	9	16	**10 9**
45	9.77 694	17	9.87 317	26	10.12 683	9.90 377	9	15	1 1.0 0.9
46	9.77 711	17	9.87 343	26	10.12 657	9.90 368	10	14	2 2.0 1.8
47	9.77 728	16	9.87 369	27	10.12 631	9.90 358	9	13	3 3.0 2.7
48	9.77 744	17	9.87 396	26	10.12 604	9.90 349	10	12	4 4.0 3.6
49	9.77 761	17	9.87 422	26	10.12 578	9.90 339	9	11	5 5.0 4.5
50	9.77 778	17	9.87 448	27	10.12 552	9.90 330	10	10	6 6.0 5.4
51	9.77 795	17	9.87 475	26	10.12 525	9.90 320	9	9	7 7.0 6.3
52	9.77 812	17	9.87 501	26	10.12 499	9.90 311	10	8	8 8.0 7.2
53	9.77 829	17	9.87 527	27	10.12 473	9.90 301	9	7	9 9.0 8.1
54	9.77 846	16	9.87 554	26	10.12 446	9.90 292	10	6	
55	9.77 862	17	9.87 580	26	10.12 420	9.90 282	9	5	
56	9.77 879	17	9.87 606	27	10.12 394	9.90 273	10	4	
57	9.77 896	17	9.87 633	26	10.12 367	9.90 263	9	3	
58	9.77 913	17	9.87 659	26	10.12 341	9.90 254	10	2	
59	9.77 930	16	9.87 685	26	10.12 315	9.90 244	9	1	
60	9.77 946		9.87 711		10.12 289	9.90 235		0	
	L Cos	d	L Cot	c d	L Tan	L Sin	d	′	Prop. Pts.

TABLE 3

53°

37°

′	L Sin	d	L Tan	c d	L Cot	L Cos	d		Prop. Pts.
0	9.77 946	17	9.87 711	27	10.12 289	9.90 235	10	60	
1	9.77 963	17	9.87 738	26	10.12 262	9.90 225	9	59	
2	9.77 980	17	9.87 764	26	10.12 236	9.90 216	10	58	
3	9.77 997	16	9.87 790	27	10.12 210	9.90 206	9	57	
4	9.78 013	17	9.87 817	26	10.12 183	9.90 197	10	56	
5	9.78 030	17	9.87 843	26	10.12 157	9.90 187	9	55	
6	9.78 047	16	9.87 869	26	10.12 131	9.90 178	10	54	
7	9.78 063	17	9.87 895	27	10.12 105	9.90 168	9	53	
8	9.78 080	17	9.87 922	26	10.12 078	9.90 159	10	52	
9	9.78 097	16	9.87 948	26	10.12 052	9.90 149	10	51	
10	9.78 113	17	9.87 974	26	10.12 026	9.90 139	9	50	
11	9.78 130	17	9.88 000	27	10.12 000	9.90 130	10	49	**27** **26**
12	9.78 147	16	9.88 027	26	10.11 973	9.90 120	9	48	1 2.7 2.6
13	9.78 163	17	9.88 053	26	10.11 947	9.90 111	10	47	2 5.4 5.2
14	9.78 180	17	9.88 079	26	10.11 921	9.90 101	10	46	3 8.1 7.8
15	9.78 197	16	9.88 105	26	10.11 895	9.90 091	9	45	4 10.8 10.4
16	9.78 213	17	9.88 131	27	10.11 869	9.90 082	10	44	5 13.5 13.0
17	9.78 230	16	9.88 158	26	10.11 842	9.90 072	9	43	6 16.2 15.6
18	9.78 246	17	9.88 184	26	10.11 816	9.90 063	10	42	7 18.9 18.2
19	9.78 263	17	9.88 210	26	10.11 790	9.90 053	10	41	8 21.6 20.8
20	9.78 280	16	9.88 236	26	10.11 764	9.90 043	9	40	9 24.3 23.4
21	9.78 296	17	9.88 262	27	10.11 738	9.90 034	10	39	
22	9.78 313	16	9.88 289	26	10.11 711	9.90 024	10	38	
23	9.78 329	17	9.88 315	26	10.11 685	9.90 014	9	37	
24	9.78 346	16	9.88 341	26	10.11 659	9.90 005	10	36	
25	9.78 362	17	9.88 367	26	10.11 633	9.89 995	10	35	
26	9.78 379	16	9.88 393	27	10.11 607	9.89 985	9	34	
27	9.78 395	17	9.88 420	26	10.11 580	9.89 976	10	33	**17** **16**
28	9.78 412	16	9.88 446	26	10.11 554	9.89 966	10	32	1 1.7 1.6
29	9.78 428	17	9.88 472	26	10.11 528	9.89 956	9	31	2 3.4 3.2
30	9.78 445	16	9.88 498	26	10.11 502	9.89 947	10	30	3 5.1 4.8
31	9.78 461	17	9.88 524	26	10.11 476	9.89 937	10	29	4 6.8 6.4
32	9.78 478	16	9.88 550	27	10.11 450	9.89 927	9	28	5 8.5 8.0
33	9.78 494	16	9.88 577	26	10.11 423	9.89 918	10	27	6 10.2 9.6
34	9.78 510	17	9.88 603	26	10.11 397	9.89 908	10	26	7 11.9 11.2
35	9.78 527	16	9.88 629	26	10.11 371	9.89 898	10	25	8 13.6 12.8
36	9.78 543	17	9.88 655	26	10.11 345	9.89 888	9	24	9 15.3 14.4
37	9.78 560	16	9.88 681	26	10.11 319	9.89 879	10	23	
38	9.78 576	16	9.88 707	26	10.11 293	9.89 869	10	22	
39	9.78 592	17	9.88 733	26	10.11 267	9.89 859	10	21	
40	9.78 609	16	9.88 759	27	10.11 241	9.89 849	9	20	
41	9.78 625	17	9.88 786	26	10.11 214	9.89 840	10	19	
42	9.78 642	16	9.88 812	26	10.11 188	9.89 830	10	18	
43	9.78 658	16	9.88 838	26	10.11 162	9.89 820	10	17	
44	9.78 674	17	9.88 864	26	10.11 136	9.89 810	9	16	**10** **9**
45	9.78 691	16	9.88 890	26	10.11 110	9.89 801	10	15	1 1.0 0.9
46	9.78 707	16	9.88 916	26	10.11 084	9.89 791	10	14	2 2.0 1.8
47	9.78 723	16	9.88 942	26	10.11 058	9.89 781	10	13	3 3.0 2.7
48	9.78 739	17	9.88 968	26	10.11 032	9.89 771	10	12	4 4.0 3.6
49	9.78 756	16	9.88 994	26	10.11 006	9.89 761	9	11	5 5.0 4.5
50	9.78 772	16	9.89 020	26	10.10 980	9.89 752	10	10	6 6.0 5.4
51	9.78 788	17	9.89 046	27	10.10 954	9.89 742	10	9	7 7.0 6.3
52	9.78 805	16	9.89 073	26	10.10 927	9.89 732	10	8	8 8.0 7.2
53	9.78 821	16	9.89 099	26	10.10 901	9.89 722	10	7	9 9.0 8.1
54	9.78 837	16	9.89 125	26	10.10 875	9.89 712	10	6	
55	9.78 853	16	9.89 151	26	10.10 849	9.89 702	9	5	
56	9.78 869	17	9.89 177	26	10.10 823	9.89 693	10	4	
57	9.78 886	16	9.89 203	26	10.10 797	9.89 683	10	3	
58	9.78 902	16	9.89 229	26	10.10 771	9.89 673	10	2	
59	9.78 918	16	9.89 255	26	10.10 745	9.89 663	10	1	
60	9.78 934		9.89 281		10.10 719	9.89 653		0	
	L Cos	d	L Cot	c d	L Tan	L Sin	d	′	Prop. Pts.

TABLE 3

52°

38°

′	L Sin	d	L Tan	c d	L Cot	L Cos	d	′	Prop. Pts.
0	9.78 934		9.89 281		10.10 719	9.89 653		60	
1	9.78 950	16	9.89 307	26	10.10 693	9.89 643	10	59	
2	9.78 967	17	9.89 333	26	10.10 667	9.89 633	10	58	
3	9.78 983	16	9.89 359	26	10.10 641	9.89 624	9	57	
4	9.78 999	16	9.89 385	26	10.10 615	9.89 614	10	56	
5	9.79 015	16	9.89 411	26	10.10 589	9.89 604	10	55	
6	9.79 031	16	9.89 437	26	10.10 563	9.89 594	10	54	
7	9.79 047	16	9.89 463	26	10.10 537	9.89 584	10	53	
8	9.79 063	16	9.89 489	26	10.10 511	9.89 574	10	52	
9	9.79 079	16	9.89 515	26	10.10 485	9.89 564	10	51	
10	9.79 095	16	9.89 541	26	10.10 459	9.89 554	10	50	
11	9.79 111	17	9.89 567	26	10.10 433	9.89 544	10	49	
12	9.79 128	16	9.89 593	26	10.10 407	9.89 534	10	48	
13	9.79 144	16	9.89 619	26	10.10 381	9.89 524	10	47	
14	9.79 160	16	9.89 645	26	10.10 355	9.89 514	10	46	
15	9.79 176	16	9.89 671	26	10.10 329	9.89 504	9	45	
16	9.79 192	16	9.89 697	26	10.10 303	9.89 495	10	44	
17	9.79 208	16	9.89 723	26	10.10 277	9.89 485	10	43	
18	9.79 224	16	9.89 749	26	10.10 251	9.89 475	10	42	
19	9.79 240	16	9.89 775	26	10.10 225	9.89 465	10	41	
20	9.79 256	16	9.89 801	26	10.10 199	9.89 455	10	40	
21	9.79 272	16	9.89 827	26	10.10 173	9.89 445	10	39	
22	9.79 288	16	9.89 853	26	10.10 147	9.89 435	10	38	
23	9.79 304	15	9.89 879	26	10.10 121	9.89 425	10	37	
24	9.79 319	16	9.89 905	26	10.10 095	9.89 415	10	36	
25	9.79 335	16	9.89 931	26	10.10 069	9.89 405	10	35	
26	9.79 351	16	9.89 957	26	10.10 043	9.89 395	10	34	
27	9.79 367	16	9.89 983	26	10.10 017	9.89 385	10	33	
28	9.79 383	16	9.90 009	26	10.09 991	9.89 375	11	32	
29	9.79 399	16	9.90 035	26	10.09 965	9.89 364	10	31	
30	9.79 415	16	9.90 061	25	10.09 939	9.89 354	10	30	
31	9.79 431	16	9.90 086	26	10.09 914	9.89 344	10	29	
32	9.79 447	16	9.90 112	26	10.09 888	9.89 334	10	28	
33	9.79 463	15	9.90 138	26	10.09 862	9.89 324	10	27	
34	9.79 478	16	9.90 164	26	10.09 836	9.89 314	10	26	
35	9.79 494	16	9.90 190	26	10.09 810	9.89 304	10	25	
36	9.79 510	16	9.90 216	26	10.09 784	9.89 294	10	24	
37	9.79 526	16	9.90 242	26	10.09 758	9.89 284	10	23	
38	9.79 542	16	9.90 268	26	10.09 732	9.89 274	10	22	
39	9.79 558	15	9.90 294	26	10.09 706	9.89 264	10	21	
40	9.79 573	16	9.90 320	26	10.09 680	9.89 254	10	20	
41	9.79 589	16	9.90 346	25	10.09 654	9.89 244	11	19	
42	9.79 605	16	9.90 371	26	10.09 629	9.89 233	10	18	
43	9.79 621	15	9.90 397	26	10.09 603	9.89 223	10	17	
44	9.79 636	16	9.90 423	26	10.09 577	9.89 213	10	16	
45	9.79 652	16	9.90 449	26	10.09 551	9.89 203	10	15	
46	9.79 668	16	9.90 475	26	10.09 525	9.89 193	10	14	
47	9.79 684	15	9.90 501	26	10.09 499	9.89 183	10	13	
48	9.79 699	16	9.90 527	26	10.09 473	9.89 173	11	12	
49	9.79 715	16	9.90 553	25	10.09 447	9.89 162	10	11	
50	9.79 731	15	9.90 578	26	10.09 422	9.89 152	10	10	
51	9.79 746	16	9.90 604	26	10.09 396	9.89 142	10	9	
52	9.79 762	16	9.90 630	26	10.09 370	9.89 132	10	8	
53	9.79 778	15	9.90 656	26	10.09 344	9.89 122	10	7	
54	9.79 793	16	9.90 682	26	10.09 318	9.89 112	11	6	
55	9.79 809	16	9.90 708	26	10.09 292	9.89 101	10	5	
56	9.79 825	15	9.90 734	25	10.09 266	9.89 091	10	4	
57	9.79 840	16	9.90 759	26	10.09 241	9.89 081	10	3	
58	9.79 856	16	9.90 785	26	10.09 215	9.89 071	11	2	
59	9.79 872	15	9.90 811	26	10.09 189	9.89 060	10	1	
60	9.79 887		9.90 837		10.09 163	9.89 050		0	
	L Cos	d	L Cot	c d	L Tan	L Sin	d	′	Prop. Pts.

Prop. Pts.

	26	25
1	2.6	2.5
2	5.2	5.0
3	7.8	7.5
4	10.4	10.0
5	13.0	12.5
6	15.6	15.0
7	18.2	17.5
8	20.8	20.0
9	23.4	22.5

	17	16	15
1	1.7	1.6	1.5
2	3.4	3.2	3.0
3	5.1	4.8	4.5
4	6.8	6.4	6.0
5	8.5	8.0	7.5
6	10.2	9.6	9.0
7	11.9	11.2	10.5
8	13.6	12.8	12.0
9	15.3	14.4	13.5

	11	10	9
1	1.1	1.0	0.9
2	2.2	2.0	1.8
3	3.3	3.0	2.7
4	4.4	4.0	3.6
5	5.5	5.0	4.5
6	6.6	6.0	5.4
7	7.7	7.0	6.3
8	8.8	8.0	7.2
9	9.9	9.0	8.1

TABLE

3

51°

39°

′	L Sin	d	L Tan	c d	L Cot	L Cos	d		Prop. Pts.
0	9.79 887		9.90 837		10.09 163	9.89 050		60	
1	9.79 903	16	9.90 863	26	10.09 137	9.89 040	10	59	
2	9.79 918	15	9.90 889	26	10.09 111	9.89 030	10	58	
3	9.79 934	16	9.90 914	25	10.09 086	9.89 020	10	57	
4	9.79 950	16	9.90 940	26	10.09 060	9.89 009	11	56	
5	9.79 965	15	9.90 966	26	10.09 034	9.88 999	10	55	
6	9.79 981	16	9.90 992	26	10.09 008	9.88 989	10	54	
7	9.79 996	15	9.91 018	26	10.08 982	9.88 978	11	53	
8	9.80 012	16	9.91 043	25	10.08 957	9.88 968	10	52			26	25
9	9.80 027	15	9.91 069	26	10.08 931	9.88 958	10	51	1	2.6	2.5
10	9.80 043	16	9.91 095	26	10.08 905	9.88 948	10	50	2	5.2	5.0
11	9.80 058	15	9.91 121	26	10.08 879	9.88 937	11	49	3	7.8	7.5
12	9.80 074	16	9.91 147	26	10.08 853	9.88 927	10	48	4	10.4	10.0
13	9.80 089	15	9.91 172	25	10.08 828	9.88 917	10	47	5	13.0	12.5
14	9.80 105	16	9.91 198	26	10.08 802	9.88 906	11	46	6	15.6	15.0
15	9.80 120	15	9.91 224	26	10.08 776	9.88 896	10	45	7	18.2	17.5
16	9.80 136	16	9.91 250	26	10.08 750	9.88 886	10	44	8	20.8	20.0
17	9.80 151	15	9.91 276	25	10.08 724	9.88 875	11	43	9	23.4	22.5
18	9.80 166	15	9.91 301	26	10.08 699	9.88 865	10	42	
19	9.80 182	16	9.91 327	26	10.08 673	9.88 855	10	41	
20	9.80 197	15	9.91 353	26	10.08 647	9.88 844	11	40	
21	9.80 213	16	9.91 379	26	10.08 621	9.88 834	10	39	
22	9.80 228	15	9.91 404	25	10.08 596	9.88 824	10	38	
23	9.80 244	16	9.91 430	26	10.08 570	9.88 813	11	37	
24	9.80 259	15	9.91 456	26	10.08 544	9.88 803	10	36	
25	9.80 274	15	9.91 482	26	10.08 518	9.88 793	10	35	
26	9.80 290	16	9.91 507	25	10.08 493	9.88 782	11	34			16	15
27	9.80 305	15	9.91 533	26	10.08 467	9.88 772	10	33	1	1.6	1.5
28	9.80 320	15	9.91 559	26	10.08 441	9.88 761	11	32	2	3.2	3.0
29	9.80 336	16	9.91 585	26	10.08 415	9.88 751	10	31	3	4.8	4.5
30	9.80 351	15	9.91 610	25	10.08 390	9.88 741	10	30	4	6.4	6.0
31	9.80 366	15	9.91 636	26	10.08 364	9.88 730	11	29	5	8.0	7.5
32	9.80 382	16	9.91 662	26	10.08 338	9.88 720	10	28	6	9.6	9.0
33	9.80 397	15	9.91 688	26	10.08 312	9.88 709	11	27	7	11.2	10.5
34	9.80 412	15	9.91 713	25	10.08 287	9.88 699	10	26	8	12.8	12.0
35	9.80 428	16	9.91 739	26	10.08 261	9.88 688	11	25	9	14.4	13.5
36	9.80 443	15	9.91 765	26	10.08 235	9.88 678	10	24	
37	9.80 458	15	9.91 791	26	10.08 209	9.88 668	10	23	
38	9.80 473	15	9.91 816	25	10.08 184	9.88 657	11	22	
39	9.80 489	16	9.91 842	26	10.08 158	9.88 647	10	21	
40	9.80 504	15	9.91 868	26	10.08 132	9.88 636	11	20	
41	9.80 519	15	9.91 893	25	10.08 107	9.88 626	10	19	
42	9.80 534	15	9.91 919	26	10.08 081	9.88 615	11	18	
43	9.80 550	16	9.91 945	26	10.08 055	9.88 605	10	17	
44	9.80 565	15	9.91 971	26	10.08 029	9.88 594	11	16			11	10
45	9.80 580	15	9.91 996	25	10.08 004	9.88 584	10	15	1	1.1	1.0
46	9.80 595	15	9.92 022	26	10.07 978	9.88 573	11	14	2	2.2	2.0
47	9.80 610	15	9.92 048	26	10.07 952	9.88 563	10	13	3	3.3	3.0
48	9.80 625	16	9.92 073	25	10.07 927	9.88 552	11	12	4	4.4	4.0
49	9.80 641	15	9.92 099	26	10.07 901	9.88 542	10	11	5	5.5	5.0
50	9.80 656	15	9.92 125	25	10.07 875	9.88 531	11	10	6	6.6	6.0
51	9.80 671	15	9.92 150	26	10.07 850	9.88 521	10	9	7	7.7	7.0
52	9.80 686	15	9.92 176	26	10.07 824	9.88 510	11	8	8	8.8	8.0
53	9.80 701	15	9.92 202	25	10.07 798	9.88 499	11	7	9	9.9	9.0
54	9.80 716	15	9.92 227	26	10.07 773	9.88 489	10	6	
55	9.80 731	15	9.92 253	26	10.07 747	9.88 478	11	5	
56	9.80 746	16	9.92 279	25	10.07 721	9.88 468	10	4	
57	9.80 762	15	9.92 304	26	10.07 696	9.88 457	11	3	
58	9.80 777	15	9.92 330	26	10.07 670	9.88 447	10	2	
59	9.80 792	15	9.92 356	25	10.07 644	9.88 436	11	1	
60	9.80 807		9.92 381		10.07 619	9.88 425	11	0	
	L Cos	d	L Cot	c d	L Tan	L Sin	d	′	Prop. Pts.

TABLE 3

50°

40°

′	L Sin	d	L Tan	c d	L Cot	L Cos	d		Prop. Pts.
0	9.80 807		9.92 381		10.07 619	9.88 425		60	
1	9.80 822	15	9.92 407	26	10.07 593	9.88 415	10	59	
2	9.80 837	15	9.92 433	26	10.07 567	9.88 404	11	58	
3	9.80 852	15	9.92 458	25	10.07 542	9.88 394	10	57	
4	9.80 867	15	9.92 484	26	10.07 516	9.88 383	11	56	
5	9.80 882	15	9.92 510	26	10.07 490	9.88 372	11	55	
6	9.80 897	15	9.92 535	25	10.07 465	9.88 362	10	54	
7	9.80 912	15	9.92 561	26	10.07 439	9.88 351	11	53	
8	9.80 927	15	9.92 587	26	10.07 413	9.88 340	11	52	
9	9.80 942	15	9.92 612	25	10.07 388	9.88 330	10	51	
10	9.80 957	15	9.92 638	26	10.07 362	9.88 319	11	50	
11	9.80 972	15	9.92 663	25	10.07 337	9.88 308	11	49	
12	9.80 987	15	9.92 689	26	10.07 311	9.88 298	10	48	
13	9.81 002	15	9.92 715	26	10.07 285	9.88 287	11	47	
14	9.81 017	15	9.92 740	25	10.07 260	9.88 276	11	46	
15	9.81 032	15	9.92 766	26	10.07 234	9.88 266	10	45	
16	9.81 047	15	9.92 792	26	10.07 208	9.88 255	11	44	
17	9.81 061	14	9.92 817	25	10.07 183	9.88 244	11	43	26 25
18	9.81 076	15	9.92 843	26	10.07 157	9.88 234	10	42	1 2.6 2.5
19	9.81 091	15	9.92 868	25	10.07 132	9.88 223	11	41	2 5.2 5.0
20	9.81 106	15	9.92 894	26	10.07 106	9.88 212	11	40	3 7.8 7.5
21	9.81 121	15	9.92 920	26	10.07 080	9.88 201	11	39	4 10.4 10.0
22	9.81 136	15	9.92 945	25	10.07 055	9.88 191	10	38	5 13.0 12.5
23	9.81 151	15	9.92 971	26	10.07 029	9.88 180	11	37	6 15.6 15.0
24	9.81 166	15	9.92 996	25	10.07 004	9.88 169	11	36	7 18.2 17.5
25	9.81 180	14	9.93 022	26	10.06 978	9.88 158	11	35	8 20.8 20.0
26	9.81 195	15	9.93 048	26	10.06 952	9.88 148	10	34	9 23.4 22.5
27	9.81 210	15	9.93 073	25	10.06 927	9.88 137	11	33	
28	9.81 225	15	9.93 099	26	10.06 901	9.88 126	11	32	
29	9.81 240	15	9.93 124	25	10.06 876	9.88 115	11	31	
30	9.81 254	14	9.93 150	26	10.06 850	9.88 105	10	30	
31	9.81 269	15	9.93 175	25	10.06 825	9.88 094	11	29	
32	9.81 284	15	9.93 201	26	10.06 799	9.88 083	11	28	
33	9.81 299	15	9.93 227	26	10.06 773	9.88 072	11	27	
34	9.81 314	15	9.93 252	25	10.06 748	9.88 061	11	26	15 14
35	9.81 328	14	9.93 278	26	10.06 722	9.88 051	10	25	1 1.5 1.4
36	9.81 343	15	9.93 303	25	10.06 697	9.88 040	11	24	2 3.0 2.8
37	9.81 358	15	9.93 329	26	10.06 671	9.88 029	11	23	3 4.5 4.2
38	9.81 372	14	9.93 354	25	10.06 646	9.88 018	11	22	4 6.0 5.6
39	9.81 387	15	9.93 380	26	10.06 620	9.88 007	11	21	5 7.5 7.0
40	9.81 402	15	9.93 406	25	10.06 594	9.87 996	11	20	6 9.0 8.4
41	9.81 417	15	9.93 431	26	10.06 569	9.87 985	10	19	7 10.5 9.8
42	9.81 431	14	9.93 457	25	10.06 543	9.87 975	11	18	8 12.0 11.2
43	9.81 446	15	9.93 482	26	10.06 518	9.87 964	11	17	9 13.5 12.6
44	9.81 461	15	9.93 508	25	10.06 492	9.87 953	11	16	
45	9.81 475	14	9.93 533	26	10.06 467	9.87 942	11	15	
46	9.81 490	15	9.93 559	25	10.06 441	9.87 931	11	14	
47	9.81 505	15	9.93 584	26	10.06 416	9.87 920	11	13	
48	9.81 519	14	9.93 610	26	10.06 390	9.87 909	11	12	
49	9.81 534	15	9.93 636	25	10.06 364	9.87 898	11	11	
50	9.81 549	15	9.93 661	26	10.06 339	9.87 887	10	10	11 10
51	9.81 563	14	9.93 687	25	10.06 313	9.87 877	11	9	1 1.1 1.0
52	9.81 578	15	9.93 712	26	10.06 288	9.87 866	11	8	2 2.2 2.0
53	9.81 592	14	9.93 738	25	10.06 262	9.87 855	11	7	3 3.3 3.0
54	9.81 607	15	9.93 763	26	10.06 237	9.87 844	11	6	4 4.4 4.0
55	9.81 622	15	9.93 789	25	10.06 211	9.87 833	11	5	5 5.5 5.0
56	9.81 636	14	9.93 814	26	10.06 186	9.87 822	11	4	6 6.6 6.0
57	9.81 651	15	9.93 840	25	10.06 160	9.87 811	11	3	7 7.7 7.0
58	9.81 665	14	9.93 865	26	10.06 135	9.87 800	11	2	8 8.8 8.0
59	9.81 680	15	9.93 891	26	10.06 109	9.87 789	11	1	9 9.9 9.0
60	9.81 694	14	9.93 916	25	10.06 084	9.87 778	11	0	
	L Cos	d	L Cot	c d	L Tan	L Sin	d	′	Prop. Pts.

49°

TABLE

3

41°

′	L Sin	d	L Tan	c d	L Cot	L Cos	d	′
0	9.81 694	15	9.93 916	26	10.06 084	9.87 778	11	60
1	9.81 709	14	9.93 942	25	10.06 058	9.87 767	11	59
2	9.81 723	15	9.93 967	26	10.06 033	9.87 756	11	58
3	9.81 738	14	9.93 993	25	10.06 007	9.87 745	11	57
4	9.81 752	15	9.94 018	26	10.05 982	9.87 734	11	56
5	9.81 767	14	9.94 044	25	10.05 956	9.87 723	11	55
6	9.81 781	15	9.94 069	26	10.05 931	9.87 712	11	54
7	9.81 796	14	9.94 095	25	10.05 905	9.87 701	11	53
8	9.81 810	15	9.94 120	26	10.05 880	9.87 690	11	52
9	9.81 825	14	9.94 146	25	10.05 854	9.87 679	11	51
10	9.81 839	15	9.94 171	26	10.05 829	9.87 668	11	50
11	9.81 854	14	9.94 197	25	10.05 803	9.87 657	11	49
12	9.81 868	14	9.94 222	26	10.05 778	9.87 646	11	48
13	9.81 882	15	9.94 248	25	10.05 752	9.87 635	11	47
14	9.81 897	14	9.94 273	26	10.05 727	9.87 624	11	46
15	9.81 911	15	9.94 299	25	10.05 701	9.87 613	12	45
16	9.81 926	14	9.94 324	26	10.05 676	9.87 601	11	44
17	9.81 940	15	9.94 350	25	10.05 650	9.87 590	11	43
18	9.81 955	14	9.94 375	26	10.05 625	9.87 579	11	42
19	9.81 969	14	9.94 401	25	10.05 599	9.87 568	11	41
20	9.81 983	15	9.94 426	26	10.05 574	9.87 557	11	40
21	9.81 998	14	9.94 452	25	10.05 548	9.87 546	11	39
22	9.82 012	14	9.94 477	26	10.05 523	9.87 535	11	38
23	9.82 026	15	9.94 503	25	10.05 497	9.87 524	11	37
24	9.82 041	14	9.94 528	26	10.05 472	9.87 513	12	36
25	9.82 055	14	9.94 554	25	10.05 446	9.87 501	11	35
26	9.82 069	15	9.94 579	25	10.05 421	9.87 490	11	34
27	9.82 084	14	9.94 604	26	10.05 396	9.87 479	11	33
28	9.82 098	14	9.94 630	25	10.05 370	9.87 468	11	32
29	9.82 112	14	9.94 655	26	10.05 345	9.87 457	11	31
30	9.82 126	15	9.94 681	25	10.05 319	9.87 446	12	30
31	9.82 141	14	9.94 706	26	10.05 294	9.87 434	11	29
32	9.82 155	14	9.94 732	25	10.05 268	9.87 423	11	28
33	9.82 169	15	9.94 757	26	10.05 243	9.87 412	11	27
34	9.82 184	14	9.94 783	25	10.05 217	9.87 401	11	26
35	9.82 198	14	9.94 808	26	10.05 192	9.87 390	12	25
36	9.82 212	14	9.94 834	25	10.05 166	9.87 378	11	24
37	9.82 226	14	9.94 859	25	10.05 141	9.87 367	11	23
38	9.82 240	15	9.94 884	26	10.05 116	9.87 356	11	22
39	9.82 255	14	9.94 910	25	10.05 090	9.87 345	11	21
40	9.82 269	14	9.94 935	26	10.05 065	9.87 334	12	20
41	9.82 283	14	9.94 961	25	10.05 039	9.87 322	11	19
42	9.82 297	14	9.94 986	26	10.05 014	9.87 311	11	18
43	9.82 311	15	9.95 012	25	10.04 988	9.87 300	12	17
44	9.82 326	14	9.95 037	25	10.04 963	9.87 288	11	16
45	9.82 340	14	9.95 062	26	10.04 938	9.87 277	11	15
46	9.82 354	14	9.95 088	25	10.04 912	9.87 266	11	14
47	9.82 368	14	9.95 113	26	10.04 887	9.87 255	12	13
48	9.82 382	14	9.95 139	25	10.04 861	9.87 243	11	12
49	9.82 396	14	9.95 164	26	10.04 836	9.87 232	11	11
50	9.82 410	14	9.95 190	25	10.04 810	9.87 221	12	10
51	9.82 424	15	9.95 215	25	10.04 785	9.87 209	11	9
52	9.82 439	14	9.95 240	26	10.04 760	9.87 198	11	8
53	9.82 453	14	9.95 266	25	10.04 734	9.87 187	12	7
54	9.82 467	14	9.95 291	26	10.04 709	9.87 175	11	6
55	9.82 481	14	9.95 317	25	10.04 683	9.87 164	11	5
56	9.82 495	14	9.95 342	26	10.04 658	9.87 153	12	4
57	9.82 509	14	9.95 368	25	10.04 632	9.87 141	11	3
58	9.82 523	14	9.95 393	25	10.04 607	9.87 130	11	2
59	9.82 537	14	9.95 418	26	10.04 582	9.87 119	12	1
60	9.82 551		9.95 444		10.04 556	9.87 107		0
′	L Cos	d	L Cot	c d	L Tan	L Sin	d	′

48°

Prop. Pts.

	26	25
1	2.6	2.5
2	5.2	5.0
3	7.8	7.5
4	10.4	10.0
5	13.0	12.5
6	15.6	15.0
7	18.2	17.5
8	20.8	20.0
9	23.4	22.5

	15	14
1	1.5	1.4
2	3.0	2.8
3	4.5	4.2
4	6.0	5.6
5	7.5	7.0
6	9.0	8.4
7	10.5	9.8
8	12.0	11.2
9	13.5	12.6

	12	11
1	1.2	1.1
2	2.4	2.2
3	3.6	3.3
4	4.8	4.4
5	6.0	5.5
6	7.2	6.6
7	8.4	7.7
8	9.6	8.8
9	10.8	9.9

TABLE 3

42°

′	L Sin	d	L Tan	c d	L Cot	L Cos	d	′
0	9.82 551	14	9.95 444	25	10.04 556	9.87 107	11	60
1	9.82 565	14	9.95 469	26	10.04 531	9.87 096	11	59
2	9.82 579	14	9.95 495	25	10.04 505	9.87 085	12	58
3	9.82 593	14	9.95 520	25	10.04 480	9.87 073	11	57
4	9.82 607	14	9.95 545	26	10.04 455	9.87 062	12	56
5	9.82 621	14	9.95 571	25	10.04 429	9.87 050	11	55
6	9.82 635	14	9.95 596	26	10.04 404	9.87 039	11	54
7	9.82 649	14	9.95 622	25	10.04 378	9.87 028	12	53
8	9.82 663	14	9.95 647	25	10.04 353	9.87 016	11	52
9	9.82 677	14	9.95 672	26	10.04 328	9.87 005	12	51
10	9.82 691	14	9.95 698	25	10.04 302	9.86 993	11	50
11	9.82 705	14	9.95 723	25	10.04 277	9.86 982	12	49
12	9.82 719	14	9.95 748	26	10.04 252	9.86 970	11	48
13	9.82 733	14	9.95 774	25	10.04 226	9.86 959	12	47
14	9.82 747	14	9.95 799	26	10.04 201	9.86 947	11	46
15	9.82 761	14	9.95 825	25	10.04 175	9.86 936	12	45
16	9.82 775	13	9.95 850	25	10.04 150	9.86 924	11	44
17	9.82 788	14	9.95 875	26	10.04 125	9.86 913	11	43
18	9.82 802	14	9.95 901	25	10.04 099	9.86 902	12	42
19	9.82 816	14	9.95 926	26	10.04 074	9.86 890	11	41
20	9.82 830	14	9.95 952	25	10.04 048	9.86 879	12	40
21	9.82 844	14	9.95 977	25	10.04 023	9.86 867	12	39
22	9.82 858	14	9.96 002	26	10.03 998	9.86 855	11	38
23	9.82 872	13	9.96 028	25	10.03 972	9.86 844	12	37
24	9.82 885	14	9.96 053	25	10.03 947	9.86 832	11	36
25	9.82 899	14	9.96 078	26	10.03 922	9.86 821	12	35
26	9.82 913	14	9.96 104	25	10.03 896	9.86 809	11	34
27	9.82 927	14	9.96 129	26	10.03 871	9.86 798	12	33
28	9.82 941	14	9.96 155	25	10.03 845	9.86 786	11	32
29	9.82 955	13	9.96 180	25	10.03 820	9.86 775	12	31
30	9.82 968	14	9.96 205	26	10.03 795	9.86 763	11	30
31	9.82 982	14	9.96 231	25	10.03 769	9.86 752	12	29
32	9.82 996	14	9.96 256	25	10.03 744	9.86 740	12	28
33	9.83 010	13	9.96 281	26	10.03 719	9.86 728	11	27
34	9.83 023	14	9.96 307	25	10.03 693	9.86 717	12	26
35	9.83 037	14	9.96 332	25	10.03 668	9.86 705	11	25
36	9.83 051	14	9.96 357	26	10.03 643	9.86 694	12	24
37	9.83 065	13	9.96 383	25	10.03 617	9.86 682	12	23
38	9.83 078	14	9.96 408	25	10.03 592	9.86 670	11	22
39	9.83 092	14	9.96 433	26	10.03 567	9.86 659	12	21
40	9.83 106	14	9.96 459	25	10.03 541	9.86 647	12	20
41	9.83 120	13	9.96 484	26	10.03 516	9.86 635	11	19
42	9.83 133	14	9.96 510	25	10.03 490	9.86 624	12	18
43	9.83 147	14	9.96 535	25	10.03 465	9.86 612	12	17
44	9.83 161	13	9.96 560	26	10.03 440	9.86 600	11	16
45	9.83 174	14	9.96 586	25	10.03 414	9.86 589	12	15
46	9.83 188	14	9.96 611	25	10.03 389	9.86 577	12	14
47	9.83 202	13	9.96 636	26	10.03 364	9.86 565	11	13
48	9.83 215	14	9.96 662	25	10.03 338	9.86 554	12	12
49	9.83 229	13	9.96 687	25	10.03 313	9.86 542	12	11
50	9.83 242	14	9.96 712	26	10.03 288	9.86 530	12	10
51	9.83 256	14	9.96 738	25	10.03 262	9.86 518	11	9
52	9.83 270	13	9.96 763	25	10.03 237	9.86 507	12	8
53	9.83 283	14	9.96 788	26	10.03 212	9.86 495	12	7
54	9.83 297	13	9.96 814	25	10.03 186	9.86 483	11	6
55	9.83 310	14	9.96 839	25	10.03 161	9.86 472	12	5
56	9.83 324	14	9.96 864	26	10.03 136	9.86 460	12	4
57	9.83 338	13	9.96 890	25	10.03 110	9.86 448	12	3
58	9.83 351	14	9.96 915	25	10.03 085	9.86 436	11	2
59	9.83 365	13	9.96 940	26	10.03 060	9.86 425	12	1
60	9.83 378		9.96 966		10.03 034	9.86 413		0
	L Cos	d	L Cot	c d	L Tan	L Sin	d	′

Prop. Pts.

	26	25
1	2.6	2.5
2	5.2	5.0
3	7.8	7.5
4	10.4	10.0
5	13.0	12.5
6	15.6	15.0
7	18.2	17.5
8	20.8	20.0
9	23.4	22.5

	14	13
1	1.4	1.3
2	2.8	2.6
3	4.2	3.9
4	5.6	5.2
5	7.0	6.5
6	8.4	7.8
7	9.8	9.1
8	11.2	10.4
9	12.6	11.7

	12	11
1	1.2	1.1
2	2.4	2.2
3	3.6	3.3
4	4.8	4.4
5	6.0	5.5
6	7.2	6.6
7	8.4	7.7
8	9.6	8.8
9	10.8	9.9

47°

TABLE 3

43°

′	L Sin	d	L Tan	c d	L Cot	L Cos	d	′
0	9.83 378		9.96 966		10.03 034	9.86 413		**60**
1	9.83 392	14	9.96 991	25	10.03 009	9.86 401	12	59
2	9.83 405	13	9.97 016	25	10.02 984	9.86 389	12	58
3	9.83 419	14	9.97 042	26	10.02 958	9.86 377	12	57
4	9.83 432	13	9.97 067	25	10.02 933	9.86 366	11	56
5	9.83 446	14	9.97 092	25	10.02 908	9.86 354	12	55
6	9.83 459	13	9.97 118	26	10.02 882	9.86 342	12	54
7	9.83 473	14	9.97 143	25	10.02 857	9.86 330	12	53
8	9.83 486	13	9.97 168	25	10.02 832	9.86 318	12	52
9	9.83 500	14	9.97 193	26	10.02 807	9.86 306	12	51
10	9.83 513	13	9.97 219	25	10.02 781	9.86 295	11	**50**
11	9.83 527	14	9.97 244	25	10.02 756	9.86 283	12	49
12	9.83 540	13	9.97 269	26	10.02 731	9.86 271	12	48
13	9.83 554	14	9.97 295	25	10.02 705	9.86 259	12	47
14	9.83 567	13	9.97 320	25	10.02 680	9.86 247	12	46
15	9.83 581	14	9.97 345	26	10.02 655	9.86 235	12	45
16	9.83 594	13	9.97 371	25	10.02 629	9.86 223	12	44
17	9.83 608	14	9.97 396	25	10.02 604	9.86 211	11	43
18	9.83 621	13	9.97 421	26	10.02 579	9.86 200	12	42
19	9.83 634	13	9.97 447	25	10.02 553	9.86 188	12	41
20	9.83 648	14	9.97 472	25	10.02 528	9.86 176	12	**40**
21	9.83 661	13	9.97 497	26	10.02 503	9.86 164	12	39
22	9.83 674	13	9.97 523	25	10.02 477	9.86 152	12	38
23	9.83 688	14	9.97 548	25	10.02 452	9.86 140	12	37
24	9.83 701	13	9.97 573	25	10.02 427	9.86 128	12	36
25	9.83 715	14	9.97 598	26	10.02 402	9.86 116	12	35
26	9.83 728	13	9.97 624	25	10.02 376	9.86 104	12	34
27	9.83 741	13	9.97 649	25	10.02 351	9.86 092	12	33
28	9.83 755	14	9.97 674	26	10.02 326	9.86 080	12	32
29	9.83 768	13	9.97 700	25	10.02 300	9.86 068	12	31
30	9.83 781	13	9.97 725	25	10.02 275	9.86 056	12	**30**
31	9.83 795	14	9.97 750	26	10.02 250	9.86 044	12	29
32	9.83 808	13	9.97 776	25	10.02 224	9.86 032	12	28
33	9.83 821	13	9.97 801	25	10.02 199	9.86 020	12	27
34	9.83 834	13	9.97 826	25	10.02 174	9.86 008	12	26
35	9.83 848	14	9.97 851	26	10.02 149	9.85 996	12	25
36	9.83 861	13	9.97 877	25	10.02 123	9.85 984	12	24
37	9.83 874	13	9.97 902	25	10.02 098	9.85 972	12	23
38	9.83 887	14	9.97 927	26	10.02 073	9.85 960	12	22
39	9.83 901	13	9.97 953	25	10.02 047	9.85 948	12	21
40	9.83 914	13	9.97 978	25	10.02 022	9.85 936	12	**20**
41	9.83 927	13	9.98 003	26	10.01 997	9.85 924	12	19
42	9.83 940	14	9.98 029	25	10.01 971	9.85 912	12	18
43	9.83 954	13	9.98 054	25	10.01 946	9.85 900	12	17
44	9.83 967	13	9.98 079	25	10.01 921	9.85 888	12	16
45	9.83 980	13	9.98 104	26	10.01 896	9.85 876	12	15
46	9.83 993	13	9.98 130	25	10.01 870	9.85 864	13	14
47	9.84 006	14	9.98 155	25	10.01 845	9.85 851	12	13
48	9.84 020	13	9.98 180	26	10.01 820	9.85 839	12	12
49	9.84 033	13	9.98 206	25	10.01 794	9.85 827	12	11
50	9.84 046	13	9.98 231	25	10.01 769	9.85 815	12	**10**
51	9.84 059	13	9.98 256	25	10.01 744	9.85 803	12	9
52	9.84 072	13	9.98 281	26	10.01 719	9.85 791	12	8
53	9.84 085	13	9.98 307	25	10.01 693	9.85 779	13	7
54	9.84 098	14	9.98 332	25	10.01 668	9.85 766	12	6
55	9.84 112	13	9.98 357	26	10.01 643	9.85 754	12	5
56	9.84 125	13	9.98 383	25	10.01 617	9.85 742	12	4
57	9.84 138	13	9.98 408	25	10.01 592	9.85 730	12	3
58	9.84 151	13	9.98 433	25	10.01 567	9.85 718	12	2
59	9.84 164	13	9.98 458	26	10.01 542	9.85 706	13	1
60	9.84 177		9.98 484		10.01 516	9.85 693		**0**
	L Cos	d	L Cot	c d	L Tan	L Sin	d	′

46°

TABLE 3

Prop. Pts.

	26	25
1	2.6	2.5
2	5.2	5.0
3	7.8	7.5
4	10.4	10.0
5	13.0	12.5
6	15.6	15.0
7	18.2	17.5
8	20.8	20.0
9	23.4	22.5

	14	13
1	1.4	1.3
2	2.8	2.6
3	4.2	3.9
4	5.6	5.2
5	7.0	6.5
6	8.4	7.8
7	9.8	9.1
8	11.2	10.4
9	12.6	11.7

	12	11
1	1.2	1.1
2	2.4	2.2
3	3.6	3.3
4	4.8	4.4
5	6.0	5.5
6	7.2	6.6
7	8.4	7.7
8	9.6	8.8
9	10.8	9.9

44°

′	L Sin	d	L Tan	c d	L Cot	L Cos	d	′
0	9.84 177	13	9.98 484	25	10.01 516	9.85 693	12	60
1	9.84 190	13	9.98 509	25	10.01 491	9.85 681	12	59
2	9.84 203	13	9.98 534	26	10.01 466	9.85 669	12	58
3	9.84 216	13	9.98 560	25	10.01 440	9.85 657	12	57
4	9.84 229	13	9.98 585	25	10.01 415	9.85 645	13	56
5	9.84 242	13	9.98 610	25	10.01 390	9.85 632	12	55
6	9.84 255	14	9.98 635	26	10.01 365	9.85 620	12	54
7	9.84 269	13	9.98 661	25	10.01 339	9.85 608	12	53
8	9.84 282	13	9.98 686	25	10.01 314	9.85 596	13	52
9	9.84 295	13	9.98 711	26	10.01 289	9.85 583	12	51
10	9.84 308	13	9.98 737	25	10.01 263	9.85 571	12	50
11	9.84 321	13	9.98 762	25	10.01 238	9.85 559	12	49
12	9.84 334	13	9.98 787	25	10.01 213	9.85 547	13	48
13	9.84 347	13	9.98 812	26	10.01 188	9.85 534	12	47
14	9.84 360	13	9.98 838	25	10.01 162	9.85 522	12	46
15	9.84 373	12	9.98 863	25	10.01 137	9.85 510	13	45
16	9.84 385	13	9.98 888	25	10.01 112	9.85 497	12	44
17	9.84 398	13	9.98 913	26	10.01 087	9.85 485	12	43
18	9.84 411	13	9.98 939	25	10.01 061	9.85 473	13	42
19	9.84 424	13	9.98 964	25	10.01 036	9.85 460	12	41
20	9.84 437	13	9.98 989	26	10.01 011	9.85 448	12	40
21	9.84 450	13	9.99 015	25	10.00 985	9.85 436	13	39
22	9.84 463	13	9.99 040	25	10.00 960	9.85 423	12	38
23	9.84 476	13	9.99 065	25	10.00 935	9.85 411	12	37
24	9.84 489	13	9.99 090	26	10.00 910	9.85 399	13	36
25	9.84 502	13	9.99 116	25	10.00 884	9.85 386	12	35
26	9.84 515	13	9.99 141	25	10.00 859	9.85 374	13	34
27	9.84 528	12	9.99 166	25	10.00 834	9.85 361	12	33
28	9.84 540	13	9.99 191	26	10.00 809	9.85 349	12	32
29	9.84 553	13	9.99 217	25	10.00 783	9.85 337	13	31
30	9.84 566	13	9.99 242	25	10.00 758	9.85 324	12	30
31	9.84 579	13	9.99 267	26	10.00 733	9.85 312	13	29
32	9.84 592	13	9.99 293	25	10.00 707	9.85 299	12	28
33	9.84 605	13	9.99 318	25	10.00 682	9.85 287	13	27
34	9.84 618	12	9.99 343	25	10.00 657	9.85 274	12	26
35	9.84 630	13	9.99 368	26	10.00 632	9.85 262	12	25
36	9.84 643	13	9.99 394	25	10.00 606	9.85 250	13	24
37	9.84 656	13	9.99 419	25	10.00 581	9.85 237	12	23
38	9.84 669	13	9.99 444	25	10.00 556	9.85 225	13	22
39	9.84 682	12	9.99 469	26	10.00 531	9.85 212	12	21
40	9.84 694	13	9.99 495	25	10.00 505	9.85 200	13	20
41	9.84 707	13	9.99 520	25	10.00 480	9.85 187	12	19
42	9.84 720	13	9.99 545	25	10.00 455	9.85 175	13	18
43	9.84 733	12	9.99 570	26	10.00 430	9.85 162	12	17
44	9.84 745	13	9.99 596	25	10.00 404	9.85 150	13	16
45	9.84 758	13	9.99 621	25	10.00 379	9.85 137	12	15
46	9.84 771	13	9.99 646	26	10.00 354	9.85 125	13	14
47	9.84 784	12	9.99 672	25	10.00 328	9.85 112	12	13
48	9.84 796	13	9.99 697	25	10.00 303	9.85 100	13	12
49	9.84 809	13	9.99 722	25	10.00 278	9.85 087	13	11
50	9.84 822	13	9.99 747	26	10.00 253	9.85 074	12	10
51	9.84 835	12	9.99 773	25	10.00 227	9.85 062	13	9
52	9.84 847	13	9.99 798	25	10.00 202	9.85 049	12	8
53	9.84 860	13	9.99 823	25	10.00 177	9.85 037	13	7
54	9.84 873	12	9.99 848	26	10.00 152	9.85 024	12	6
55	9.84 885	13	9.99 874	25	10.00 126	9.85 012	13	5
56	9.84 898	13	9.99 899	25	10.00 101	9.84 999	13	4
57	9.84 911	12	9.99 924	25	10.00 076	9.84 986	12	3
58	9.84 923	13	9.99 949	26	10.00 051	9.84 974	13	2
59	9.84 936	13	9.99 975	25	10.00 025	9.84 961	12	1
60	9.84 949		10.00 000		10.00 000	9.84 949		0
	L Cos	d	L Cot	c d	L Tan	L Sin	d	′

45°

Prop. Pts.

	26	25
1	2.6	2.5
2	5.2	5.0
3	7.8	7.5
4	10.4	10.0
5	13.0	12.5
6	15.6	15.0
7	18.2	17.5
8	20.8	20.0
9	23.4	22.5

	14	13	12
1	1.4	1.3	1.2
2	2.8	2.6	2.4
3	4.2	3.9	3.6
4	5.6	5.2	4.8
5	7.0	6.5	6.0
6	8.4	7.8	7.2
7	9.8	9.1	8.4
8	11.2	10.4	9.6
9	12.6	11.7	10.8

TABLE

3

TABLE **4**

SQUARES, CUBES, SQUARE AND CUBE ROOTS, AND
RECIPROCALS OF THE NUMBERS 1 TO 100

1-50

No.	Square	Cube	Sq. Root	Cu. Root	Reciprocal
1	1	1	1.0000	1.0000	1.000000000
2	4	8	1.4142	1.2599	.500000000
3	9	27	1.7321	1.4422	.333333333
4	16	64	2.0000	1.5874	.250000000
5	25	125	2.2361	1.7100	.200000000
6	36	216	2.4495	1.8171	.166666667
7	49	343	2.6458	1.9129	.142857143
8	64	512	2.8284	2.0000	.125000000
9	81	729	3.0000	2.0801	.111111111
10	100	1,000	3.1623	2.1544	.100000000
11	121	1,331	3.3166	2.2240	.090909091
12	144	1,728	3.4641	2.2894	.083333333
13	169	2,197	3.6056	2.3513	.076923077
14	196	2,744	3.7417	2.4101	.071428571
15	225	3,375	3.8730	2.4662	.066666667
16	256	4,096	4.0000	2.5198	.062500000
17	289	4,913	4.1231	2.5713	.058823529
18	324	5,832	4.2426	2.6207	.055555556
19	361	6,859	4.3589	2.6684	.052631579
20	400	8,000	4.4721	2.7144	.050000000
21	441	9,261	4.5826	2.7589	.047619048
22	484	10,648	4.6904	2.8020	.045454545
23	529	12,167	4.7958	2.8439	.043478261
24	576	13,824	4.8990	2.8845	.041666667
25	625	15,625	5.0000	2.9240	.040000000
26	676	17,576	5.0990	2.9625	.038461538
27	729	19,683	5.1962	3.0000	.037037037
28	784	21,952	5.2915	3.0366	.035714286
29	841	24,389	5.3852	3.0723	.034482759
30	900	27,000	5.4772	3.1072	.033333333
31	961	29,791	5.5678	3.1414	.032258065
32	1,024	32,768	5.6569	3.1748	.031250000
33	1,089	35,937	5.7446	3.2075	.030303030
34	1,156	39,304	5.8310	3.2396	.029411765
35	1,225	42,875	5.9161	3.2711	.028571429
36	1,296	46,656	6.0000	3.3019	.027777778
37	1,369	50,653	6.0828	3.3322	.027027027
38	1,444	54,872	6.1644	3.3620	.026315789
39	1,521	59,319	6.2450	3.3912	.025641026
40	1,600	64,000	6.3246	3.4200	.025000000
41	1,681	68,921	6.4031	3.4482	.024390244
42	1,764	74,088	6.4807	3.4760	.023809524
43	1,849	79,507	6.5574	3.5034	.023255814
44	1,936	85,184	6.6332	3.5303	.022727273
45	2,025	91,125	6.7082	3.5569	.022222222
46	2,116	97,336	6.7823	3.5830	.021739130
47	2,209	103,823	6.8557	3.6088	.021276596
48	2,304	110,592	6.9282	3.6342	.020833333
49	2,401	117,649	7.0000	3.6593	.020408163
50	2,500	125,000	7.0711	3.6840	.020000000

TABLE

4

50-100

No.	Square	Cube	Sq. Root	Cu. Root	Reciprocal
50	2,500	125,000	7.0711	3.6840	.020000000
51	2,601	132,651	7.1414	3.7084	.019607843
52	2,704	140,608	7.2111	3.7325	.019230769
53	2,809	148,877	7.2801	3.7563	.018867925
54	2,916	157,464	7.3485	3.7798	.018518519
55	3,025	166,375	7.4162	3.8030	.018181818
56	3,136	175,616	7.4833	3.8259	.017857143
57	3,249	185,193	7.5498	3.8485	.017543860
58	3,364	195,112	7.6158	3.8709	.017241379
59	3,481	205,379	7.6811	3.8930	.016949153
60	3,600	216,000	7.7460	3.9149	.016666667
61	3,721	226,981	7.8102	3.9365	.016393443
62	3,844	238,328	7.8740	3.9579	.016129032
63	3,969	250,047	7.9373	3.9791	.015873016
64	4,096	262,144	8.0000	4.0000	.015625000
65	4,225	274,625	8.0623	4.0207	.015384615
66	4,356	287,496	8.1240	4.0412	.015151515
67	4,489	300,763	8.1854	4.0615	.014925373
68	4,624	314,432	8.2462	4.0817	.014705882
69	4,761	328,509	8.3066	4.1016	.014492754
70	4,900	343,000	8.3666	4.1213	.014285714
71	5,041	357,911	8.4261	4.1408	.014084507
72	5,184	373,248	8.4853	4.1602	.013888889
73	5,329	389,017	8.5440	4.1793	.013698630
74	5,476	405,224	8.6023	4.1983	.013513514
75	5,625	421,875	8.6603	4.2172	.013333333
76	5,776	438,976	8.7178	4.2358	.013157895
77	5,929	456,533	8.7750	4.2543	.012987013
78	6,084	474,552	8.8318	4.2727	.012820513
79	6,241	493,039	8.8882	4.2908	.012658228
80	6,400	512,000	8.9443	4.3089	.012500000
81	6,561	531,441	9.0000	4.3267	.012345679
82	6,724	551,368	9.0554	4.3445	.012195122
83	6,889	571,787	9.1104	4.3621	.012048193
84	7,056	592,704	9.1652	4.3795	.011904762
85	7,225	614,125	9.2195	4.3968	.011764706
86	7,396	636,056	9.2736	4.4140	.011627907
87	7,569	658,503	9.3274	4.4310	.011494253
88	7,744	681,472	9.3808	4.4480	.011363636
89	7,921	704,969	9.4340	4.4647	.011235955
90	8,100	729,000	9.4868	4.4814	.011111111
91	8,281	753,571	9.5394	4.4979	.010989011
92	8,464	778,688	9.5917	4.5144	.010869565
93	8,649	804,357	9.6437	4.5307	.010752688
94	8,836	830,584	9.6954	4.5468	.010638298
95	9,025	857,375	9.7468	4.5629	.010526316
96	9,216	884,736	9.7980	4.5789	.010416667
97	9,409	912,673	9.8489	4.5947	.010309278
98	9,604	941,192	9.8995	4.6104	.010204082
99	9,801	970,299	9.9499	4.6261	.010101010
100	10,000	1,000,000	10.0000	4.6416	.010000000

TABLE

4

Index